Maritime & Coastguard Agency

Code of Safe Working Practices for Merchant Seafarers

2024 Edition

tso
a Williams Lea company

Published by TSO (The Stationery Office), part of Williams Lea, and available from:

Online
www.tsoshop.co.uk

Mail, Telephone & E-mail
TSO
PO Box 29, Norwich, NR3 1GN
Telephone orders/General enquiries: 0333 202 5070
E-mail: customer.services@tso.co.uk
Textphone: 0333 202 5077

© Crown Copyright 2024.

Published by The Stationery Office Limited, for the Maritime and Coastguard Agency, under licence from the Controller of His Majesty's Stationery Office.

You may re-use the content in this document/publication free of charge in any format or medium, under the terms of the Open Government Licence. The design, cover design, images, photos, logos, and typographical arrangements are NOT subject to open government licence terms and cannot be reused. To view this licence, visit www.nationalarchives.gov.uk/doc/open-government-licence/version/3/, or write to the Information Policy Team, The National Archives, Kew, Richmond, Surrey, TW9 4DU, and/or email: psi@nationalarchives.gov.uk.

ISBN 978 0 11 554132 2

Printed in the United Kingdom for The Stationery Office Limited.
SD000175 6/24

Contents

	About this Code	v
	How to use this document	x
1	Managing occupational health and safety	1
2	Safety induction for personnel working on ships	29
3	Living on board	35
4	Emergency drills and procedures	53
5	Fire precautions	71
6	Security on board	75
7	Workplace health surveillance	79
8	Personal protective equipment	85
9	Safety signs and their use	99
10	Manual handling	117
11	Safe movement on board ship	129
12	Noise, vibration and other physical agents	141
13	Safety officials	159
14	Permit to work systems	181
15	Entering enclosed spaces	201
16	Hatch covers and access lids	223
17	Work at height	229
18	Provision, care and use of work equipment	245
19	Lifting equipment and operations	275
20	Work on machinery and power systems	305
21	Hazardous substances and mixtures	333
22	Boarding arrangements	345
23	Food preparation and handling in the catering department	361
24	Hot work	373
25	Painting	389

26	Anchoring, mooring and towing operations	393
27	Roll-on/roll-off ferries	411
28	Dry cargo	425
29	Tankers and other ships carrying bulk liquid cargoes	437
30	Port towage industry	443
31	Ships serving offshore oil and gas installations	447
32	Ships serving offshore renewables installations	461
33	Ergonomics	467
Appendix 1	Regulations, marine notices and guidance issued by the Maritime and Coastguard Agency	475
Appendix 2	Other sources of information	489
Appendix 3	Standards and specifications referred to in this Code	499
Appendix 4	Acknowledgements	507
Glossary		511
Index		515

About this Code

General

1. This Code is published by the Maritime and Coastguard Agency (MCA) and endorsed by the National Maritime Occupational Health and Safety Committee, UK Chamber of Shipping, Nautilus International and the National Union of Rail, Maritime and Transport Workers (RMT) as best practice guidance for improving health and safety on board ship.

2. It is intended primarily for merchant seafarers on UK-registered ships. The Code is addressed to everyone on a ship regardless of rank or rating, and to those ashore responsible for safety, because the recommendations can be effective only if they are understood by all, and if everyone cooperates in their implementation. Those not actually engaged in a job in hand should be aware of what is being done, so that they may avoid putting themselves at risk or causing risk to others by impeding or needlessly interfering with the conduct of their work.

3. The Code covers the regulatory framework and provides best practice guidance for health and safety on board ship. It also gives guidance on safety management, identifies statutory duties underlying the advice and includes practical information for safe working on board.

4. From 2024 onwards, this document reflects a large-scale review of the Code, which aimed to improve the functionality of the Code as a reference document for seafarers. As a result of this review, the MCA has redesigned and modernised the Code to improve its structure, ensure consistency and simplify its language. The use of recurring design features will enable quicker reference and generally aid understanding, making it clearer and easier to follow. You may wish to refer to 'How to use this document', which outlines key features to help you make the most of this document. In addition, there is a change to the physical format of the document from loose-leaf to bound, to improve the end-user experience, remove the yearly manual updating process and seek a more cost-effective and sustainable solution.

5. This updated Code was developed with support from industry (see acknowledgements, Appendix 4), including an online survey, several focus groups, a stakeholder engagement exercise and a public consultation which was held from August to November 2023. As part of this review, individuals working in the maritime industry, MCA survey and inspection teams and MCA policy teams contributed to the new content, structure and design features for each chapter of the Code.

Living on board: occupational health and safety risks

6. Occupational health and safety risks may lead to death, permanent disability, temporary disability or reduced work capability. Occupational health and safety risks may arise from work-related hazards or from the general living and working conditions on board, sometimes referred to as ambient factors. In cases where some risks are unavoidable, appropriate control measures should be implemented to minimise exposure to hazards that may cause injury, disease or death. Harmful exposure may have short-term or long-term adverse health effects.

7. Risks inherent in the working environment must be identified and evaluated ('risk assessment'), and measures must be taken to remove or minimise those risks, to protect seafarers and others from harm, so far as is reasonably practicable.

8. These risks include, but are not limited to:

    - ambient factors, such as noise, vibration, lighting, ultra-violet light, non-ionising radiation and extreme temperatures
    - inherent hazards, such as the vessel's structure, means of access, ergonomic hazards and hazardous materials such as asbestos
    - hazards arising from work activities, such as work in enclosed spaces, use of equipment and machinery, working on and below deck in adverse weather, dangerous cargo and ballast operations, and exposure to biological hazards or chemicals
    - health risks, such as fatigue and impacts on mental occupational health
    - the emergency and accident response.

9. In addition, there are risks from violence in the workplace, tobacco smoking, drug abuse, alcohol misuse and drug or alcohol dependence.

10. Each of these risks is covered in this Code.

The status of the Code of Safe Working Practices for Merchant Seafarers

11. In the UK, the Merchant Shipping Act 1995 allows the Secretary of State to make regulations to secure the safety of ships and those on them. Much of the Code relates to matters that are the subject of such regulations. In such cases, the Code is intended to give guidance as to how the statutory obligations should be fulfilled.

12. Many regulations lay down specific requirements for standards of safety, equipment or operations, which must be satisfied to comply with the law. Where there are no specific requirements, the MCA generally considers compliance with the Code as demonstrating that the company, employer or seafarer did what was reasonable to comply with the regulations. Each situation will be considered and evaluated on an individual basis. The guidance must never be regarded as superseding or amending regulation, and risk assessment should always be used to ensure that all risks are addressed.

13. References to British Standards (BS) or European Norms (EN) contained in this Code are made with the understanding that 'an alternative Standard which provides, in use, equivalent levels of safety, suitability and fitness for purpose' is equally acceptable.

14. The Code provides guidance on safe working practices for many situations that commonly arise on ships, and the basic principles can be applied to many other work situations that are not specifically covered. However, it should not be considered a comprehensive guide to safety: the advice it contains should always be considered in conjunction with the findings of the company's or employer's risk assessment, and any information, procedures or working instructions provided by the manufacturer, supplier or any other source should be followed.

15. It is a statutory requirement that seafarers are provided with the information necessary to ensure their health and safety. The MCA considers that on UK-registered ships this means that all those with specific responsibilities for safety should have immediate access to this Code, and that it should be readily available to all seafarers on board; for example, a copy should be kept in the mess room. It should be provided in appropriate formats (e.g. electronic and hard copy) in sufficient quantity to ensure easy access. The Code should be supplemented by safety manuals, work instructions and other guidance issued by shipping companies for their particular ships, as appropriate.

16. Non-UK-registered ships are not subject to all UK health and safety regulations, although failure to meet international standards of safety enshrined in those regulations may result in enforcement action while the ship is in UK waters.

International Management Code for the Safe Operation of Ships and for Pollution Prevention (International Safety Management (ISM) Code)

17. All ships of 500 gross tonnage (GT) and over are required to operate a safety management system in compliance with the ISM Code. The ISM Code provides for safety management on board the ships to which it applies. The safety management system may not in itself cover all aspects of seafarer safety and health as required by the Maritime Labour Convention 2006 (MLC 2006); for example, with respect to disease prevention. However, a shipowner may develop that system to do so. Duplication should be avoided.

18. Compliance with the ISM Code complements existing health and safety regulations and use of the guidance in this Code. For example:

    - The ISM Code requires that the Company's safety management system should 'ensure that applicable codes, guidelines and standards recommended by the … Administration' are taken into account. This Code is one such 'applicable code', and an ISM audit may consider how the guidance it contains has been implemented.
    - The ISM Code requires that the 'safety management objectives of the Company should, inter alia, … establish safeguards against all identified risks …'. This Code will assist the company in identifying risks and establishing safe practices to safeguard against them.
    - The ISM Code requires the company to 'define and document the responsibility, authority and interrelation of all personnel who manage, perform and verify work relating to and affecting safety and pollution prevention'. This Code gives advice on the roles of those with particular safety responsibilities, and highlights work areas where specific responsibilities should be allocated to a **competent person**. 🔍

Merchant Shipping and Fishing Vessels (Health and Safety at Work) Regulations

Duties of shipowners

19. It is the duty of shipowners and employers to protect the health and safety of seafarers and others so far as is reasonably practicable. The principles that should underpin health and safety measures are:

    📑 *S.I. 1997/2962, Reg. 5*

    - the avoidance of risks, which among other things includes the combating of risks at source and the replacement of dangerous practices, substances or equipment by non-dangerous or less dangerous practices, substances or equipment
    - the evaluation of unavoidable risks and the taking of action to reduce them
    - the adoption of work patterns and procedures that take account of the capacity of the individual, especially in respect of the design of the workplace and the choice of work equipment, with a view in particular to alleviating monotonous work and to reducing any consequent adverse effect on workers' health and safety
    - the adaptation of procedures to take account of new technology and other changes in working practices, equipment, the working environment and any other factors that may affect health and safety
    - the adoption of a coherent approach to management of the vessel or undertaking, taking account of health and safety at every level of the organisation
    - giving collective protective measures priority over individual protective measures
    - the provision of appropriate and relevant information and instruction for workers.

20. The company and other employers owe a duty of care to other workers on board who may be affected. Where passengers are also covered, this will normally be stated.

   MGN 492 (M+F) Amendment 1

21. The company is also responsible for ensuring that seafarers have the appropriate information, training and instruction to enable them to work safely, making arrangements for consultation with seafarers about health and safety matters, and having systems for recording and investigating safety incidents and accidents on board. Further information about each of these aspects is contained in the following chapters.

22. The master is the representative of the company.

Duties of seafarers

23. Seafarers are required to:

    - take reasonable care for their own health and safety and that of others on board who may be affected by their acts or omissions
    - cooperate with anyone else carrying out health and safety duties, including compliance with control measures identified during the employer's or company's risk assessment
    - report any identified serious hazards or deficiencies immediately to the appropriate officer or other responsible person
    - make proper use of plant and machinery, and treat any hazard to health or safety (such as a dangerous substance) with due caution.

24. Under the regulations, it is also an offence for any person intentionally or recklessly to interfere with or misuse anything provided in the interests of health and safety.

Terms used in this Code

25. In this Code, unless otherwise defined in the specific chapter:

    'Company' is used in the sense that it is used in the ISM Code, as the person responsible for the operation of the ship. (This is often the same organisation as the 'shipowner' referred to in health and safety regulations.)

    'Competent person' means someone who has sufficient training and experience or knowledge and other qualities that allow them to carry out the work in hand effectively and safely. The level of competence required will depend on the complexity of the situation and the particular work involved.

    'Responsible person' means the person designated to take responsibility for a particular work activity. There may be particular competency requirements attached to that work activity.

    'Seafarer' means anyone whose normal place of work is on board the ship, whether or not they are employed.

    'Thorough examination' means a systematic and detailed examination of the equipment and safety-critical parts, carried out at specified intervals by a competent person, who must then complete a written report.

'Inspection': the purpose of an inspection is to identify whether work equipment can be operated, adjusted and maintained safely, with any deterioration detected and remedied before it results in a health and safety risk. The need for inspection and inspection frequencies should be determined through risk assessment. In many cases, a quick visual check before use will be sufficient. However, inspection is necessary for any equipment where significant risks to health and safety may arise from incorrect installation, reinstallation, deterioration or any other circumstances.

'Safety management system' means the safety management system for the time being in place on the ship.

Regulations, standards, documents and other sources of information referred to in the Code

26. Where chapters of the Code refer to other documents, these are referenced in the margin, and further details, including how to obtain them, are contained in the appendices:

Appendix 1 Regulations, marine notices and guidance issued by the Maritime and Coastguard Agency

Appendix 2 Other sources of information

Appendix 3 Standards and specifications referred to in this Code

Appendix 4 Acknowledgements.

Keeping the Code up to date

27. The MCA intends to issue regular updates to the Code to ensure that it remains relevant and reflects changes in standards and in working practices. Updates will be considered by the industry working group and will be subject to wider consultation before final agreement. If you notice anything that requires updating, please notify the MCA at seafarersafety@mcga.gov.uk

28. The Code will be produced in digital form in due course.

How to use this document

This is a brief guide to the new features and what they mean.

Boxes and sidebars

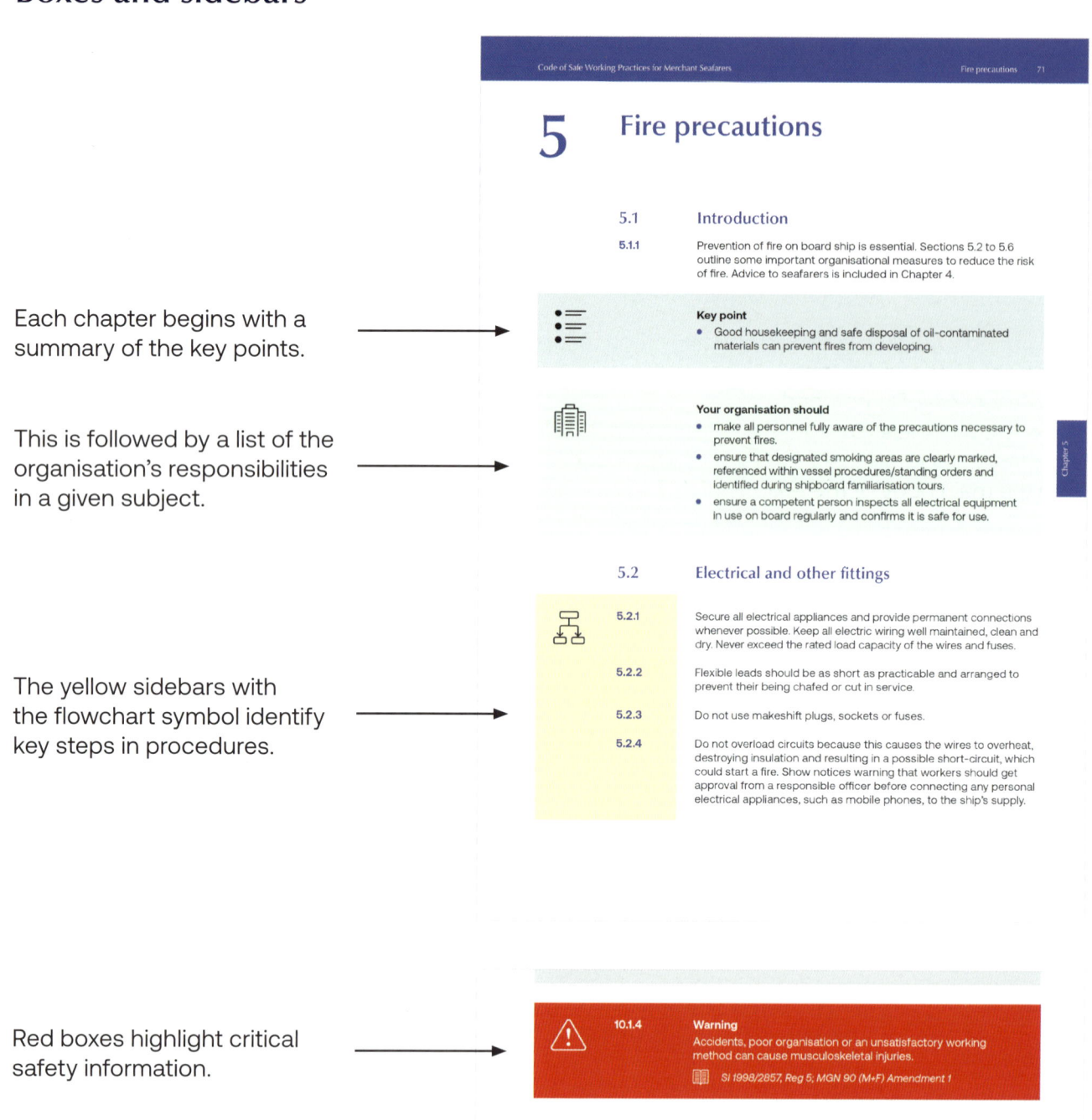

Each chapter begins with a summary of the key points.

This is followed by a list of the organisation's responsibilities in a given subject.

The yellow sidebars with the flowchart symbol identify key steps in procedures.

Red boxes highlight critical safety information.

This box highlights important points to remember.

These sidebars show information that may help you create a checklist.

These boxes shows key tips for safe working. Do not read these in isolation as you could miss important information from the main body of the text.

Symbols in the text

The book symbol means further reading. This could be the legislation on which the guidance is based, or another publication on the same subject.

The magnifying glass symbol means the word or phrase is explained in the glossary.

1 Managing occupational health and safety

1.1 Introduction

Seafarers, like shore workers, have the right and expectation to remain safe at work.

Seafarers have a duty to cooperate with their employer and the company in health, safety and welfare matters.

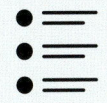

Key points

Every person on board has a responsibility for their own occupational health and safety and that of others, including:

- complying with instructions, safety procedures and any other measures in place for their own or others' safety
- reporting any defects in equipment or unsafe conditions to a responsible person
- not interfering with or altering any safety device provided on board.

Every task carried out on board the vessel should be risk assessed. However, a risk assessment does not necessarily need writing every time a simple task is done.

As well as following procedures identified by risk assessment, seafarers should proactively speak up about any concerns they have.

SI 1997/2962

Your organisation should

- ensure the health, safety and welfare of all seafarers and other workers on board
- create a culture where everyone takes responsibility for a safe working environment, and seafarers are enabled to speak up about any safety concerns.

For more information on topics in this chapter see MCA's Wellbeing at Sea: A Guide for Organisations.

1.2 What does a safe working culture look like?

1.2.1

The following elements contribute to a safe working culture:

- risk awareness
- clearly defined expectations
- good communications
- good planning
- accountability
- clear leadership
- good safety culture
- effective knowledge management.

These elements should be implemented at all levels within the safety management system and actioned on board by the master and crew.

It is important to involve the entire workforce, from the most junior crew members to the senior managers ashore, in the development of these elements. Many may already be present within management systems. But if any are missing, there will be weaknesses in the management system.

A gap analysis can identify any elements that are missing or weak, and can be used to amend the systems accordingly. The more developed and comprehensive the systems, the more effective they can be.

Guidance on these elements follows, with some examples. Although the details may differ between companies and vessels, the principles remain the same.

1.2.2 Clearly defined expectations

It is important that seafarers at all levels of the organisation understand what is expected of them, and have a clear and accurate job description.

Seafarers should feel confident to stop work if they feel unsafe; this is known as 'stop work authority'.

The company should carry out a comprehensive and clear induction process for every joining member of the crew, with respect to company- and vessel-specific requirements. The inductions should explain requirements and expectations in a way that crew members can easily understand. They should include an overview of the rules and how to apply them, along with information on where to find further information; for example:

- the company handbook
- the vessel guidebook
- pocket cards.

See Chapter 2 for further information.

There should be clear and concise policies, procedures and safety rules within the safety management system and associated documentation. Review these regularly to ensure that they are appropriate, remain valid and can be communicated to the crew in various ways including:

- during the company and vessel inductions
- as part of the on-board and external training programmes
- through on-board supervision and monitoring
- in safety committee meetings.

Seafarers need to know what will happen if they do not follow rules. This can be achieved through a 'just culture' policy (see section 1.2.7), and ensuring that all seafarers are aware of the Code of Conduct for the Merchant Navy.

Improvement plans with clear, achievable targets and goals are useful in managing continuous and sustainable improvement. It is important to communicate these plans well and to involve all seafarers in both their development and implementation. Improvement plans can be standalone or incorporated into other planning tools. Use them to set priorities and measure progress.

1.2.3 Good communications

Effective communication and workforce involvement are crucial in ensuring a safe living and working environment. Communication is a two-way process: to gain information, and to act upon it. Systems need to be in place to facilitate this at all levels in the organisation. Some examples include:

- ensuring everyone understands their roles and responsibilities
- ensuring everyone understands, acknowledges and acts upon orders and instructions
- passing safety-critical information between watchkeepers and changing crews
- ensuring there is appropriate communication between workers on the ship and ashore
- ensuring information posters, signs and instructions are clear and understandable
- ensuring safety alerts, memos and newsletters are clear and understandable
- encouraging feedback, improvement suggestions and safety observations, and acting on that information
- taking minutes of safety meetings, distributing the reports and acting upon them where appropriate
- ensuring a good, clear and reliable system of emergency response communications is in place.

Formal arrangements for consultation and communication (through the safety committee) are described in Chapter 13. However, communication should go beyond workers with a formal role under those arrangements. Individual seafarers should report any issues for the purpose of learning, and companies should encourage this behaviour.

There should be a clear and simple system for reporting problems and suggesting solutions. This would typically use an improvement suggestion system and a proactive reporting system for unsafe acts and conditions. These are most effective when developed in consultation with the workforce.

Use clear, unambiguous language at all times. Avoid jargon and acronyms unless everyone understands what they mean. While it may be reasonable to believe that all seafarers understand common nautical terminology, it is not reasonable to expect them to understand terms found in local slang or dialects. Use the designated working language of the vessel. On ships with multicultural crews, take care to avoid misunderstanding as a result of different body language or cultural norms.

Actively encourage face-to-face communications and use techniques to confirm understanding. This can be particularly effective during visits by senior and line management, and can indicate how the company's values and safety procedures are being implemented.

Discuss change and actively seek input from everyone involved. Clear information regarding the reasons and need for the change should be provided and discussed. Give prompt feedback on any issues raised, both positive and negative. This will ensure that all concerned are part of the process and help them to be fully engaged and committed to any necessary changes.

There should be an open-door policy that encourages and enables people to discuss any concerns and issues they may have. Consider any issues they may raise and give feedback.

Company magazines, newsletters and regular sharing of learning bulletins are all good additions to safety alerts and other official communications in getting the safety message across in an accessible and understandable manner. You should give credit to any contributing seafarers.

1.2.4 Clear leadership

Leadership has a significant impact on the safety of maritime operations. The effectiveness of the International Safety Management (ISM) Code depends heavily on how leaders approach its implementation, and this in turn depends on the skills and qualities of leaders – at sea, at the ship–shore interface, and on shore.

Despite your best endeavours to work safely, sometimes real life makes things difficult. Time pressures, economic constraints and everyday circumstances sometimes seem to conspire against good safety leadership. What counts is how leaders behave in everyday situations. Seafarers will draw inferences about safety leadership based on what they see their leaders do and hear them say, far more than what they hear in formal spoken or written communication.

There are many models of leadership, and some companies run their own leadership programmes. However, there are ten core qualities that provide the foundation for safety leadership in the maritime industry.

Leading for Safety: A Practical Guide for Leaders in the Maritime Industry (MCA)

The ten core safety leadership qualities

1. Instil respect and command authority

The ability to instil respect in, and command authority over, seafarers is probably the first thing that comes to mind when people think of leadership. In many ways, it happens on its own when everything else is right. Leaders get respect and command authority when crews believe that they:

- are willing to exercise the power vested in their position
- have the necessary knowledge and competence
- understand their situation and care about their welfare
- can communicate clearly
- are prepared to act confidently and decisively
- listen.

2. Lead the team by example

Leading the team by example is a combination of two things: being seen to comply with the safety procedures, and working as a key part of the team. This includes being willing, where necessary, to get involved in subordinates' tasks.

3. Draw on knowledge and experience

Adequate knowledge and experience are prerequisites for effective leadership. In the context of safety leadership, this means in particular:

- good knowledge of safety-related regulations, codes and standards
- experience and skills, not only in technical and operational issues but also in people management.

4. Remain calm in a crisis

People need strong, clear leadership in a crisis and rely more on their leaders than they would otherwise. Calmness in a crisis is a core requirement and people will rely on many of the other leadership qualities described, including commanding authority and drawing on knowledge and experience. It is particularly important to have confidence and trust in the crew's abilities and emergency preparedness. Attendance at safety training and at response drilling is essential for all seafarers.

5. Practise 'tough empathy'

Empathy means identifying with, and understanding, another's situation, feelings and motives. It requires the capacity to put oneself in another's place, and the cultivation of good listening skills. Good leaders empathise realistically with others and care intensely about the work they do – but this does not mean that they always agree with them or join in with concerns and grumbles. Instead they practise 'tough empathy', which means giving people what they need, rather than necessarily what they want. Another way of looking at this is 'care with detachment'; for example, providing staff with comfortable, safe footwear, rather than spending more money on a more fashionable style.

6. Be sensitive to different cultures

Crews of mixed nationalities are the norm. Good leaders are sensitive to differences in the social and behavioural norms of national cultures, yet at the same time value all seafarers equally irrespective of their nationality. They know how to interpret different behavioural signals, and how best to react to exert the strongest influence.

7. Recognise seafarers' limitations

Good leaders have a clear understanding of how operational and other demands can be realistically met by seafarers, and can judge whether fatigue levels are such that they need to take action.

8. Motivate a sense of community

Research has shown that people in work are typically motivated by satisfaction or pride in completing a good job, and feeling like part of a team – not just by money. Leaders have an important role to play in creating the conditions to encourage and maintain these 'healthy' motivators. Demonstrating respect for staff is often an essential part of this. Meeting someone's basic needs is often the key to keeping their motivation high.

9. Place the safety of crew and passengers above everything

It is universally accepted that commitment from the leader is essential for good safety. Leaders need to demonstrate this commitment clearly to their staff through their actions, rather than just through formal declarations or policy statements. In practice, this means showing that the safety of the crew and passengers is placed above everything else – 'nothing we do is worth getting hurt for'.

10. Communicate clearly

The ability to communicate clearly is important at all levels in an organisation. For a master, the key issue is most often how to encourage better two-way rather than one-way communication, balancing authority and approachability. Being open to criticism is a part of this.

1.2.5 Good planning

Good planning is essential in ensuring occupational health and safety at work. You can only control risks adequately by ensuring that everyone involved is aware, activities are coordinated and good communication is maintained.

You should carefully consider what you want to achieve, what actions are necessary, how these will be carried out and what effect they may have on seafarers' health and safety at work, taking into account that there may be indirect and unintended consequences. Consideration should include:

- what might cause harm to people and whether you are doing enough to prevent it
- how to prioritise improvements
- who will be responsible for occupational health and safety tasks; what they should do; when; and with what results
- how to measure achievements against objectives and review them.

Workers should participate in the planning process, so consider those who may be affected. Provide clear instructions for the required activities, and adequate time and resources. Confirm that all workers fully understand the instructions (this is known as 'closed-loop communication'). Use **permit to work** systems where appropriate (see Chapter 14), capture learning and apply it to future work.

Management of change

Most effective change management on board is adequately controlled through pre-existing processes such as handover procedures, safe systems of work and sound navigational practices. However, some changes introduce new factors that existing controls may not cover. These could include unexpected changes to personnel, fatigue, adverse weather, a change to the operation while it is already underway, or more complex changes, such as fitting new equipment or a change in operations.

Changes can become necessary for a variety of reasons. It is important to manage them effectively to ensure that:

- they are necessary
- they are realistic and achievable
- they are planned and systematically managed
- any impact on operations, both negative and positive, is understood and managed
- they are effectively communicated
- they are effectively implemented
- workers affected are consulted.

The appropriate level of change management required will vary according to circumstances. Some companies have formal procedures in place that define the level of change management necessary. Annex 1.1 gives an example of such a procedure.

1.2.6 Risk awareness and risk assessment

Risk awareness

If seafarers are fully informed and aware of the risks to their health, safety and welfare, they are much more likely to ensure they avoid the risks and remain safe. We get this knowledge through risk assessment and in other ways throughout our lives, including training in theory and practical application, information, observation, instructions, supervision and personal experience. We can improve the quality and usefulness of the information available through effective knowledge management (see section 1.2.9).

Key terms

A hazard is a source of potential injury, harm or damage. It may come from many sources, such as situations, the environment or a human element.

Risk has two elements:

- the likelihood that harm or damage may occur
- the potential severity of the harm or damage.

A toolbox talk before the work begins is key in ensuring that all workers involved in the work understand and are aware of any hazards and their associated risks.

Our values, beliefs, attitudes and behaviours, and views of others influence how we apply the knowledge in the workplace. A safe working culture will help facilitate this (see section 1.2.8).

Risk assessment

The risk assessment process identifies the hazards in a work undertaking, analyses the level of risk, considers workers in danger and evaluates whether hazards are adequately controlled, taking into account any measures already in place.

There are many ways to do a risk assessment. It will depend on the type of ship, the nature of the operation, and the type and extent of the hazards and risks. The process should be simple but meaningful.

Any risk assessment must address risks to the occupational health and safety of seafarers. Advice on assessment in relation to using personal protective equipment, **manual handling** operations and using work equipment is given in Chapters 8, 10 and 18. In addition, specific areas of work involving significant risk, and recommended measures to address that risk, are covered in more detail in later chapters of this Code.

Although the risk assessment must be 'suitable and sufficient' the process should be simple. The amount of effort that is put into an assessment should depend on the level of risks identified and whether those risks are already controlled by satisfactory precautions or procedures to ensure that they are as low as reasonably practicable. The assessment is not expected to cover risks that are not reasonably foreseeable.

Refer to the relevant legislation regarding risk assessments when deciding on what methodology to use.

Seafarers must be informed of any significant findings of the assessment and measures for their protection, and of any subsequent revisions made. It is recommended that each vessel carries copies on board and that there is a process for carrying out regular revisions. In particular, review the risk assessment and update it as necessary, to ensure that it reflects any significant changes of equipment or procedure or the circumstances at the time; for example, the weather or level of expertise of workers carrying out the task.

Take into account potential language difficulties. Where relevant consider temporary staff, or workers who are new to the ship or the company and are not fully familiar with the safety management system or other operational details. Seafarers who need special consideration include young persons and pregnant women.

MSN 1838 (M) Amendment 1; MSN 1890 (M+F) Amendment 2; Regs 7(1) and (6)

Risk assessment is a continuous process. In practice, the risks in the workplace should be assessed before work begins on any task for which no valid risk assessment exists.

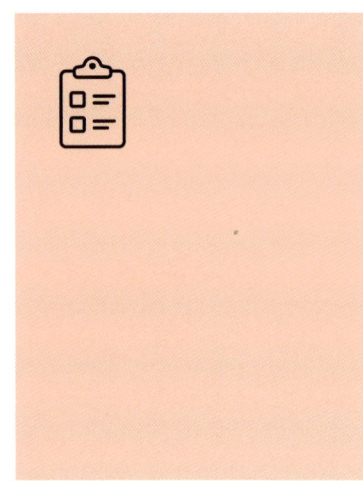

Effective risk assessments should:

- correctly and accurately identify all hazards
- identify who may be harmed and how
- determine the likelihood of harm arising
- quantify the severity of the harm
- identify and disregard inconsequential risks
- record the significant findings
- provide the basis for implementing or improving control measures
- provide a basis for regular review and updating.

Annex 1.2 provides a simple guide for small businesses.

The company may benefit from adopting a four-level process to risk assessment, as outlined below.

Risk assessment level 1: generic

The ISM Code states that the safety management objectives of the company should assess the risks associated with all identified hazards, in respect to its ships, personnel and the environment, and establish appropriate safeguards.

These risk assessments should therefore be carried out at a high level in the company by appropriately knowledgeable and experienced personnel. The results should be used to ensure that appropriate safeguards and control measures are contained within the company's safety management system in the form of policies, procedures and work instructions.

Risk assessment level 2: task based

In addition to the general requirements under the ISM Code, The Merchant Shipping and Fishing Vessels (Health and Safety at Work) Regulations 1997 require that a suitable and sufficient assessment shall be made of the risks to the occupational health and safety of seafarers arising in the normal course of their activities or duties.

Any generic risk assessments are used in context, and are not suitable for specific tasks. For this, people involved in the work carry out **task-based risk assessments** (TBRAs) on board each vessel.

You can develop a range of vessel-specific generic TBRAs for all routine and low-risk tasks. These should be periodically reviewed, but frequency would depend on the particular circumstances on the vessel and the level of risk.

You would use a different type of TBRA for specific high-risk jobs that are not routine, such as working aloft or enclosed space entry. These should relate to the specific people who will be doing the work and are valid only for the duration of that job.

In both cases, the assessments should be carried out by a **competent person** 🔍 or persons who understand the work being assessed. Seafarers who will be involved in the work should also be involved in the assessment process.

Risk assessment level 3: toolbox talk

A toolbox talk is another form of risk assessment carried out in support of a TBRA. Its prime purpose is to talk through the procedures of the job in hand and the findings of the TBRA with the seafarers involved.

When carrying out a toolbox talk, it is important to actively involve workers doing the work and others who may be at risk; in other words, seafarers, sub-contractors and others on board ship who may be affected. Encourage full and active participation, and discuss any questions or concerns. Once finished, confirm that everyone fully understands their role in the task and the precautions in place (this is known as 'closed-loop communication'). Record this along with details of any relevant risk assessment to which you have referred.

Give a toolbox talk before any work is carried out that involves more than one person and where there is significant risk to persons or assets.

Risk assessment stage 4: personal assessment of risk

This is an informal assessment of day-to-day risks carried out as you are working and in life generally. It is used to ensure that we perform even the most mundane of tasks without getting hurt. It is also used to stay aware of our environment and to help to identify and control immediate hazards as we work. Personal assessment of risk should be developed and encouraged.

This is about taking a few minutes to step back, look at the job to be done, consider what could go wrong and how it may occur, and what steps you can personally take to avoid any incident occurring. As the work is proceeding, you should also monitor the worksite for any change in conditions that might alter the hazards and controls in place. If there is any concern, stop the work, reassess the controls and, if necessary, replan and reassess the task.

This approach may also be called a **dynamic risk assessment**. 🔍 If the person does not believe that it is sufficient move back to stage 2.

Every task carried out on board the vessel should be risk-assessed. This does not mean writing a risk assessment every time you do a simple task, but the existing risk assessment must be referred to as part of a toolbox talk (stage 3) before beginning the task to ensure that the hazards and controls are fully understood, still relevant and appropriate.

Once the task commences, monitor the work site for any changes in conditions that might alter the hazards and controls in place. If there is any concern, use stop work authority.

Once the task is completed it is important to record or feed back any lessons learned and make improvements for next time including, where appropriate, reviewing and updating existing risk assessments. Everyone should be encouraged to contribute.

It is recommended that a proactive hazard-reporting system with empowerment and expectation for immediate corrective action is also in place and that information on hazards and risks is shared as widely as possible.

1.2.7 Accountability

Maintaining a safe living and working environment on a vessel is a shared responsibility of everyone on board and ashore. All personnel have a role to play and they can adversely affect others on board by their acts and/or omissions. For these reasons, it is important that:

- there are well-defined rules and guidelines, which are clearly understood
- responsibilities are clearly defined for everyone on board and ashore
- consequences of unacceptable (safety) behaviour are made clear
- there is a fair, transparent and consistent response to unacceptable safety behaviour; this is commonly referred to as a **just culture**. 🔍

Points 1 and 2 above have been covered under 'Clearly defined expectations' (section 1.2.2) and 'Good communications' (section 1.2.3).

Just culture

A just culture policy is an important part of a positive health and safety culture. It clearly sets out the expectations for adherence to procedures in the workplace and provides a context for enforcing them. It recognises behaviours that exceed company expectations as well as those that fall below expectation, but are not always the fault of the seafarer.

A just culture places responsibilities on management to provide support, training and resources such that seafarers will be competent to undertake their tasks to the required standard.

The just culture policy provides a process (with appropriate support) for managing behaviours that fall below expectations in a transparent and fair manner. A just culture seeks to improve the organisational culture and the performance of the organisation by modifying behaviour, encouraging seafarers to take greater personal responsibility for their actions, and rewarding behaviour exceeding expectations. It also recognises that firm action may be needed in circumstances where, despite management having carried out their responsibilities, inappropriate behaviours are still evident.

The just culture decision tree (see Figure 1.1) is a guide for ensuring consistent management for workers who exceed or deviate from company standards. The model presents a simple, yet robust, means of dealing with both exemplary and inappropriate behaviours, linked with a structure for an appropriate management response. It also recognises that there are overlaps between the areas of any given established disciplinary response. However, before managers or supervisors apply the decision tree it is important that they fully understand the causal factors and root causes of an event. Where incorrect causes have been identified and applied to the model, there is a danger of taking inappropriate action.

The decision tree operates on an increasing personal accountability baseline:

- On the proactive side, the baseline covers a range from expected to exemplary behaviour.
- On the reactive side, the baseline covers a range from initiating actions that were malevolent or reckless (at the most extreme end) through to a no-blame error.

The decision tree is linked to a company action model:

- On the proactive side, company actions range from those for management to encourage behaviour through to rewarding seafarers for their exemplary work.
- On the reactive side, company actions range from dismissal (at the most extreme response end) to coaching/mentoring (at the least extreme response end).

This recognises that both seafarer and company have responsibilities for achieving improvements in behaviour and increasing the company's safety culture.

Substitution test

The substitution test asks a reasonable person: 'Given the circumstances at the time of the event, could you be sure that you would not have committed the same, or similar, breach of procedures, standards, unsafe act, etc.?' Several people should do this independently and everyone involved should review it to gain agreement and consensus.

Management of supervisory interventions

Management of supervisory interventions following breaches of procedures/codes of practice/standards or any formalised company/vessel rules can be an effective and powerful way of modifying individual behaviour.

However, it is essential that the type of management response is appropriate. Figure 1.1 provides a framework to guide management in identifying an appropriate and common response.

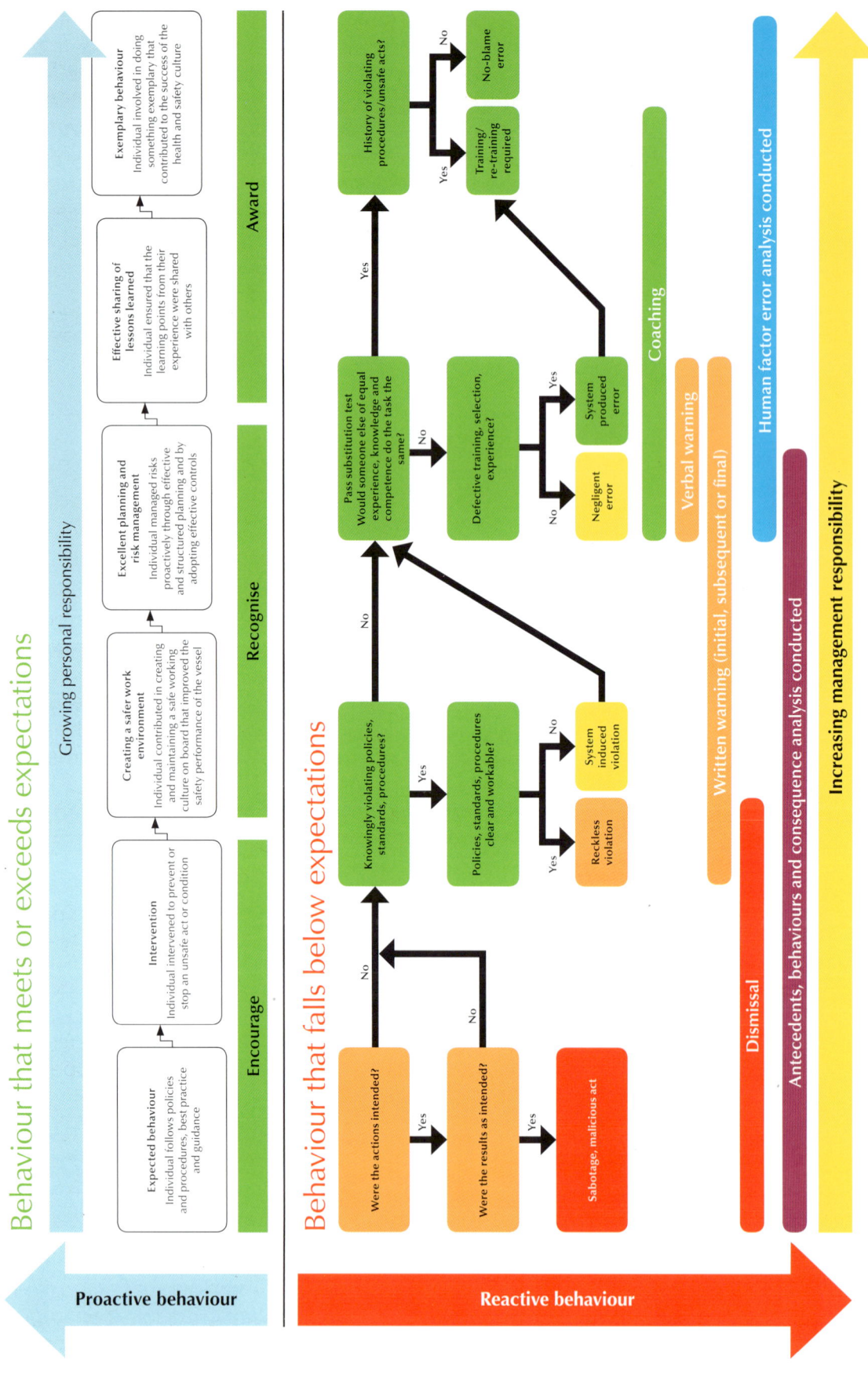

Figure 1.1 Just culture decision tree

1.2.8 Good safety culture

A good safety culture is one where safety is an integral part of everything that is planned, discussed, done and documented. Everyone in the company should think about safety and new ways of improving it as a matter of course. They should be on the alert for any unsafe acts or unsafe conditions; look out for each other; intervene to prevent accidents and incidents; actively share good ideas; and always seek to improve.

To achieve a good safety culture, there are certain key components to encourage. This begins with ensuring that all seafarers fully understand their roles and responsibilities; not just what they have to do, but also why it is important. They need to be informed and share their knowledge to help inform others.

All personnel, at every level of the company, need to be fully engaged and committed to nurturing and developing the safety culture. Establish compliance with safety rules as a core company requirement – good safety behaviours should be the norm.

Another key aspect of developing a good safety culture is continual improvement: the company should be a learning organisation. This should be a personal commitment and responsibility of everyone in the company. Systems and infrastructure need to be in place to facilitate this process. A proactive reporting system for unsafe acts and conditions, and improvement suggestions, should be in place. Investigate all accidents and incidents and disseminate the findings widely. See section 1.2.9 on effective knowledge management.

There needs to be an open and just culture that recognises that it is normal for human beings to make mistakes. It also needs to recognise that there are wider organisational factors that affect our behaviours and can create barriers to safe behaviours. It is vital that all workers are empowered and feel comfortable in reporting unsafe acts, unsafe conditions, accidents and incidents without fear of unjust reprisals.

Often most of the component parts are already in place in some form or another. However, for any culture to be truly safe, all the elements discussed in this chapter should be fully developed.

> *Further information on promoting good safety culture is available in the National Maritime Occupational Health and Safety Committee's Guidelines to Shipping Companies on Behavioural Safety Systems.*

1.2.9 Effective knowledge management

From an occupational health and safety perspective, efficient management of knowledge can significantly improve learning and understanding and prevent accidents and incidents from happening again. This is particularly useful in the maritime industry where similar high-risk activities are being carried out on numerous autonomous units, such as a fleet of ships.

It has been said: 'A person learns from their mistakes, but a wise person learns from the mistakes of others.' By effectively collecting relevant information, organising it so it can be understood and distributing it to workers who can use it, we can share experiences and increase our knowledge. By applying this knowledge to our own working environment we can reduce the likelihood of the same type of accident or incident reoccurring on our vessel.

Knowledge management is about:

- **Getting the right information** Understand what information and knowledge has value, can improve safety, operations or services, or is necessary for fast and effective decision making.
- **Making it easy to understand** Convert the information into a format that can be easily understood and acted upon at all levels in the company.
- **Getting it to the people who need it, when they need it** Create the necessary technical and cultural 'delivery systems' and organise information and knowledge so it is useful and available.
- **Encouraging people to use it** Develop an organisational structure and culture that encourages seafarers to take what they know, apply it effectively for both continuous improvement and innovation, and share it with others.

Knowledge management does not have to be complicated or difficult. Most companies will have many of the elements in place already; it is often just a case of ensuring that they are all working together.

Getting the right information

We gather information from the data we retrieve, both internally and externally, as Figure 1.2 shows. The sources include accident and incident investigations, Marine Accident Investigation Branch reports, safety alerts, audits and inspections, maintenance records, trip reports, safety meeting reports, masters' reviews, vessel visits, safety observations and improvement suggestions. You will probably need to do some form of data analysis. You can do this in several ways, including spreadsheets to create statistics. It is important to involve all personnel at all levels in gathering this information.

Making it easy to understand

Different levels of the organisation may need different approaches. For example, statistics on a spreadsheet may be appropriate for senior management but safety alerts, amendments to procedures, bulletins and learning points memos may be more effective in introducing any lessons to the crew from the accidents and incidents depicted in the statistics. It is important to convert the data into useful information that makes sense to the end user. Ask for feedback from the end user on how useful the information is.

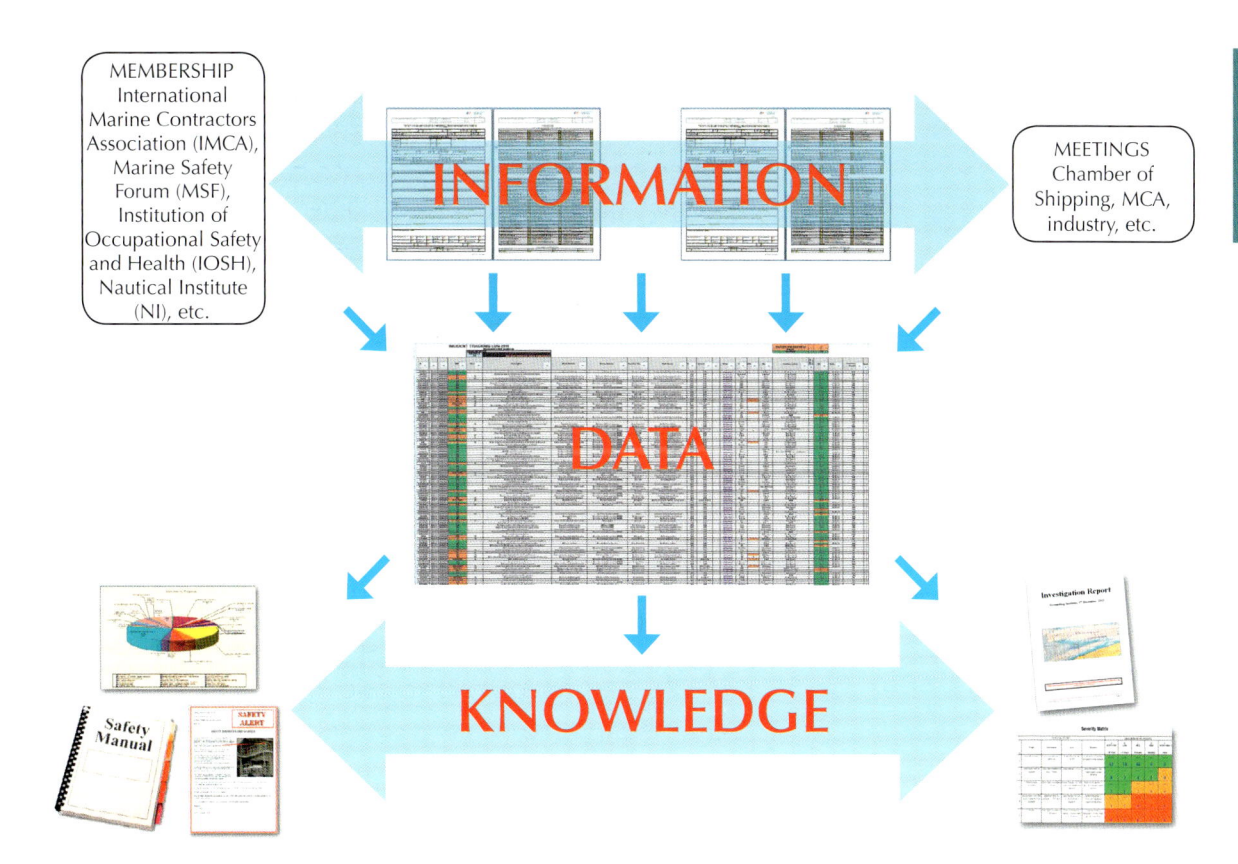

Figure 1.2   Effective knowledge management

Getting it to workers who need it, when they need it

Present this information so the end user can understand it and it is clear, useful and available. There are many ways to do this, such as posters, memos, video, computer-based training, amendments to the safety management system, and safety alerts. The choice of the best medium for the information will vary for each company. A company newsletter can be very effective and easy to understand.

Encouraging them to use it

Shared knowledge will not be useful unless workers receiving it are empowered and feel comfortable using it. It is essential to have an open and honest safety culture that encourages all seafarers to share the same high values and beliefs in healthy and safe working. Everyone should be encouraged to use the knowledge and to gather useful information to share.

In real terms, therefore, the basis of good knowledge management lies in having effective systems to gather, process, distribute, learn and review throughout the company and industry. This will improve understanding of what can cause harm and lead to accidents and incidents, and will encourage everyone to be fully engaged.

Learning from incidents

📖 *MGN 484(M) Amendment 3*

The ISM Code requires that a safety management system includes procedures for reporting, investigating and analysing every non-conformity, accident and hazardous situation, to improve safety and pollution prevention. This should then lead to the implementation of corrective actions.

The safety officer will often lead on this work, and guidance is provided in Chapter 13. However, on ships with no safety officer, the company must make other arrangements to carry out this function. Record any accident or incident so that it can be investigated to find out what went wrong and to see what can be done to prevent it happening again.

Every seafarer has a responsibility to:

- report deficiencies, conditions that are causing concern and things that could be improved so that workers with specific safety responsibilities can put things right
- share their views on how to make things safer.

Lessons can also be learned from accidents and incidents on other ships and even in other sectors. Some industry organisations publish accident statistics and safety information, and these may help to identify likely risks and suitable safety measures. Information is available in Marine Guidance Note MGN 484(M) Amendment 3.

Tips for managing occupational health and safety

- Remember that safety culture is integral to how seafarers view health and safety on board. It means that all are empowered and feel comfortable in reporting unsafe acts, unsafe conditions, accidents and incidents without fear of unjust reprisals.
- Use the just culture decision tree (see Figure 1.1) to ensure consistent handling of deviations from acceptable standards of behaviour.
- Do a risk assessment for every activity on board. This does not necessarily have to be formal or written down.

- Occupational health and safety is everyone's responsibility.
- In the most effective knowledge management systems, all accidents, near misses, unsafe acts, unsafe conditions and non-conformities are investigated for the purpose of learning and improving.

Annex 1.1   Management of change

Sample procedure covering simple and complex change

Simple change

To make a simple change the procedure is as follows.

This procedure is for task-based changes only. For routine tasks it should be used only if the ship's master or superintendent operations personnel consider it necessary, or if it involves a complex change.

Stop the job if a simple, task-based change develops or is recognised while a task is under way (eg during adverse weather, reduced time to carry out the task, unexpected changes to personnel, operational changes). To manage the simple change, redo the toolbox talk for the task, make any required alterations to working practices or control measures, and keep records.

This will reduce the risk to as low a level as is reasonably practicable. If this is not achievable, stop the job until the level of risk is acceptable, then the task may proceed.

If the task is being carried out under a permit to work, notify the permit controller as soon as possible; they should suspend the permit until the change has been assessed.

Once the change has been assessed and recorded, the permit (where applicable) may be reinstated and the task can be restarted.

To make a change it is important to revisit the risk assessment and identify any changes needed through the management of change process. You should add these to the risk assessment for future use and consideration.

Complex change

To make a complex change take the following steps.

1. You can identify a change through:

   - an improvement suggestion form
   - the safety officer's recommendation
   - the master's/chief engineer's management review
   - the manager's review
   - internal or external audit findings.

2. Clearly identify the owner of the change. The company should allocate a responsible manager or superintendent (responsible person) to investigate how applicable, suitable and practical the change is. Bring any concerns related to the change to the attention of the responsible person. Allocate the ownership of the change to the most appropriate available manager/superintendent for the task at the discretion of the company.

3. The responsible person should begin documenting the change using the management of change form.

4. The responsible person should then evaluate the cost versus the benefits, assessing the impact of the change on seafarers, processes, materials and plans. Report the results of this process to the appropriate company for a decision.

5. If the change is approved, complete a plan and risk assessment. The responsible person, in consultation with everyone involved, should assess whether there are any factors that would render the task unsafe. If so, put risk reduction measures in place to lower any risk as far as reasonably practicable. If this is not achievable stop the job until the level of risk is acceptable and the job can be resumed, or permanently stopped.

6. Where the change is a temporary or interim measure:

    - Make seafarers who have recently joined the vessel aware of the change(s) during their induction and continual training/assessment.
    - Any locally produced procedures should be printed, signed by the master and/or chief engineer, laminated and prominently displayed. If the change involves machinery use, display the procedure prominently at or near the machinery, and update any corresponding risk assessments.

7. If any procedures in the safety management system need amending, notify the designated person ashore/responsible person immediately so they can make the change.

8. All people associated with the change shall verify that they are aware and have understood it. Record this on the management of change form; the responsible person should ensure that it is completed.

9. The management of change form should then be sent to the responsible person or, in their absence, a member of their department, to verify all steps have been followed and sufficiently recorded before starting the task.

10. If the change is long term, the responsible person or, in their absence, a member of their department, should, where necessary, ensure that the change process is concluded once the task is completed, or reviewed at a defined period within the change plan. This process or review should also include a review of the associated risk assessment, including making any amendments identified and communicating these changes to workers concerned.

11. Once the change has been carried out, and all temporary modifications have been removed, the responsible person should sign the management of change form to conclude the process.

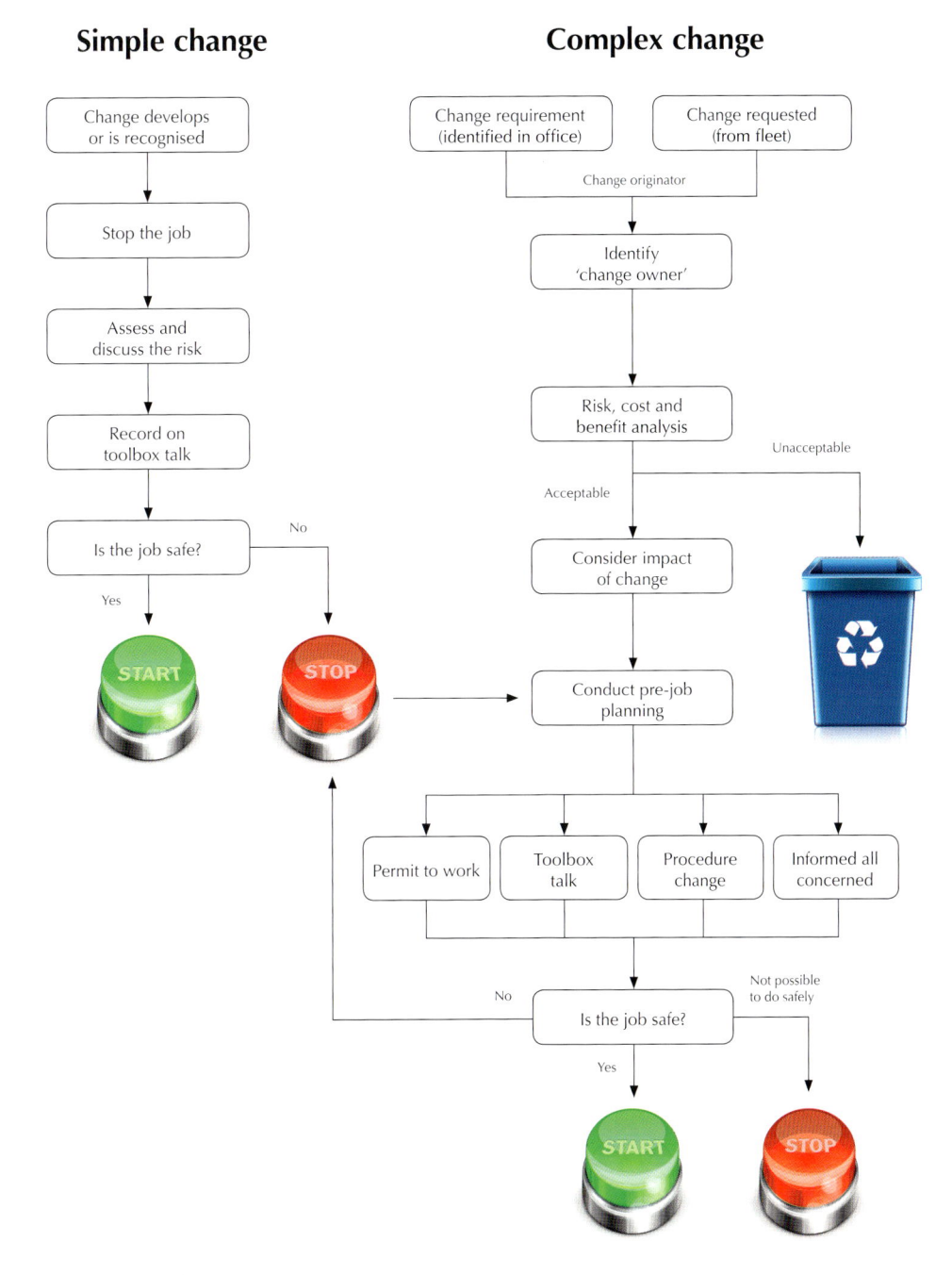

Figure 1.3   Procedures for simple and complex changes

Annex 1.2 Five steps to risk assessment

The information here is based on health and safety guidance published at http://www.hse.gov.uk. Alternative text is available in MGN 636(M) Amendment 2 and MGN 587(F) Amendment 1.

📖 *MGN 636(M) Amendment 2; MGN 587(F) Amendment 1*

Step 1: Identify the hazards

First you need to work out how people could be harmed. When you work in a place every day, it is easy to overlook hazards, so here are some tips to help you identify those that matter:

- Walk around your workplace and look at what could reasonably be expected to cause harm.
- Ask your employees or their representatives what they think. They may have noticed things that are not immediately obvious to you.
- Consider published information on accidents and near misses on ships, which will highlight common hazards and high-risk activities.

📖 *MGN 484(M) Amendment 3*

- If you are a member of a trade association or protection and indemnity insurance (P&I) club, contact them. Many produce very helpful guidance.
- Check manufacturers' instructions or data sheets for chemicals and equipment because they can be very helpful in spelling out the hazards and putting them in their true perspective.
- Look back at your accident and ill-health records – these often help to identify less obvious hazards.
- Remember to think about long-term hazards to health (eg high levels of noise or exposure to harmful substances) as well as safety hazards.
- Consider people who may be particularly vulnerable (eg young persons or pregnant seafarers).

Step 2: Decide who might be harmed and how

For each hazard, you need to be clear about who might be harmed, because this will help you to identify the best way of managing the risk. This means identifying groups of people (eg 'people working in the storeroom' or 'passers-by') rather than listing everyone by name.

Remember:

- Some seafarers need particular consideration. New and young seafarers, those for whom the working language of the ship is not their first language, or those new to the ship who may not be familiar with company or ship safety procedures may be at particular risk. Some hazards need extra thought.
- Stevedores, contractors and surveyors may not be in the workplace all the time.
- Members of the public could be hurt by your activities.
- If you share your workplace with others, think about how your work affects them as well as vice versa; talk to them.
- Ask your crew if they can think of anyone you may have missed.
- In each case, identify how they might be harmed; in other words, what type of injury or ill health might occur. For example, crew on roll-on/roll-off ferry car decks may be at risk from excess fumes.

Step 3: Evaluate the risks and decide on precautions

Once you have spotted the hazards, decide what to do about them. The law requires you to do everything 'reasonably practicable' to protect people from harm. You can work this out for yourself, but the easiest way is to compare what you are doing with good practice.

First, look at what you're already doing; think about what controls you have in place and how the work is organised. Then compare this with the good practice and see if there is more you should be doing to bring yourself up to standard. Ask yourself the following:

- Can I get rid of the hazard altogether?
- If not, how can I control the risks so that harm is unlikely?

When controlling risks, apply the principles below, if possible in the following order:

- Try a less risky option (eg switch to using a less hazardous chemical).
- Prevent access to the hazard (eg by guarding).
- Organise work to reduce exposure to the hazard (eg put barriers between pedestrians and traffic).
- Issue personal protective equipment (eg clothing, footwear, goggles).
- Provide welfare facilities (eg first-aid and washing facilities for removal of contamination).

Improving occupational safety and health need not cost a lot. For instance, placing a mirror on a dangerous blind corner to help prevent vehicle accidents is a low-cost precaution considering the risks. Failure to take simple precautions can cost you a lot more if an accident does happen.

Involve staff to make sure that what you propose to do will work in practice and will not introduce any new hazards.

Step 4: Record your findings and implement them

Putting the results of your risk assessment into practice will make a difference when looking after people and your operation.

Writing down the results of your risk assessment, and sharing them with your staff, helps you to do this. Annexes 1.3 and 1.4 show two examples of risk assessment forms. When writing down your results, keep it simple; for example, 'Tripping over rubbish: bins provided, staff instructed, weekly housekeeping checks', or 'Fume from welding: local exhaust ventilation used and regularly checked'.

A risk assessment does not have to perfect, but it must be suitable and sufficient. You need to be able to show that:

- a proper check was made
- you asked who might be affected
- you dealt with all the obvious significant hazards, taking into account the number of people who could be involved
- the precautions are reasonable, and the remaining risk is low
- you involved your staff or their representatives in the process.

If, like many businesses, you find that you could make many improvements, big and small, don't try to do everything at once. Make a plan of action to deal with the most important things first. Occupational safety and health inspectors acknowledge the efforts of businesses that are clearly trying to improve.

A good plan of action often includes a mixture of things such as:

- cheap or easy improvements that can be done quickly, perhaps as a temporary solution until more reliable controls are in place
- long-term solutions to risks that are most likely to cause accidents or ill health
- long-term solutions to risks with the worst potential consequences
- arrangements for training employees on the main risks that remain and how to control them
- regular checks to make sure that the control measures stay in place
- clear responsibilities – who will lead on what action and by when.

Remember: prioritise and tackle the most important things first. As you complete each action, tick it off your plan.

Step 5: Review your risk assessment and update if necessary

Few workplaces stay the same. Sooner or later, you will bring in new equipment, substances and procedures that could lead to new hazards. It makes sense, therefore, to review what you are doing on an ongoing basis.

Look at your risk assessment and think about whether there have been any changes. Are there any improvements you still need to make? Have your seafarers spotted a problem? Have you learned anything from accidents or near misses? Keep your risk assessment up to date.

When you are running a business it is easy to forget about reviewing your risk assessment – until something has gone wrong and it is too late.

If there is a significant change check your risk assessment straight away and, where necessary, amend it. If possible, think about the risk assessment when you are planning your change – that way you leave yourself more flexibility.

Annex 1.3 Risk assessment form: Example 1

Name of ship				Record no.	
Work area being assessed					
Task ID no.	Work process/ action undertaken in area	Hazards associated with activity	Controls already in place	Significant risks identified	Further assessment required? (Y/N)

Declaration
Where no significant risk has been listed, we as assessors have judged that the only risks identified were of an inconsequential nature and therefore do not require a more detailed assessment.

Signed _____

Print name _____ Date (DD/MM/YY) _____

Annex 1.4 Risk assessment form: Example 2

Ship name _____

Record number _____

Current assessment date (DD/MM/YY) _____

Last assessment date (DD/MM/YY) _____

Work activity being assessed _____
_____

Section 1: Analysis of the intended work activity

Hazard no.	Description of identified hazards	Existing control measures to protect personnel from harm
1		(a) (b) (c)
2		(a) (b) (c)
3		(a) (b) (c)
4		(a) (b) (c)
5		(a) (b) (c)
6		(a) (b) (c)
7		(a) (b) (c)
8		(a) (b) (c)
9		(a) (b) (c)
10		(a) (b) (c)

Section 2: Assessment of risk factor

Likelihood of harm	Severity of harm			Hazard no.	Likelihood of harm	Severity of harm	Risk factor
	Slight harm	Moderate harm	Extreme harm	1			
Very unlikely	Very low risk	Very low risk	High risk	2			
				3			
				4			
Unlikely	Very low risk	Medium risk	Very high risk	5			
				6			
Likely	Low risk	High risk	Very high risk	7			
				8			
Very likely	Low risk	Very high risk	Very high risk	9			
				10			

To assess the risk factor arising from the hazard:
1. From the left-hand column select the expression for likelihood which most applies to the hazard.
2. From the next three columns select the expression for severity of harm which most applies to the hazard.
3. Cross-reference using the shaded area to determine the level of risk.
4. If the risk factor is medium (yellow), high (orange) or very high (red), additional control measures should be implemented and recorded in section 3.

Section 3: Additional control measures to reduce the risk of harm

Hazard no.	Further risk control measures	Remedial action date	Review date (MM/DD/YY)
1			
2			
3			
4			
5			
6			
7			
8			
9			
10			

Additional comments _____

_____

_____

Assessment review date (DD/MM/YY) _____

2 Safety induction for personnel working on ships

2.1 Introduction

2.1.1 Companies should design and implement a standard induction programme for each vessel, covering the Standards of Training, Certification and Watchkeeping (STCW) and Maritime Labour Convention (MLC) requirements, and incorporating any expanded detail specific to that vessel's particular needs. This chapter gives guidance on the subjects to cover.

> *International Convention on Standards of Training, Certification and Watchkeeping for Seafarers, 1978, as amended (STCW); Maritime Labour Convention, 2006 (MLC 2006)*

2.1.2 Once the new personnel have completed the standard safety induction, they should have the appropriate security training and departmental induction covering safe working practices, areas of responsibility, departmental standing orders, and training/certification requirements to operate specific machinery or undertake specific tasks.

> *See the Merchant Shipping and Fishing Vessels (Health and Safety at Work) Regulations 1997 and MGN 636 Amendment 2.*
>
> *SI 1997/2962 Reg 5 and Reg 12; MGN 636 Amendment 2*

Key points
- Arrange the safety induction to start as soon as anyone employed or engaged in any capacity joins on board.
- Run familiarisation training to make sure everyone on board knows the vessel's layout and important areas.

Your organisation should
- Inform personnel of the company's and personal duties regarding health and safety. Give enough information and instruction for seafarers to be able to do their job safely.
- Ensure all personnel know the action to take in cases of accidents or a medical emergency on board.
- Carry out health and safety risk assessments of activities on board.

2.2 Vessel familiarisation training

2.2.1 Before they are assigned to shipboard duties, all personnel employed or engaged on a ship, other than passengers, must have familiarisation training on board and be given enough information and instruction to be able to:

- communicate with other people on board on elementary safety matters and understand safety information symbols, signs and alarm signals
- know what to do if:
  - a person falls overboard
  - fire or smoke is detected
  - the fire or abandon ship alarm is sounded
- identify alarm points, muster and embarkation stations, and emergency escape routes
- locate and don lifejackets
- know how to use portable fire extinguishers
- take immediate action upon encountering an accident or other medical emergency before seeking further medical assistance on board
- close and open the fire, weathertight and watertight doors fitted in the particular ship other than those for hull openings.

For more information on topics covered in this chapter, see MCA's Wellbeing at Sea: A Guide for Organisations, section 2.1.9.

2.3 Basic training in standards of training, certification and watchkeeping

2.3.1 Anyone employed or engaged on board a vessel in any capacity with designated safety or pollution prevention duties should, before being assigned to any of those duties, receive appropriate basic training as listed below (from the tables in the STCW Code) relevant to those duties, and relevant refresher training as required:

- personal survival techniques as set out in Table A-VI/1-1
- fire prevention and firefighting as set out in Table A-VI/1-2
- elementary first aid as set out in Table A-VI/1-3
- personal safety and social responsibilities as set out in Table A-VI/1-4.

2.4 Emergency procedures and fire precautions

2.4.1 All new personnel should be given a clear explanation of the vessel's alarm signals, and instruction on the emergency assembly stations, lifeboat stations and fire drill/team requirements (see Chapter 4).

　　SI 1999/2722; MGN 71 (M)

2.4.2 Smoking regulations on the vessel should be strictly observed. Safe and correct disposal of cigarette ends is essential. Smoking or non-smoking areas, as appropriate, should be identified and clearly marked.

2.4.3 Rules concerning smoking should be strictly obeyed. E-cigarettes are a source of ignition and should not be used in hazardous areas.

2.4.4 Fire aboard a vessel can be disastrous. Common causes are:

- faulty electrical appliances/circuitry
- overloading of electrical circuitry
- careless disposal of cigarette ends
- spontaneous combustion of damp or dirty waste/rags, especially if contaminated with oil
- damp storage of linen/materials
- spillage/leakage in machinery spaces
- galley fires due to overheating of cooking oils
- carelessness with hand-pressing irons
- incorrect methods of drying laundry.

2.4.5 All personnel should be aware of these risks and ensure that fire risks are removed wherever possible; for example, through good housekeeping, and regular inspection and maintenance of electrical circuitry and appliances.

2.5 Accidents and medical emergencies

2.5.1 All personnel should know the action to take in cases of accident or medical casualty on board. For example, as a minimum they will need to know how to raise the alarm and seek assistance.

2.6 Health and hygiene

2.6.1 It is the responsibility of individuals to ensure high standards of personal hygiene and to look after their own health. Personnel should pay attention to:

- personal cleanliness
- sensible diet
- adequate sleep during rest periods
- regular exercise
- any cuts/abrasions
- keeping working clothes and protective equipment clean
- appropriate dress for the work and climate
- avoidance of excess alcohol/tobacco
- avoidance of recreational drugs.

2.6.2 On international voyages, any vaccinations/inoculations required should be fully updated. Medications for the prevention of illness (eg anti-malarial tablets) should be taken as and when required.

MGN 652 (M+F) Amendment 1 Infectious diseases at sea

2.6.3 In hot climates, it is important to protect the skin from strong sunlight and drink plenty of salt-containing liquids to replace the body fluids lost through perspiration (see Chapter 3).

2.7 Good housekeeping

2.7.1 All ships move in a seaway and as space is very limited on board any vessel, good housekeeping is essential for safe working/access and hygiene control. Particular attention should be paid to the:

- safe and secure stowage of loose items
- proper securing of doors
- good maintenance of fittings and fixtures
- adequate lighting of all work/transit areas
- avoidance of overloading of electrical circuits
- clear and legible signs/operational notices
- proper clearance and disposal of garbage/waste materials.

2.8 Environmental responsibilities

2.8.1 The maintenance of good standards to protect the environment, whether local (i.e. accommodation/work areas) or the wider environment, is important and the responsibility of all personnel. Many aspects are covered by international legislation and it is the duty of all personnel to ensure strict compliance with such legislation.

MGN 632 (M+F) Amendment 1; SI 2020/621

2.8.2 The handling and storage of garbage can present health and safety hazards to crews and ships. The requirements of the garbage management plan should be observed.

2.8.3 Particular attention should be paid to the correct methods of disposal of waste oils (bilge or other), chemicals, galley waste (including used cooking oil), garbage (especially plastics, glass, drums and other non-biodegradable items) and redundant items (eg moorings, dunnage or cargo cleanings) in line with the vessel's garbage management plan.

2.8.4 Incinerators and compactors should always be operated by competent personnel, and operating instructions should be strictly followed.

2.9 Occupational health and safety

2.9.1 All new personnel should be made aware of the company's procedures governing occupational health and safety on board, including activity-specific requirements, such as those governing the use of lifting equipment or means of access.

2.9.2 Where there are no specific regulations, the general duties contained in the Merchant Shipping and Fishing Vessels (Health and Safety at Work) Regulations 1997 apply. The main principle of these regulations is that all safety measures should be based on an assessment of the risks involved in a particular task, and the identification of the most effective measures to limit that risk. Guidance on risk assessment is in Chapter 1.

2.10 Worker responsibilities

2.10.1 It is important that personnel are reminded to follow any training, oral or written instructions that they have been given, and know to whom they should report any deficiencies in equipment or unsafe practices that they may notice.

2.10.2 Personnel who find any defects in any equipment, or a condition that they believe to be hazardous or unsafe, should immediately report it to a responsible person, who should take appropriate action.

2.11 Consultation procedures

2.11.1 New personnel must be told about the procedures for consultation on health and safety matters, including who their safety representatives are, and should be encouraged to contribute towards continuous improvement.

- Be aware of health and safety responsibilities when working on board.
- Follow company procedures on occupational health and any specific requirements when carrying out particular work activities.

- Personnel employed or working in any capacity on board must have induction and familiarisation training.
- Follow instructions for good housekeeping and personal hygiene.
- Follow instructions in the event of an accident or emergency.

3 Living on board

3.1 Introduction

3.1.1 This Code provides information and guidance on improving the health and safety of people living and working on board ship. This chapter gives some more specific advice for the individual seafarer.

> *For more information on topics covered in this chapter see Wellbeing at Sea: A Guide for Organisations and Wellbeing at Sea: A Pocket Guide for Seafarers.*

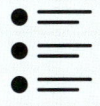

Key points
- All seafarers must have a medical fitness certificate (ENG1 on UK-registered ships).
- Seafarers should understand why health, wellbeing and cleanliness are important; not only to perform well but also to help reduce instances of illness on board.

Your organisation should
- organise work activities to limit the effects of seafarer fatigue
- have health, fitness and wellbeing policies that encourage seafarers by providing dedicated spaces and facilities. Smoking policies should allow for smokers while protecting non-smokers
- apply risk assessment to activities depending on the climate the vessel is operating in and provide appropriate personal protective equipment (PPE)
- ensure medically trained personnel on board are equipped to provide initial medical care for a range of health problems. If a seafarer develops a serious health problem or injury contact telemedical services. Where necessary, arrange to transport the sick or injured seafarer ashore for medical treatment. For further advice on medical care see the *Ship Captain's Medical Guide*.

3.2 Fitness, health and hygiene

3.2.1 Seafarers are responsible for looking after their own health and fitness; their work requires both to be of a high standard and they must hold a valid certificate of medical fitness (an ENG1 or recognised equivalent on a UK-registered ship) to join a ship. This confirms that at the time of the medical examination:

- the seafarer's hearing and sight, and where relevant colour vision, met the appropriate standards for their role on board
- the seafarer had no conditions likely to be made worse by service at sea or to make the seafarer unfit for their duties or endanger other persons on board.

MSN 1886 (M+F) Amendment 1; MSN 1815 (M+F) Amendment 6

3.2.2 If a change in health affects a seafarer's fitness for duty they should seek advice so a doctor can reassess the validity of their medical certificate. If they do not do this their medical certificate may be invalid and they may also place themselves or their colleagues at risk.

3.2.3 Seafarers should keep high standards of personal cleanliness and hygiene. On board ship infections can easily spread so preventive measures, as well as simple, effective treatment, are essential.

3.2.4 Good health depends on sensible diet, adequate sleep and regular exercise. Guidance on healthy eating is available from the National Health Service (NHS) website. Seafarers should avoid recreational drugs, substance or drug misuse, and excesses of alcohol and tobacco. They should seek treatment straight away for minor injuries, clean cuts and abrasions and give first-aid treatment to protect against infection. Barrier creams can help to protect exposed skin against dermatitis and make thorough cleansing easier.

3.2.5 Rats and other rodents may carry infection so never touch them with bare hands, whether dead or alive.

3.3 Smoking

3.3.1 Smoking damages health and can expose others to second-hand smoke. Seafarers must follow company smoking policies on board. The policy should recognise that the ship is also the seafarers' home and place of recreation, but will usually prioritise protecting non-smokers from second-hand smoke.

3.3.2 The smoking policy should limit the places on the ship where seafarers may smoke. It should include education on the health benefits of giving up smoking by promoting schemes to help seafarers to quit, including advice available on the NHS website.

3.3.3 In addition to the health risks, smoking may create a fire risk if matches and cigarettes are not carefully extinguished and disposed of safely in the designated places. Do not throw matches and cigarette ends overboard since there is a danger that they may be blown back on board.

E-cigarettes

3.3.4 E-cigarettes can cause fires and faulty lithium ion batteries can explode, both of which may cause injuries. As with all rechargeable electrical equipment, to avoid the risk of fire always use the correct charger. Never leave e-cigarettes charging unattended or overnight. Buy e-cigarette products from a reputable retailer to ensure that they comply with UK safety regulations.

3.3.5 In line with section 3.15, or where companies have introduced policies for the vetting/portable appliance testing of electrical appliances being brought on board for personal use, e-cigarettes and their chargers may fall under the scope of these requirements. Seafarers should therefore get approval from a responsible officer before using an e-cigarette.

3.4 Medication

3.4.1 Anyone taking medication must follow the relevant medical directions outlined on the label or by a doctor. These may include not drinking alcohol or not doing certain activities when alertness is affected. Seafarers should discuss the possible side-effects with the approved doctor during the medical examination and then tell the responsible officer on board. This is to ensure that the correct information is available in a medical emergency, and so allowances can be made when allocating tasks.

> *MSN 1886 (M+F) Amendment 1*

3.4.2 Avoid drinking alcohol while under treatment with certain medications. Even common remedies such as aspirin, seasickness tablets, anti-malarial tablets and codeine may be dangerous when taken with alcohol. Seafarers are responsible for keeping their inoculations and vaccinations for international voyages up to date and for taking medications to prevent illness, such as anti-malarial tablets, when required.

3.5 Malaria

3.5.1 Seafarers must start preventive medication for malaria before they arrive in an affected area. The length of time may vary according to treatment, but around one to three weeks is normal. Medication should continue for four weeks after leaving the area. The company will need to take medical advice on the best medication for particular areas.

> *MGN 652 (M+F) Amendment 1*

3.5.2 While in infected areas, do the following to minimise the risk of insect bites:

- Wear long-sleeved tops and trousers when going on deck or ashore.
- Use mosquito wire-screening and nets.
- Keep openings closed.
- Use anti-mosquito preparations or insecticides.

3.5.3 Anyone who falls ill after being in a malarial area should tell a doctor immediately that they are at risk of malarial infection.

> *More detailed guidance on prevention is available in Marine Guidance Note MGN 652 (M+F) Amendment 1 and on prevention and care in The Ship Captain's Medical Guide.*

3.6 Avoiding the effects of fatigue (tiredness)

3.6.1 The International Maritime Organization (IMO) defines fatigue as: 'A reduction in physical and/or mental capability as the result of physical, mental or emotional exertion which may impair nearly all physical abilities including: strength; speed; reaction time; coordination; decision making; or balance.'

> *IMO; MSC/Circ 813*

3.6.2 Fatigue among seafarers is a serious issue affecting maritime safety. There is evidence that fatigue contributes to accidents, injuries, death, long-term ill health, major damage to and loss of vessels, leading to environmental harm.

3.6.3 The company and the master should organise work so as to minimise fatigue, but seafarers also have a duty to take care of their own health and safety and that of their fellow workers.

3.6.4 To prevent fatigue:

- Arrive on board well rested at the start of a period of work.
- Take your scheduled rest periods.
- Use rest periods to gain adequate, uninterrupted sleep as far as possible (about eight hours of sleep in each 24 hours).
- Eat regular, well balanced meals, but eat lightly before sleep.
- Avoid alcohol and caffeine before sleep.
- Record your hours of rest accurately, so management are aware if there are workload pressures at particular times.
- Avoid using electronic devices shortly before sleep. The blue light from screens can interfere with the body's natural rhythms and production of hormones; particularly melatonin, which is key in helping us sleep.

Further information about fatigue, making the most of sleep patterns and ways to maintain alertness is available in MGN 505 (M) Amendment 1.

MGN 505 (M) Amendment 1

3.7 Working in hot or sunny climates and hot environments

3.7.1 High humidity and high temperatures can lead to heat exhaustion and heat stroke. Perspiration is the body's best heat-control mechanism but sweat consists mainly of salt and water, which must be replaced. When working in hot and/or humid conditions drink at least 4.5 litres (8 pints) of cool (but not iced) water daily. Drink small quantities frequently to keep hydrated. You can get salt in food and also have salt-containing drinks to prevent heat cramps. Avoid drinking alcohol.

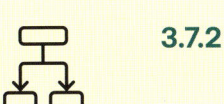

3.7.2 Limit the length of time that seafarers are exposed to hot conditions. Provide breaks in the shade or in fresh air. Mechanical aids to make physically demanding work easier will help to reduce the impacts of hot environments or when seafarers are wearing a lot of clothing or equipment. See guidance from the Health and Safety Executive (HSE) on thermal comfort.

3.7.3 If seafarers are working in an enclosed space ensure that it is as well ventilated as possible. They should wear light clothing to allow the largest possible surface for free evaporation of sweat.

3.7.4 Seafarers should avoid exposure to the sun, especially in tropical areas and during the hottest part of the day. When it is necessary to work in exceptionally hot or humid conditions, seafarers should wear appropriate clothing (including a hat) protecting both body and head. Light cotton clothing will reflect the heat and help to keep the body temperature down. Keep the upper body covered, especially around midday. Skin that has not been exposed to the sun for several months burns very easily.

3.7.5 Sunscreen can add useful protection for the seafarer's body, which it is not easy to shade from UVA and UVB rays and sunlight. In European climatic regions use a sun protection factor (SPF) rating of at least 15. In tropical regions or other areas of high risk, use an SPF of at least 30 (or higher for those with fair skin).

To avoid sunburn:

- The best protection is to shade the skin from direct sunlight.
- Get to know your skin. This will help decide what precautions you need to take. Getting burnt now might increase the chances of developing skin cancer in later years.
- Try to avoid reddening – it is the first sign of skin damage as well as being an early sign of burning.
- A suntan may give some protection against burning but does not eliminate the long-term cancer risk; nor will it protect against premature ageing.
- When on leave, continue to take care as the skin remembers every exposure.
- Use sunscreen generously and reapply regularly.

3.7.6 When seafarers are working in exceptionally hot and/or humid conditions or when they are wearing respiratory equipment, breaks at intervals in the fresh air or in the shade may be necessary. They should take off protective clothing and equipment during breaks to allow the body to cool down, but they must put it on again before restarting work.

3.8 Working in cold climates and environments

3.8.1 Working in cold climates can impair the seafarer's ability to carry out simple tasks, as the cold temperatures can severely affect dexterity. At even colder temperatures, deeper muscles are affected, which results in reduced muscular strength and stiffened joints. See the Canadian Centre for Occupational Health and Safety's guidelines for working in cold weather (https://www.ccohs.ca/).

3.8.2 Early signs that the body is under stress from the cold include:

- persistent shivering
- poor coordination
- blue lips and fingers
- irrational or confused behaviour
- reduced mental alertness.

3.8.3 Seafarers should wear appropriate clothing including gloves, hat and warm socks. Ensure that this clothing is compatible with any PPE needed for the work.

Cold weather-related injuries/conditions

3.8.4 Frostbite/frostnip can damage the skin and the tissue of the parts of the body that are left exposed to freezing temperatures. Extremities, specifically hands, feet, ears, nose and lips, are particularly vulnerable. Seafarers should wear clothing that protects the extremities.

3.8.5 Hypothermia happens when the body's core temperature falls below 35°C (95°F) and it can be life threatening. It is usually caused by being outdoors in cold conditions for a long time or falling into cold water. See the National Health Service (NHS) Choices website for information on frostbite and hypothermia.

3.9 Working clothes

3.9.1 Clothing should be appropriate for the working conditions. Clothes should be close-fitting with no loose flaps, pockets or ties, which could become caught up in moving parts of machinery or on obstructions or projections. Where there is a risk of burning or scalding, as in galleys, clothing and shoes should adequately cover the body. Material should be of low flammability, such as cotton.

3.9.2 Long-sleeved shirts or overalls provide better protection than short sleeves and sleeves should not be rolled up. Long hair should be tied back and covered. Industrial or safety footwear should be worn when appropriate.

3.10 Shipboard housekeeping

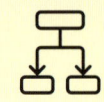

3.10.1 Good housekeeping is essential in promoting health and safety on board:

- Store equipment and other items safely and securely. This ensures not only that defects are discovered but also that you can find articles when required.
- Maintain fixtures and fittings properly.
- Provide adequate lighting for all work and transit areas.
- Do not overload electrical circuits, particularly in cabins.
- Clear up garbage and waste and dispose of it correctly and promptly.
- Secure doors and drawers properly.
- Keep emergency signage and fire and life-saving equipment clear at all times.
- Keep instruction plates, notices and operating indicators clean and legible, and do not allow other items to obstruct them.

3.10.2 Aerosols may have volatile and inflammable contents. Never use or place them near naked flames or other heat sources even when empty. Dispose of empty canisters properly.

3.10.3 Some fumigating or insecticidal sprays contain ingredients which, though perhaps themselves harmless to human beings, may be decomposed when heated. Smoking may therefore be dangerous in sprayed atmospheres until the spray has dissipated and the area has been ventilated.

3.11 Substances hazardous to health

3.11.1 Many substances found on ships can damage the health of people exposed to them. They include not only recognised hazardous substances, such as dangerous goods cargoes and asbestos, but also some maintenance and cleaning substances. For example, caustic soda and bleaching powders or liquids can burn or penetrate the skin. They may react dangerously with other substances and should never be mixed.

3.11.2 Where personnel are working near hazardous substances take appropriate safety measures to remove, control or minimise the risk of exposure. Packaged cargoes and stores should carry hazard warning labels, where appropriate. Do a risk assessment to identify other hazardous substances and give seafarers information about the hazards and the measures in place to protect them.

3.11.3 Read all labels on chemical containers carefully before opening them to find out about any hazards from the contents. Do not use a chemical from an unlabelled container unless you can clearly find out what it is. Further advice is in Chapter 21.

3.11.4 Older ships may have asbestos-containing products in panels, cladding or insulation. Report any damage to such materials during a voyage immediately to the departmental head. If possible seal off the area until the damage can be repaired properly and insulate or cover the exposed edges or surfaces. This will prevent asbestos fibres from being released and dispersed in the air.

3.11.5 Long exposure to mineral oils and detergents may cause skin problems. Seafarers should wash all traces of oil thoroughly from the skin using a skin cleaner designed for oil removal. They should not use chemical solvents as they may damage the skin. If a seafarer inadvertently touches toxic chemicals or other harmful substances they should report this immediately and take the appropriate remedial action. Launder working clothes frequently and do not put oil-soaked rags in pockets.

3.11.6 Breathing irritant dust can cause coughs and lung damage. The risk is usually much greater for a smoker than for a non-smoker.

3.11.7 Seafarers should be aware of and understand the risks arising from their work, the precautions to take and the results of any monitoring of exposure.

3.11.8 Seafarers should always comply with any control measures in place, and wear any protective clothing and equipment supplied.

3.11.9 In cases where failure of the control measures could result in serious risks to health, or where their adequacy or efficiency is in doubt, this should be reported so that health surveillance can be undertaken.

3.12 Common personal injuries

3.12.1 Chapter 8 includes advice on suitable PPE that will help to prevent the following injuries.

Hand injuries

3.12.2 Wear gloves when handling sharp or hot objects. However, gloves may get trapped on drum ends or on machinery. Although loose-fitting gloves allow hands to slip out readily, they do not give a good grip on ladders. Wet or oily gloves may be slippery so take great care when working in them. Wearing gloves for long periods may make skin hot and sweaty, leading to damage. Wearing separate cotton gloves inside protective gloves will help to prevent this.

Foot injuries

3.12.3 Unsuitable footwear (such as sandals, plimsolls and flip-flops) gives little protection if there is a risk of burning or scalding and may lead to trips and falls. Keep feet away from moving machinery, bights of ropes and hawsers.

Eye injuries

3.12.4 Take care to protect the eyes. Seafarers should wear appropriate protective goggles for any work involving sparks, chips of wood, paint or metal, and dangerous substances.

Head injuries

3.12.5 Remember to duck when stepping over coamings, etc. to avoid hitting your head on the door frame. Seafarers should wear head protection where appropriate.

Cuts

3.12.6 To avoid cuts, handle sharp implements and objects with care. Do not leave them lying around where someone may accidentally cut themselves. In the galley, do not mix sharp knives and choppers with other items for washing up but clean them individually and store them in a safe place. Sweep up broken glass carefully; do not pick it up by hand.

Burns and scalds

3.12.7 Burns and scalds are commonly caused by hot pipelines and stoves, as well as by fires. Think of every hot machine and container of scalding liquid as a hazard that can cause injury, and take adequate precautions.

3.12.8 Faulty electrical equipment can cause severe burns as well as an electric shock. Check equipment before use. If something appears wrong report it.

Misuse of tools

3.12.9 Misusing tools can cause injury. Always use the correct tool for the job and use it in the right way. Never leave tools lying around where they can fall on someone, or be tripped over. After finishing a job put them away in a safe place.

Manual handling

3.12.10 It is easy to strain muscles when **manual handling**. You may avoid pulled muscles by using proper lifting techniques. Chapter 10 gives guidance on handling loads.

Mooring

3.12.11 Mooring and unmooring operations can cause serious accidents. Personnel should never stand in the bight of a rope or near a rope under tension, and they should treat ropes on drums and bollards with the utmost care.

Electrical hazards

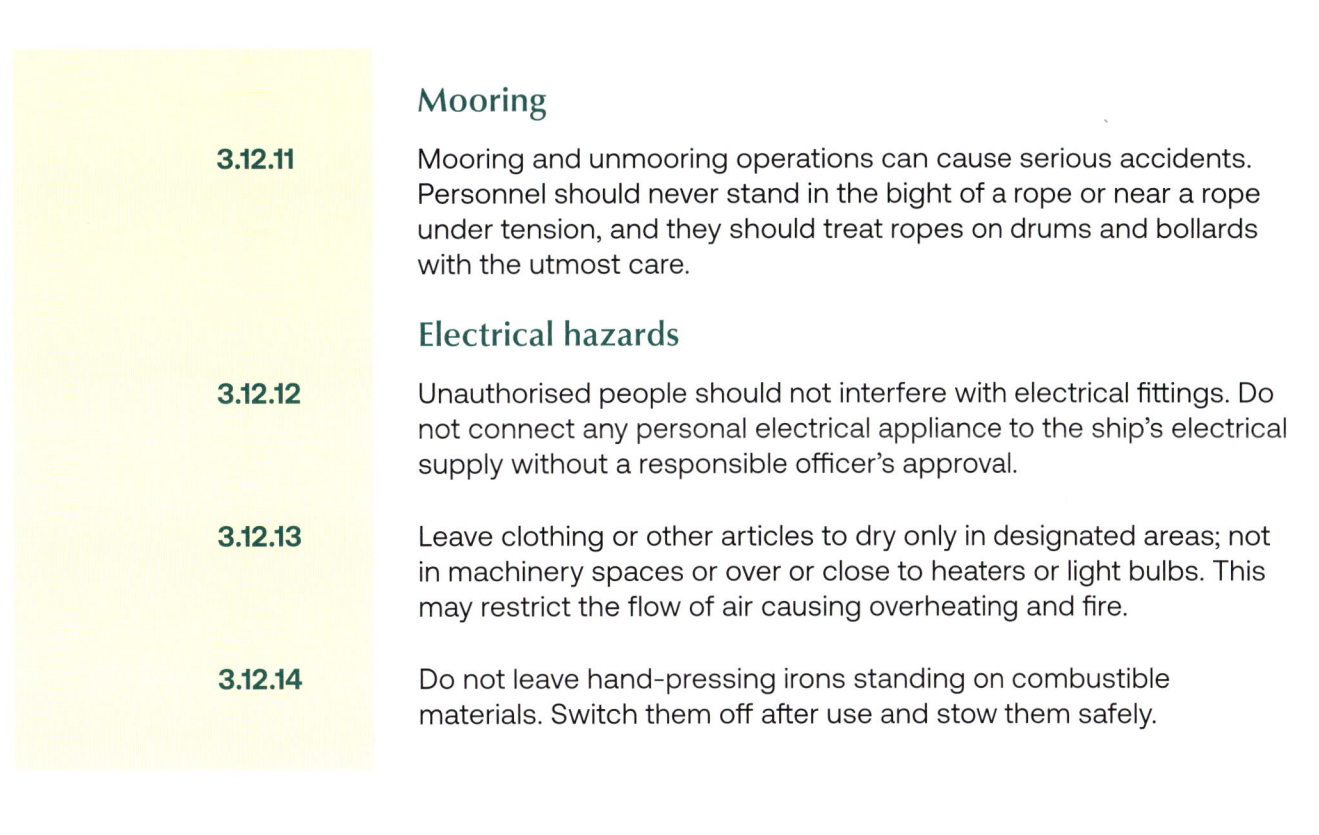

3.12.12 Unauthorised people should not interfere with electrical fittings. Do not connect any personal electrical appliance to the ship's electrical supply without a responsible officer's approval.

3.12.13 Leave clothing or other articles to dry only in designated areas; not in machinery spaces or over or close to heaters or light bulbs. This may restrict the flow of air causing overheating and fire.

3.12.14 Do not leave hand-pressing irons standing on combustible materials. Switch them off after use and stow them safely.

3.13 Sunglasses

3.13.1 The bright light from the sun reflecting off the surface of a calm sea or from ice caps in the polar regions, or from the vessel itself, can dazzle the seafarer and damage the eyes.

3.13.2 Seafarers working on the bridge or on the open deck in sunshine should wear sunglasses. These are an important piece of protective equipment and more than just tinted eye protection. For protection on the bridge, consider collective protection systems such as sunblinds.

3.13.3 Photochromic lenses react with UV radiation by darkening. Seafarers should not wear sunglasses with photochromic lenses during the hours of darkness as they can significantly reduce night vision. For further information see MGN 357 (M+F).

📖 *MGN 357 (M+F)*

3.13.4 Polarised lenses reduce the amount of light passing through the lens by selectively filtering certain electromagnetic spectral planes. Do not use polarised lenses when viewing instrument panels as some images may be unclear. In some situations, for example, when navigating in shallow water, these lenses may help as they reduce the reflected glare from the surrounding water. Seafarers should be aware of the benefits and limitations of polarised lenses.

3.13.5 All frames should be well fitting and large enough to allow sufficient protection from oblique sunlight. All seafarers who need a spectacle prescription must have a clear pair of correcting lenses but they can also have prescription sunglasses as their second pair. They must not wear non-prescription sunglasses on top of prescription glasses. For navigational watches during the hours of darkness they should not wear any type of sunglasses.

3.13.6 Follow this guidance when buying sunglasses:

- The lens tint should be neutral – ideally either grey or brown as these cause the least colour distortion.
- The lens tint should be no darker than 80% absorption.
- A graduated tint may be useful, with the darkest at the top of the lens, lightening towards the bottom.
- Glasses should be CE marked and to the British Standard BS EN ISO 12312-1:2013+A1:2015. BS EN ISO 12311:2013 ensures that the sunglasses offer a safe level of ultraviolet protection.

BS EN ISO 12311:2013; BS EN ISO 12312-1:2013+A1:2015

3.14 Risk from sharps

Introduction

3.14.1 The term 'sharps' includes needles, syringes and razor blades.

3.14.2 Sharps may be used for the treatment of medical conditions, for recreational drug use or for wet shaving. Housekeeping staff may therefore find them in bed linen, on surfaces or in bins. Take care to avoid injury and the risk of contamination with blood-borne viruses (BBVs). The main BBVs of concern are:

- hepatitis B (HBV)
- hepatitis C (HCV)
- human immunodeficiency virus (HIV).

There is a risk of bacterial or viral infection from used sharps.

3.14.3 As there is always a risk of finding sharps unexpectedly, get advice from a medical practitioner about whether seafarers exposed to this risk should have a tetanus or hepatitis B vaccination as a precaution. For UK residents, these are provided free of charge on the NHS.

3.14.4 To reduce the risk from sharps:

- Train all housekeeping staff in safe systems of work, and what to do if they find a sharp.
- Ensure that supervisors know the safe systems of work and what to do in the event of injury.

Rubbish collection

3.14.5 Never remove items by hand from the bin.

3.14.6 Before removing a bin liner full of rubbish check that it is not too heavy to lift. Remove the liner fully from the bin and place it in a sturdy rubbish collection sack. To avoid being accidentally stabbed with a discarded needle or razor, never put your hands inside a sack or a bin when emptying the contents.

3.14.7 Where no bin liner is in use, check that the bin is not too heavy to lift. Empty the contents directly into a sturdy rubbish collection sack and do not overfill it. If you need to compress the contents do this with a brush or similar object, not with your hands.

3.14.8 Carry rubbish sacks as far from the body as possible to prevent any unseen sharp objects causing injury. If you notice sharp objects in the bag, put the bag on the floor and get someone to help.

Cleaning/housekeeping

3.14.9 Avoid putting your hands into obstructed/hidden areas such as toilet U-bends or under sheets or pillows.

3.14.10 Dispose of all needles/syringes in a sharps container (or rigid-sided container) following the safe system of work (see section 3.14.15). If you find a sharp assume it is infectious. Close off the area immediately to all personnel. Report the incident to a supervisor so they can keep a record of the location, date and time as a hazardous occurrence.

   BS EN ISO 23907-1:2019

3.14.11 Handle broken glass and crockery carefully and wrap it in several sheets of paper before putting it in the bin.

Removal of sharps: safe working procedure

3.14.12 Only trained staff should remove sharps.

3.14.13 Never carry a sharp to a container for disposal; instead take the container to the sharp and place it near the sharp.

3.14.14 Never pick up sharps with bare hands or pass them from hand to hand. Do one of the following:

- Put on appropriate protective clothing (eg stout rigger, rubber gloves or specialist anti-needle gloves) to remove the sharp with a small pair of tongs.
- Use a long-handled litter-picking device to pick up the sharp. You do not need gloves as the distance between you and the sharp reduces the risk of contact.

3.14.15 Do not put sharps in a normal waste bin; always use a sharps container. Do not fill the sharps container beyond the level indicated on the side.

3.14.16 Do not try to resheath or bend the needle.

If an injury occurs

3.14.17 If a needle or razor blade pierces the skin:

- gently encourage the wound to bleed under running water but do not scrub or suck it
- wash the wound with soap and water
- report the incident immediately to a supervisor
- get telemedical advice (unless there is a doctor on board).

3.15 Mobile phones and other personal electronic devices

3.15.1 Using mobile phones, personal electronic devices and other communication devices inappropriately causes distraction and loss of awareness. In a safety-critical environment, this has led to death, injury and serious damage.

3.15.2 Operating ships and their equipment demands a lot of attention; the same is true of electronic devices. The term 'personal electronic device' means a range of devices which may be used for either personal or operational reasons, or both.

3.15.3 Using any devices while working on vessels puts increased demands on the human brain, which can lead to cognitive overload and impairment causing reduced performance; for example:

- reduced situational awareness
- failure to recognise vessels or navigational hazards
- slower reaction times
- impaired decisions in risk assessment
- loss of concentration
- greater stress and fatigue
- inattentional blindness.

3.15.4 **Warning**

As a principle of best practice, the MCA recommends putting in place robust measures and restrictions on the use of electronic devices (whether for personal or operational reasons) in safety-critical areas and during safety-critical operations. These include the navigational bridge, at mooring stations and when bunkering. There should be appropriate signage to show when and where mobile phone use is prohibited and 'safe zones' where device usage is allowed. There should also be company policies that encourage the responsible use of devices for both personal and operational reasons.

3.15.5 Devices such as mobile phones and two-way radios are legitimately used for communication during ships' operations. However, recognise the potential for distraction and loss of awareness and encourage people using them to do so safely. The most effective safeguard is to prohibit the use of personal electronic devices when they could cause distraction, as far as practicable. Where this is not possible include these aspects in the risk assessments and put mitigations in place to reduce the potential hazards.

3.15.6 Seafarers using devices for either personal or work-related communications should follow this advice:

- Ensure you are in a safe place away from moving vehicles, machinery, equipment and cargo working.
- Keep still; do not wander about.
- If possible, stand/sit with your back to a solid wall/bulkhead, enabling you to face and see any operations nearby, and to prevent the risk of vehicles or machinery approaching you from behind.
- Tell nearby colleagues you are taking a call, but do not rely on them to watch your surroundings for you.
- Alert others who may be distracted by personal devices and putting themselves at risk.
- Tell the person you are talking to about your location/situation and that you may have to break the conversation at short notice.
- Keep the conversation as short as possible. If possible reschedule calls for a more appropriate time.
- If it is necessary to receive an urgent call while on duty in the bridge or the engine room send for a replacement officer while you are taking the call.

Devices for personal use

3.15.7 As a principle of best practice, the MCA recommends that seafarers should not carry personal mobile phones and other personal electronic devices anywhere in the workplace, especially at safety-critical places, except in line with company policies. The company policies should explain the safety benefits of using devices in compliance with the policy, balancing this against the need for many seafarers to use personal devices off duty. Given the wide range of operations on ships, a consultative approach to defining appropriate device usage may be helpful.

3.15.8 With regard to mobile phones and other devices for personal use:

- only use them when off duty
- use them in a safe environment such as your cabin, rest area, or another non-working part of the ship
- do not wander into working areas of the ship
- do not distract or interrupt other crew members with your call.

If you need to take/make a personal call when on duty:

- ensure your colleagues know you are making a call and can cover for you in your absence
- do not distract or interrupt other crew members with your call.

3.15.9 MGN 299 (M+F) highlights the dangers of distraction and interference caused by the use of mobile phones at safety-critical locations and times on board. Incoming communications from personal devices, even if you ignore them, can still cause dangerous distraction in safety-critical environments.

> See further guidance on the use of mobile phones and other devices in MGN 520 (M), MGN 638 (M+F) and MGN 299 (M+F).

- Encourage seafarers to maintain a high level of health and fitness.
- Be aware of how personal devices can cause distraction.

In a medical emergency see the *Ship Captain's Medical Guide* and seek medical advice from telemedical services.

4 Emergency drills and procedures

4.1 Introduction

4.1.1 Musters and drills must take place in line with the statutory requirements that apply to the size and type of vessel. This chapter, together with the relevant marine guidance notices, advises on what to do. Annex 4.1 summarises the requirements.

 📖 *MGN 71 (M); MSN 1579 (M); SI 1999/2722; SOLAS II-1, II-2, III and V*

4.1.2 Keep the muster arrangements up to date with any changes in the ship's seafarers with designated safety and pollution prevention duties and any changes in the ship's function.

4.1.3 Musters and drills are to:

- train seafarers to respond in an organised way to dangerous situations
- allow seafarers to practise in conditions as similar as possible to a real emergency while ensuring that they are safe
- prevent loss of life and property at sea
- protect the marine environment.

4.1.4 Varying the timing of drills within the required frequency means that anyone who misses a drill can take part in the next one.

4.1.5 Correct any problems or areas for improvement identified during drills as soon as possible and keep a record of what has been done.

4.1.6 When a ship enters service for the first time, after a major character of the ship has been modified or when a new crew is engaged, undertake the drills in this chapter before sailing.

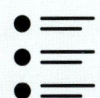

Key points

- Drills prepare seafarers on board for emergency situations, prevent loss of life and property at sea, and also protect the marine environment.
- This chapter contains guidance on drills for fire, abandoning ship, man overboard, enclosed space entry and rescue, emergency steering, leakage and spillage, and damage control.

Your organisation should
- carry out drills and record them in line with regulations
- use drills to ensure seafarers know the safe procedures and are familiar with the equipment required for ordinary and emergency circumstances.

4.1.7 Display the muster list in a visible place before the ship sails. On international voyages and for passenger ships of classes IIA and III, the muster list should also provide emergency instructions for each seafarer. Instructions could be handed out on a card or posted on all berths and bunks. They should explain:

- the allocated assembly station
- the survival craft station and emergency duty
- all emergency signals and what to do on hearing them.

SI 1999/2722 Reg 8(4), Reg 12(2) and 12(3)

4.1.8 Each crew member should take part in at least one abandon ship and one fire drill every month.

4.1.9 If more than 25% of the crew have not taken part in drills on board the ship in the previous month the master must hold an abandon ship drill and a fire drill within 24 hours of leaving port. On passenger vessels of class I, do this drill before or immediately on departure.

SOLAS III 19.3.2; MGN 71

4.1.10 On passenger ships of class I, II, II(A) and III, an abandon ship drill and fire drill must be held weekly with as many crew as practicable taking part. In any case, each crew member must take part in at least one abandon ship drill and one fire drill each month.

SOLAS III 30; MGN 71

4.1.11 Musters of newly embarked passengers who are scheduled to be on board for more than 24 hours must take place before or immediately upon departure.

SOLAS III 19.2.2

4.1.12 Seafarers should have the following onboard training as soon as possible after joining a ship:

- how to use the ship's life-saving appliances, including survival craft, evacuation systems and firefighting equipment
- their emergency duties and what the various alarm systems mean
- where their lifeboat station is
- where their life-saving and firefighting equipment is.

4.1.13 All the ship's personnel should wear properly secured lifejackets during the drill. They should keep them on during lifeboat drills and launchings. However, the master may ask them to remove lifejackets if they affect the next drill, as long as they keep them to hand.

4.2 Fire drills

4.2.1 Seafarers should have training in firefighting and the maintenance of equipment by taking part in regular drills in line with regulatory requirements. This ensures firefighting teams are ready to operate; it also helps maintain seafarers' competency in firefighting skills and to identify areas for improvement.

 SOLAS II-2 15.2.2.3

4.2.2 Keep access to firefighting equipment clear at all times and never block emergency escapes and passageways.

4.2.3 Effective firefighting needs everyone on the ship to work together. Hold a fire drill at the same time as the first stage of the abandon ship drill. The firefighting parties should assemble at their designated stations.

4.2.4 Train each member of the firefighting party in how to use breathing apparatus as part of the drill. Do fire search and rescue exercises in various parts of the ship. Clean the equipment and check that it is in good order before stowing it. Recharge self-contained breathing apparatus (SCBA) air cylinders in open and well-ventilated areas, and train seafarers in how to do this. Alternatively, carry enough spare cylinders for this purpose.

4.2.5 Start the fire pumps in the machinery spaces and check that full pressure is on fire mains. Also start and operate any emergency fire pumps outside machinery spaces. Make sure all nominated seafarers know how to operate the fire pumps.

4.2.6 The firefighting parties should go immediately from their designated stations to the place where the assumed fire is. They should take all emergency equipment, such as firefighter suits, radios, axes, lamps and breathing apparatus.

4.2.7 Use an adequate but realistic number of charged hoses to deal with the assumed fire. At some stage in the drill, operate charged hoses at appropriate locations using the general service and emergency fire pumps.

4.2.8 Test and demonstrate how to use the remote controls for ventilating fans, fuel pumps and fuel tank valves. Also test how to close the openings and **isolate** electrical equipment.

4.2.9   Test any fixed fire extinguishing installations as far as practicable.

4.2.10 Demonstrate how to use portable fire extinguishers and which types are suitable for which kinds of fire (see section 9.8.4).

4.2.11 Demonstrate how to use the fire blanket, and for what types of fire it is appropriate.

4.2.12 Demonstrate any specialist firefighting equipment carried for specific types of fire, such as vehicle fires.

4.2.13 Change the drill locations to enable seafarers to practise in different conditions and with different types and causes of fire. Make sure that accommodation, machinery spaces, storerooms, galleys, cargo areas/holds and high-risk fire areas are all covered regularly.

4.2.14 At each drill, one or more rechargeable portable fire extinguishers should be operated by a different member of the firefighting party each time. As far as practicable, use all the types of extinguishers on rotation. This can also form part of the test discharge criteria found under the periodic testing requirements for portable fire extinguishers.

📖   *MGN 71; MGN 276*

4.2.15 Explain which extinguishers cannot be charged on board and how to operate them. Recharge used extinguishers before putting them back in their normal place. Alternatively carry enough spares to use in demonstrations. You must carry enough spare charges, with instructions for recharging, for fire extinguishers on board in line with regulations.

📖   *SI 1998/2514; SOLAS II-2 10.3.3*

4.2.16 Inspect fire appliances, fire doors and watertight doors, other closing appliances, and fire detection and alarm systems that have not been used in the drill in line with the ship's safety management system. Where smoke generating equipment is available, carry out drills under controlled ventilation to remove smoke that may hinder fire control or rescue operations.

4.3 What to do if there is a fire

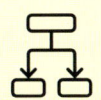

4.3.1 Raise the alarm and immediately tell the bridge or designated control centre, as appropriate, that there is a fire. Then do the following:

- If the ship is in port, tell the local fire and port authority.
- If a space is filling with smoke and fumes, tell anyone who is not wearing breathing apparatus to leave the space immediately.
- Escape from a smoke-filled space by crawling on hands and knees because the air close to deck level is likely to be clearer. Use emergency escape breathing devices if available.
- Only if safe, try to put out the fire using a suitable portable extinguisher.
- If the fire is caused by fat or oil (eg in the galley) use a fire blanket to try to smother it.
- Generally do not use water extinguishers on oil or electrical fires and do not use foam extinguishers on electrical fires (see section 9.8.4).
- Lithium ion battery fires on electric vehicles are very difficult to extinguish but you can control them with continuous boundary cooling to stop fire spread. Apply boundary cooling through fire hoses or fixed water spray systems. Monitor lithium ion batteries continually after a fire event as the fire could restart long after it appears to have been extinguished. Lithium ion fires can generate explosive and toxic gas clouds so keep the space well ventilated to prevent possible explosive gas build up.

   MGN 653 Amendment 1

- Shut openings to the space to reduce the supply of air to the fire and to prevent it spreading.
- Isolate any fuel lines that are feeding the fire or in danger from it.
- If practicable, remove any combustible materials near the fire.
- Consider using boundary cooling of compartments nearby. Monitor the temperatures if spaces are not accessible.

4.3.2 Once a fire has been extinguished, take precautions to prevent it restarting.

4.3.3 It is easier to extinguish a fire in its early stages than after it has spread, so you must act promptly and correctly.

4.3.4 Always wear breathing apparatus to re-enter a space where there has been a fire, until it has been fully ventilated.

4.3.5 There will always be a risk of fire on board a ship but its effects will be much reduced if seafarers follow the advice in this chapter.

4.4 Abandon ship drills

4.4.1 Check the prevailing weather conditions before preparing for drills. Carry out drills on board the ship. You could do this either at anchor or alongside, where there is a suitable relative movement between ship and water. When using a free-fall lifeboat you need a suitable depth of water for the launch and a clear, unobstructed launch area. If that is not possible, carry out drills in a suitable place on shore with similar conditions. Alternatively, if the master allows, hold the drill on board a ship when making headway in sheltered waters.

4.4.2 Seafarers taking part in life raft or lifeboat drills should wear clothing appropriate to the prevalent environmental conditions, and properly secured lifejackets.

4.4.3 Ensure that wires of the launch and recovery system are in good condition, and that sheaves and working parts are working properly and well lubricated. Wear a safety harness if there is any risk of falling from the davit or boat.

4.4.4 When launching lifeboats, seafarers should practise manoeuvring the boat using engines or oars, as appropriate.

4.4.5 Train seafarers how to launch, handle and operate life rafts. Explain how to board the life rafts and use any equipment and stores on them.

4.4.6 Assess all health and safety risks before doing drills. Keep lifebuoys and lines readily available at the embarkation point for survival crafts.

4.4.7 Always maintain life-saving appliances. Replace equipment immediately if it gets damaged during drills.

 📖 *MSC 1/Circ 1206/Rev 1, Annex 2*

Side-launch lifeboats

4.4.8 If the handle of the lifeboat winch would rotate while operating the winch, remove it before lowering the boat on the brake or raising the boat with an electric motor. If the handle is not removable tell seafarers to keep well clear of it.

4.4.9 When turning out davits or bringing boats or rafts inboard, seafarers should always keep clear of any moving parts.

4.4.10 Start the engines on motor lifeboats and run them ahead and astern. Do not allow the engine and the propeller shaft's stern gland to overheat. Ensure that all seafarers know how to start the engine.

4.4.11 Examine and test radio life-saving appliances, and train seafarers in how to use them.

4.4.12	Test water spray systems, where fitted, following the lifeboat manufacturer's instructions. 📖 *SI 1999/2722 Reg 10; MGN 71*
4.4.13	When carrying out a drill in port, clear and swing out as many lifeboats as possible. Launch each lifeboat and manoeuvre it in the water at least once every three months. Follow current advice and recommendations in the relevant M notices. It is better if no-one is in the boat when it is lowered or raised from the water. However, the launching crew may need to be on board. 📖 *SOLAS III 19.3.4.3, MGN 560 (M) Amendment 24.4.10*
4.4.14	When carrying out launching drills while making headway through water, you must have lifeboats and rescue boats that you can launch safely with the ship making headway at speeds of up to 5 knots in calm water. You could use any ship that has a lifeboat or rescue boat, or both, with on-load release gear adequately protected against accidental or premature release.
4.4.15	The Merchant Shipping (Musters, Training and Decision Support System) Regulations 1999 do not require you to carry out training in launching lifeboats and rescue boats from ships making headway through the water. However, follow these guidelines if giving this type of training. 📖 *MSN 1722 (M+F)*
4.4.16	When training it is not recommended to exercise at the maximum design 5 knots headway launching capability of the equipment. For safety reasons carry out drills with a low relative water speed, particularly where personnel are inexperienced. When planning the drill, ensure that, as far as practicable, the relative water speed will be at its slowest when recovering the boat.
4.4.17	When planning for and doing launching drills from ships making headway, take the precautions set out in Annex 4.2.
4.4.18	Seafarers should keep their fingers and hands clear of the long-link when unhooking or securing blocks onto lifting hooks while the boat is in the water, particularly if there is a swell.
4.4.19	Check that limit switches or similar devices are working before recovering lifeboats with gravity davits by power.

Free-fall lifeboats

4.4.20	Do the monthly drills with free-fall lifeboats following the manufacturer's instructions. Train seafarers who are going to enter the boat in an emergency in how to embark, take their seats correctly and use the safety harnesses. Also instruct them in how to act when launching into the sea.

4.4.21 Do not do a full launch of a free-fall lifeboat with seafarers during drills because of the high risk to personal safety. Where SOLAS allows, use a secondary means of launching or simulated launching, including a simulated release with restraining and/or recovery devices, following the manufacturer's instructions.

> SOLAS III 19.3.4.4; MSC 1/Circ 1206/Rev 1/Annex 2 and Appendix; MGN 560 (M) Amendment 2

Rescue boats

4.4.22 If your ship carries fast-rescue boats or rescue boats that are not lifeboats, practise launching and manoeuvring them in the water with assigned crew once a month as far as is reasonable and practicable. In any case, hold at least one such drill every three months.

4.4.23 Always check that hooks are fully engaged before launching, before recovery and after stowage, if simultaneous off-load/on-load release arrangements are provided. Use fall preventer devices as appropriate in line with the relevant M notice.

> SI 1999/2722; Reg 10(7); MGN 540 (M+F)

4.4.24 Personnel in a fast-rescue boat/rescue boat or survival craft should remain seated while it is being lowered. They should keep their hands inside the craft to avoid crushing them against the ship's side. They should wear lifejackets and, in totally enclosed lifeboats, fasten their seatbelts.

Davit launch life rafts

4.4.25 Hold on-board training in the use of davit-launched life rafts at least every four months on every ship that has them. Whenever practicable, practise inflating and lowering a life raft. If this life raft is used only for training and is not part of the ship's life-saving equipment, clearly mark it as such.

4.4.26 Do not cock the release mechanism of a davit-launched life raft until just before the raft lands on the water.

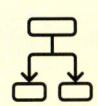

4.5 What to do when abandoning ship

4.5.1 Only the master can verbally give the order to abandon ship. Follow these instructions:

- Go to your assembly/muster point wearing appropriate clothing for the environmental conditions and a lifejacket.
- Bring any additional provisions, water or equipment and place them in the lifeboat.
- Collect electronic location beacons, such as search and rescue transponders and emergency position-indicating radio beacons, and take them to the lifeboat(s).
- Take a roll call to make sure that everyone is present before the person in charge orders the lifeboat to be lowered.

Abandon ship drills: dos and don'ts

Do	Don't
Choose a suitable place for the drill.	Allow anyone in the boat when lowering it.
Replace any damaged equipment.	Allow the engine to overheat.
Carry out launching drills monthly.	Hold drills in stormy weather.

4.6 Man overboard drills

4.6.1 Before organising man overboard drills read all relevant documentation.

> *Man Overboard! Guidelines to Shipping Companies on Procedures in Cases of Man Overboard by the National Maritime Occupational Health and Safety Committee*
>
> *Bridge Procedures Guide by the International Chamber of Shipping.*

4.6.2 Carry out man overboard drills as required, including manoeuvring of the ship.

4.6.3 If the ship carries a fast-rescue craft launch it once a month and test it in the water in a harbour or safe anchorage.

4.6.4 Ensure that all launch and recovery wires are in good condition, and that sheaves and working parts are functioning and well lubricated, as well as carrying out the statutory inspection. Wear a safety harness if there is any risk of falling.

4.6.5 Check communications with the deck and bridge before launching the rescue boat.

4.6.6 There should be as few people as possible in the boat when launching and recovering it.

4.6.7	Wear working lifejackets that do not restrict free movement.
4.6.8	Wear survival suits whenever there is a risk of hypothermia.
4.6.9	Wear other necessary protective equipment including helmet, gloves, safety shoes and suitable clothing.
4.6.10	Drills in harbours or anchorages should be as realistic as possible. Practise manoeuvring the lifeboat and recovering a training dummy from the water.
4.6.11	Train rescue boat crews in the correct techniques for getting people out of the water without injuring anybody.

4.7 What to do if someone falls overboard

4.7.1	If a person is seen falling over the side of the ship, notify the officer of the watch or bridge team immediately (see further reading in section 4.6.1).
	If a person is reported missing, take the appropriate action in line with the further reading in section 4.6.1.
4.7.2	The master, on taking charge, should consider all the ways to recover the person. This could include directly from the sea via a bunker or pilot door; ladder or gangway; crane or davit; fast-rescue craft or other boat.
4.7.3	The master should ensure that the ship is ready to receive the casualty (see further reading in section 4.6.1).

4.8 Drills for enclosed spaces

4.8.1	Read these guidelines together with Chapter 15.
4.8.2	By law you must carry out enclosed space entry and rescue drills at least every two months on vessels to which the Merchant Shipping and Fishing Vessels (Entry into Enclosed Spaces) Regulations 2022 apply.
	📖 *SI 2022/96; SOLAS III 19.3.3*
4.8.3	Train seafarers in how to identify and use the correct equipment before entry, and how to use a portable oxygen meter/multimeter.
4.8.4	Train seafarers in how to enter an enclosed space and rescue a person(s) from an enclosed space. Vary the drills to include the different types of spaces listed in Chapter 15.
4.8.5	Carry out regular drills to test whether the ship's rescue plan works under different and difficult circumstances. You can use a non-enclosed space as long as it has realistic conditions for an actual rescue. If that is not possible, make an enclosed space safe (by ventilation) before use.
4.8.6	Record each drill in the official logbook.

4.9 What to do in an enclosed space emergency

4.9.1 Put into use a vessel-specific pre-arranged plan for rescuing a person who has collapsed within an enclosed space. The plan should consider the design of the ship and the risk assessment. Base any actual rescues on this plan. Also consider the need to prepare for relief or back-up personnel.

4.9.2 If the person in the space seems to be showing signs of difficulty when exposed to the atmosphere inside the space, the person overseeing the work outside the space should immediately raise the alarm.

4.9.3 **Warning**
The person at the entrance to the space should never try to enter it before additional help has arrived. Do not attempt a rescue unless you are wearing breathing apparatus, a rescue harness and, whenever possible, communication equipment, a lifeline, and an oxygen meter/multimeter.

4.9.4 **Warning**
Never use emergency escape breathing devices to enter an enclosed space in a rescue.

4.10 Helping a casualty

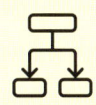

4.10.1 Anyone on board ship may find a casualty and everyone should know what to do first – for example, how to position an unconscious casualty and how to give artificial respiration. The following actions may save life until more qualified help arrives:

- If you find a casualty, first ensure that you are not at risk.
- Remove the casualty from danger if necessary, or remove the danger from the casualty. However, see below on casualties in an enclosed space.
- If there is only **one** unconscious casualty (irrespective of the total number of casualties):
    - Give immediate basic treatment to the unconscious casualty.
    - Call for help.

- If there are **two or more** unconscious casualties:
  - Call for help first.
  - Give appropriate treatment. First treat any casualty who is not breathing or whose heart has stopped.
- If the unconscious casualty is in an enclosed space:
  - Do not enter the enclosed space, but raise the alarm.
  - Assume that the atmosphere in the space is unsafe. The rescue team must not enter unless they are wearing breathing apparatus.
  - Use separate breathing apparatus or resuscitation equipment on the casualty as soon as possible.
- Remove the casualty as quickly as possible to the nearest safe area outside the enclosed space.

4.10.2 **Warning**

Never use your own breathing apparatus on the casualty while you are still in the enclosed space.

4.10.3 If you need to remove injured people from a hold or tank use the best available method. Where practicable, open all access openings and use the following equipment:

- a manually operated davit, suitably secured over the access point
- a cage or stretcher fitted with controlling lines at the lower end.

4.10.4 If a casualty has been injured, is unconscious and/or has been exposed to a hazardous chemical get advice from ashore.

4.11 Drills for leakage and spillage

4.11.1 Carry out drills to simulate a leakage or spillage of a dangerous or hazardous substance.

4.11.2 Train seafarers so they know where emergency equipment is and how to use it. Also make them aware of the potential dangers and the precautions to take.

> *Information on emergency procedures is available in the International Maritime Dangerous Goods (IMDG) Code, the IMO Emergency Procedures for Ships Carrying Dangerous Goods (EmS) Guide and the vessel's shipboard oil pollution emergency plan.*

4.12 What to do if there is a leakage or spillage

4.12.1 Wear appropriate personal protective equipment (PPE) during any clean-up operation. Information is available in the IMDG Code and/or MSDS. PPE may include eye protection (such as goggles), respirators, a dust mask and protective clothing (see Chapter 8).

4.12.2 Consider the hazards of all substances on board, including whether they are toxic, corrosive, flammable or may produce dangerous vapours:

- If the substance is a significant toxic hazard, use self-contained breathing apparatus.
- If incidents involve flammable gases or flammable liquids, avoid all sources of ignition (eg naked lights, unprotected light bulbs, electric hand tools).
- Water is generally the best firefighting medium for most dangerous goods at sea, but check the relevant EmS schedules.
- Where an EmS schedule advises against the use of foam, use the right type of extinguisher for the type of dangerous goods, such as dry powder or carbon dioxide (CO_2).
- Remove any packages of dangerous goods that are close to the fire.
- Use dry powder or CO_2 extinguishers for substances that are highly reactive with water. If there is no alternative, be aware that using water could cause a dangerous reaction.

4.12.3 There are different recommendations on emergency action depending on the type of vessel, where the goods are stowed and whether a substance is a solid, liquid or gas:

- Any disposal of dangerous goods overboard must be under the master's authority, considering that the safety of seafarers is more important than avoiding pollution of the sea. If safe to do so, collect spillages and leakages of substances, articles and materials identified in the IMDG Code as marine pollutants in receptacles for safe disposal ashore. Use absorbent material for liquids.
- Stow any collected spillages in a safe place for disposal ashore. It might not be effective to collect spillages with absorbent material in a space. You should carry out a risk assessment to decide what precautions to take when entering an enclosed space, even if the space would not normally be considered an enclosed space.
- Check carefully for structural damage after dealing with spillages of highly corrosive substances.

- Keep dangerous goods cool to reduce the risk of heat causing a chemical or physical change in a substance, or damaging packaging and potentially causing breakage and spillage. Cooling may be necessary for a long time.
- If there is an incident under deck, batten down the hatch, stop all ventilation and use the fixed firefighting installation.
- Use self-contained breathing apparatus if there is a risk of exposure to smoke, fumes or toxic gases.
- It is recommended to jettison dangerous goods if they are likely to catch fire. This may be impractical for full or nearly full container loads or other units, so do everything possible to prevent the spread of fire to those containers. If they are still liable to catch fire and explode, move people away from the area as appropriate.

4.13 Damage control drills

4.13.1 Hold a drill at least every three months for crew on passenger ships who are responsible for damage control. In each drill the relevant crew members must report to stations and prepare for their duties, as detailed on the ship's muster list.

SOLAS II-1 19-1; SOLAS III 30.3

4.13.2 Vary the drills to simulate emergency scenarios involving different damage conditions. Carry them out as if there were a real emergency.

4.13.3 If the ship has a damage stability computer on board, use it to assess the simulated damage. Set up a communications link to shore-based support, if this is provided.

4.13.4 At least once a year include a drill to activate the shore-based support, if provided (to comply with SOLAS regulation II-1/8-1.3), to assess the stability of the simulated damage conditions.

4.13.5 For each drill also include:

- operating watertight doors
- demonstrating competence in using the flooding detection system and any cross-flooding and equalisation systems
- checking bilge alarms and automatic bilge pump starting systems
- damage survey instructions and using the ship's damage control systems
- using damage control information and the damage stability computer, if fitted.

SOLAS II-1 19-1

4.14 Emergency steering drills

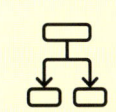

4.14.1 Carry out emergency steering drills at least once every three months to practise emergency steering procedures.

Emergency steering drills should include:

- direct control within the steering gear compartment
- communication procedure with the bridge
- operation of alternative power supplies, where applicable.

📖 *SOLAS V 26*

4.15 MES drills

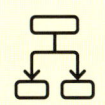

4.15.1 Crew assigned to marine evacuation systems (MES) should undertake drills regularly to practise procedures up to the point before actual deployment of the system.

Use on-board training aids such as manufacturers' instructions and training guidelines for MES to support the drills.

MES crew should take part in a wet deployment on board or at a shore-based facility recognised by the MES manufacturer, at least once every two to three years.

Wet deployment training could form part of the six-yearly rotational wet deployment of MES under SOLAS.

📖 *SOLAS III 19.3.4.8; MGN 558 Amendment 1*

Annex 4.1   How often to hold emergency drills

Drill	Class of vessel requiring specific frequency	Frequency to be carried out	Section reference
Fire drill		Monthly and within 24 hours of leaving port if more than 25% of crew have not taken part in the previous month	4.2
	Passenger vessel: class I	Before or immediately on departure, if more than 25% of crew have not taken part in the previous month	4.2
	Passenger vessel: class I, II, II(A) and III	Weekly, such that each crew member participates in one drill each month	4.2
Abandon ship		Monthly and within 24 hours of leaving port if more than 25% of crew have not taken part in the previous month	4.4
	Passenger vessel: class I	Before or immediately on departure, if more than 25% of crew have not taken part in the previous month	4.4
	Passenger vessel: class I, II, II(A) and III	Weekly, such that each crew member participates in one drill each month	4.4
Person overboard		Regularly	4.6
	Where a fast rescue craft is carried	Launched and tested monthly, if practicable. In any case launched at least once in three months	4.6.3, 4.6.11
Enclosed spaces		Every two months	4.8
Leakage and spillage	Depends on type of vessel and cargo		4.11; 4.11.2 for further information
Damage control	Passenger ships: classes I, II and II(A) and EU classes A to D	Every three months	4.13
Emergency steering		Every three months	4.14
MES		Regularly. Wet deployment training at least once every two to three years	4.15

Annex 4.2 Precautions to take when carrying out launching drills

- Ensure that an experienced officer supervises all drills, that the water is calm and weather conditions are clear.
- Prepare to send help to the boat to be used in the drill in case something goes wrong. For example, where practicable, have a rescue boat or second lifeboat ready for launching.
- When practicable, carry out the drill when the ship has minimal freeboard. When launching free-fall lifeboats, make sure there is a clear and unobstructed launch pathway and an adequate depth of water for the proposed launch.
- The officer in charge should give instructions to the boat's crew on what to do before the drill starts.
- The minimum number of crew members should be in the boat, compatible with the type of boat and type of training.
- Crew must wear lifejackets, and immersion suits where appropriate.
- Crew must wear head protection unless the boat is totally enclosed.
- For the purposes of the drill, remove skates (where fitted) unless they are designed to be retained under all launch conditions.
- Close all openings of totally enclosed boats, except the helmsman's hatch, which may stay open to provide a better view for launching. For free-fall lifeboats all openings should be closed, and any test launch restraints removed.
- Set up two-way radiotelephone communications between the officer in charge of lowering, the bridge and the boat before starting lowering, and maintain them throughout the exercise.
- During lowering, or in the case of free-fall lifeboats during the launch, and then for recovery and while the boat is close to the ship, ensure that the ship's propeller is not turning, if practicable.
- Start up the boat's engine before it enters the water.
- Hold a debriefing session after the launching and recovery to consolidate the lessons that the crew have learned.

5 Fire precautions

5.1 Introduction

5.1.1 Prevention of fire on board ship is essential. Sections 5.2 to 5.6 outline some important organisational measures to reduce the risk of fire. Advice to seafarers is included in Chapter 4.

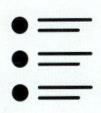

Key point
- Good housekeeping and safe disposal of oil-contaminated materials can prevent fires from developing.

Your organisation should
- make all personnel fully aware of the precautions necessary to prevent fires.
- ensure that designated smoking areas are clearly marked, referenced within vessel procedures/standing orders and identified during shipboard familiarisation tours.
- ensure a competent person inspects all electrical equipment in use on board regularly and confirms it is safe for use.

5.2 Electrical and other fittings

5.2.1 Secure all electrical appliances and provide permanent connections whenever possible. Keep all electric wiring well maintained, clean and dry. Never exceed the rated load capacity of the wires and fuses.

5.2.2 Flexible leads should be as short as practicable and arranged to prevent their being chafed or cut in service.

5.2.3 Do not use makeshift plugs, sockets or fuses.

5.2.4 Do not overload circuits because this causes the wires to overheat, destroying insulation and resulting in a possible short-circuit, which could start a fire. Show notices warning that workers should get approval from a responsible officer before connecting any personal electrical appliances, such as mobile phones, to the ship's supply.

5.2.5	Take care with devices containing rechargeable lithium-ion batteries, including personal devices such as laptops. These devices must not be left on charge unattended due to the fire hazards that can arise if their batteries are overcharged or damaged.
5.2.6	Inspect portable electrical appliances, including lights, before every use and **isolate** them from their source of electrical supply after use. Consider measuring their insulation resistance before first use and regularly thereafter, depending on the location of use/risk of damage.
5.2.7	Electrical equipment used in any cargo area should be of an approved design.
5.2.8	Ensure that all fixed electric heaters are fitted with suitable guards securely attached to the heater and that the guards remain in position.
5.2.9	When using drying cabinets or similar appliances, do not allow the ventilation apertures to be covered up by overfilling the drying space. Inspect any screens or fine mesh covers around the ventilation apertures and clean them regularly so that they do not become blocked by accumulated fluff from clothing.
5.2.10	Avoid using portable heaters except as temporary heating in port, during repairs or adverse weather. Stand the heater on a protective sheet made of a non-combustible material to protect wooden floors or bulkheads, carpets or linoleum. Provide suitable guards for portable heaters and position heaters away from furniture and other fittings.
5.2.11	Drying clothing on or above any type of heater is a fire risk. Personnel must follow instructions and use only appropriate equipment, and only in designated areas.
5.2.12	Electric heaters should be installed in accordance with the relevant regulations and instructions or the manufacturer's guidance.

5.3 Spontaneous combustion

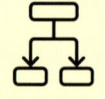

5.3.1 Dirty or damp waste, rags, sawdust and other rubbish, especially if contaminated with oil, may generate heat spontaneously, which may be sufficient to ignite flammable mixtures or set the rubbish itself on fire. Store such waste and rubbish as directed until it can be safely disposed of.

| | 5.3.2 | Materials in ships' stores, including linen, blankets and similar absorbent materials, are also liable to ignite by spontaneous combustion if damp or contaminated by oil. Strict vigilance, careful stowage and suitable ventilation are necessary to prevent this. If such materials become damp, dry them before stowing them away. If oil has soaked into them, either clean and dry them or destroy them. Do not stow them close to oil or paints, or on or near to steam pipes. |

5.4 Smoking

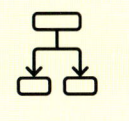

| | 5.4.1 | Display 'no smoking' signs conspicuously in any part of the ship where smoking is forbidden, and strictly enforce this rule. Provide ashtrays or other suitable containers in designated places where smoking is allowed and ensure that workers use them. E-cigarettes are also a source of ignition and should not be used in hazardous areas. |

5.5 Machinery spaces

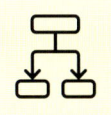

	5.5.1	Make all personnel fully aware of the precautions necessary to prevent fire in machinery spaces. In particular, maintain clean conditions, prevent oil leakage and remove all combustible materials from vulnerable positions (see Chapter 20 for more information).
	5.5.2	Provide suitable metal containers with an integral cover for the storage of cotton waste, cleaning rags or similar materials after use. Empty these containers frequently and dispose of the contents safely.
	5.5.3	Do not store wood, paints, solvents, oil and other flammable materials in boiler rooms or machinery spaces, including steering gear compartments. Store them in a designated location on board with fixed firefighting arrangements.

5.5.4 Carry out routine inspection and maintenance of insulated hot surfaces and equipment with associated piping and fittings that handle flammable liquids.

Pay particular attention to the following (and, where appropriate, the manufacturer's instructions):

- flexible hose installations
- bellows
- filters and strainers
- thermal insulation
- gauge piping
- pipe fittings.

More information and guidance is available in IMO MSC 1/Circ 1321.

5.6 Galleys

5.6.1 Galleys and pantries present particular fire risks (see Chapter 23). Particularly take care to avoid overheating or spilling fat or oil and ensure that burners or heating plates are shut off after cooking. Always keep extractor flues and ranges clean.

5.6.2 Keep a means to smother fat or cooking oil fires, such as a fire blanket, readily available close to stoves. Mark remote cut-offs and stops clearly. Ensure that galley staff know where these are and how to use them.

- Do not use personal portable space-heating appliances of any sort at sea.
- Follow instructions and operating procedures for the correct disposal of waste and to maintain a clear working environment.

- Do not place clothing on, or hang anything above, heaters to air or dry.
- Routinely inspect and maintain electronic equipment.

6 Security on board

6.1 Introduction

6.1.1 Shipboard security is essential to reduce the risks of theft, terrorism, armed robbery, stowaways, piracy and drug smuggling. The *International Ship and Port Facility Security (ISPS) Code*, published by the International Maritime Organization (IMO), was introduced on 1 July 2004 and provides a framework through which ships and port facilities can cooperate to detect and deter acts that threaten security in the maritime transport sector.

> *For more information on topics covered in this chapter, see MCA's Wellbeing at Sea: A Guide for Organisations, section 3.3, and Wellbeing at Sea: A Pocket Guide for Seafarers, section 2.5.*

Key points
- The ship's security officer is responsible for enhancing security and security awareness on board.
- To reduce the likelihood of stowaways search the vessel thoroughly before departure.

Your organisation should
- ensure all personnel are aware of the vessel's security procedures by conducting drills and training, to ensure they take correct action as required
- have the security level set according to the requirements of the port, or according to the flag administration if a higher level is required.

6.2 Ship security plans

6.2.1 The ISPS Code and its parent requirement (SOLAS XI-2) apply to the following types of ships engaged on international voyages:

- passenger ships, including high-speed passenger craft
- cargo ships, including high-speed craft, of 500 gross tonnage and upwards
- mobile offshore drilling units (MODUs).

For UK and EU ships the scope of compliance is extended to include:

- domestic 'Class A' passenger ships (domestic ships which travel more than 20 miles from a place of refuge)
- Class B passenger ships which operate domestic services within United Kingdom waters and are certified to carry more than 250 passengers, and tankers
- port facilities serving any of the types of ships detailed above.

📖 EC 725/2004

6.2.2 The ISPS Code requires a ship security plan (SSP), which is kept up to date and is relevant to the particular ship. The SSP covers, amongst other criteria, the procedures required at different security levels:

- to prevent unauthorised weapons, dangerous substances and devices intended for use against persons, ships or ports from being taken aboard
- to prevent unauthorised access to the ship
- to respond to security threats or breaches
- for the use of the ship's security alert system
- to maintain the ship's security infrastructure.

6.2.3 The SSP is protected from unauthorised access or disclosure, which may mean restricting the distribution of copies amongst the ship's crew. The SSP shall specify the requirement for training drills and exercises. The SSP shall also include the requirement for facilitating shore leave for the ship's personnel or personnel changes, as well as access for visitors to the ship.

6.2.4 The ship security officer is responsible for enhancing security awareness and vigilance on board and for ensuring that adequate training is provided to personnel with security responsibilities.

6.3 Security levels

6.3.1 Governments are required to set one of three security levels for ships flying their flag, and for ports under the government's control. The ship is required to maintain the security level set by the government of the port it is entering unless the ship's government requires a higher security level to be maintained. For UK and Red Ensign ships, the Maritime Security Division (MSD) of the Department for Transport is responsible for setting the security levels and communicates changes direct to company security officers (CSOs) for onward transmission to ships.

📖 ISPS Code A/7

6.4 Precautions

6.4.1 In port, take the appropriate security precautions set out in the SSP. These may include ensuring adequate lighting at night and maintaining a gangway watch.

6.4.2 At sea, take appropriate precautions as set out in the SSP; where appropriate, post additional lookouts and security rounds. Lookouts should be alert to the approach of lit or unlit craft.

6.4.3 At anchor, take appropriate precautions as set out in the SSP, which may include adequate lighting at night and security patrols on deck. Lookouts should be alert to the approach of lit or unlit craft.

6.5 Terrorism

6.5.1 To discourage people from trying to smuggle weapons and explosives on board, display an appropriate sign at all access points stating that 'All items brought on board this ship are liable to be searched.'

6.6 Stowaways

6.6.1 If there is any likelihood of stowaways search the vessel thoroughly before departure. This should include all accommodation, engine room, storerooms, accessible below-deck spaces, lifeboats and any other spaces where a person could hide.

6.7 Piracy and armed robbery

6.7.1 In areas of high risk of piracy or armed robbery take additional precautions in accordance with the ship's latest best management practice (BMP) publication. This should include measures to prevent people boarding the vessel at sea, at anchor or in port.

6.7.2 Ships should maintain anti-piracy watches while transiting areas of high risk. They should report all piratical and armed robbery incidents, including suspicious movements of boats and skiffs, to the 24-hour-manned IMB Piracy Reporting Centre.

6.8 Smuggling

6.8.1 Personnel should be alert to the possibility that people may attempt to smuggle drugs or other contraband on board the vessel. Make them aware of the procedures to follow if such items are found or the activity is suspected.

6.9 Personnel joining and leaving the vessel

6.9.1 Information on personal safety is available through the Foreign, Commonwealth and Development Office (FCDO), British embassies, high commissions and consulates in the area concerned. Strictly follow their advice.

All personnel should know:

- the security level in place on board and should stay vigilant
- their security duties, and have familiarisation training with them
- the security reporting procedures
- the security-related contingency plans as appropriate for their post on board.

- Follow the SSP, policies and procedures.
- Participate in security training and drills as planned in the SSP.

7 Workplace health surveillance

7.1 Introduction

7.1.1 Health surveillance enables employers to identify signs of ill health caused by hazards at work so they can take action to protect employees from further harm.

   SI 1997/2962 as amended

Key points
- It is the employer's responsibility to assess the need for health surveillance based on their risk assessment.
- Health surveillance should be introduced where risk assessment identifies that:
    - a particular work activity may cause ill health
    - an identifiable disease or adverse health condition is related to the work
    - recognised testing methods are available for early detection of an occupational disease or condition; for example, audiometry, lung function tests or skin inspection (where dermatitis is a hazard)
    - it is likely that a disease or condition may occur in relation to particular working conditions or hazards
    - surveillance is likely to protect seafarers' health.

  SI 2007/3100; SI 2007/3075; SI 2007/3077; SI 2010/323; SI 2010/330; SI 2010/332; SI 2010/2984; SI 2010/2987

Your organisation should
- identify where health surveillance is required and relevant
- explain what health surveillance is for and consult seafarers or their safety representatives on how often the procedures should take place
- act promptly on the results to protect people whose health may be harmed and to organise further assessment
- use the results to revise any risk assessments and controls
- consider how to deal with seafarers who may no longer be medically fit to be exposed, or who have restrictions placed on exposure; refer this to an occupational health practitioner

- act on results where potential workplace problems are found
- report the findings of health surveillance to management
- keep records (seafarer's details, where they work, the hazards they have been exposed to and their fitness to continue to be exposed to those hazards) for as long as the seafarer remains employed and exposed to the hazard at work
- offer any seafarer a copy of their health record when they leave their employment or the company ceases trading
- report any occupational disease to the MCA.

Records

7.1.2 Keep health surveillance records in a suitable form for up to 40 years from the date of last entry. There could be a long period between exposure and onset of ill health.

> See the Health and Safety Executive (HSE) website https://www.hse.gov.uk/health-surveillance/record-keeping.htm for information on record keeping.

7.1.3 Health surveillance of seafarers may involve one or more of the following:

- a trained, experienced person inspecting for readily detectable conditions (eg skin damage)
- asking the seafarer if they have any symptoms
- doing hearing checks (audiometry)
- doing medical examinations or company health checks
- testing blood or urine samples.

7.1.4 Decide how often to do these checks based either on suitable general guidance (eg skin inspection for skin damage) or on the advice of a qualified occupational health practitioner.

7.1.5 If it is necessary to take samples or record other personal information, respect the confidentiality of individual health records containing clinical information.

Where the employer identifies no risks to health through their risk assessment, health assessment may not be appropriate.

7.1.6 Health surveillance may be useful for:

- checking whether health control measures are effective
- early identification of changes in exposure patterns or trends through changes in marker chemicals or biological indicators in individuals or groups of people
- providing feedback on the accuracy of health risk assessment
- identifying and protecting individuals who are at increased risk.

7.1.7 Health surveillance is not a substitute for controlling risks to health and safety; control measures should always be the priority to reduce risk. Nor is it the same as medical examinations (eg pre-employment, on resumption of work after sickness, periodic examinations) to assess seafarers' fitness to work. However, wherever possible health surveillance should be conducted at the pre-employment assessment to set a baseline reference.

Where a seafarer's exposure to a hazard is close to the agreed limit advised by an occupational health practitioner, remove them from exposure before any harm is done.

If symptoms of minor ailments (eg skin rash) are detected, act to prevent them from becoming major health problems.

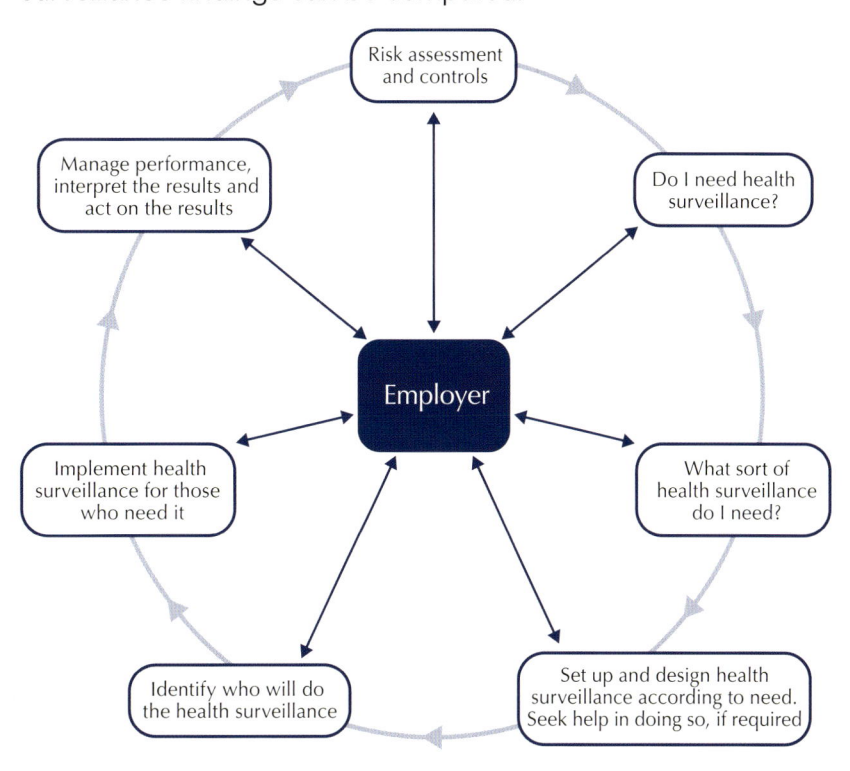

The health surveillance cycle

7.1.8 Figure 7.1 shows the recommended health surveillance cycle. If a risk assessment identifies risks to your health and safety your employer should put control measures in place to remove or reduce risk so far as is reasonably practicable. Employers may also use health surveillance and other records such as sickness reporting to help identify and monitor the impact of any remaining risks. Wherever possible, and where an appropriate health surveillance methodology is in place, your employer should do this when you start working for them, to establish a baseline reference against which future surveillance findings can be compared.

Figure 7.1 The health surveillance cycle

7.1.9 As an employee, you may be entitled to health surveillance if your work activities include:

- exposure to hazardous substances, such as chemicals or biological agents including carcinogens and mutagens
- working with vibrating tools
- exposure to high levels of noise
- use of substances known to cause dermatitis (eg solvents)
- exposure to certain dusts and fumes (eg asbestos, silica, lead)
- ionising and some non-ionising radiations.

7.2 Additional practical guidance on health surveillance for exposure to biological agents

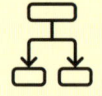

7.2.1 The doctor and/or the authority responsible for the health surveillance of seafarers exposed to biological agents must be familiar with the exposure conditions or circumstances of each seafarer.

7.2.2 Health surveillance must be done in line with the principles and practices of occupational medicine and must include at least the following measures:

- records of the seafarer's medical and occupational history
- a personalised assessment of the seafarer's state of health
- where appropriate, biological monitoring, as well as detection of early and reversible effects.

Further tests may be needed for each seafarer when they are having health surveillance, in the light of the most recent knowledge available to occupational medicine.

7.2.3 Where a seafarer has been exposed to biological agents identified in group 3 or higher (see section 21.9 and Annex 21.1), keep a record for at least ten years following the end of exposure. Where the effect of a disease may be long term, records may need to be kept for 40 years. Get medical advice from the health surveillance provider on how long to keep records.

SI 2010/323; MSN 1889 (M+F) Amendment 3

7.3 Reporting of occupational diseases

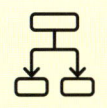

7.3.1 The Merchant Shipping (Maritime Labour Convention) (Health and Safety) (Amendment) Regulations 2014 require employers to report any occupational disease to the MCA once they have received a written report from a doctor.

SI 2014/1616; SI 1997/2962

7.3.2 Guidance on the reporting of occupational diseases is available in Merchant Shipping Notice MSN 1850 (M) Amendment 1.

7.3.3 MSN 1850 (M) Amendment 1 contains a list of reportable diseases.

7.3.4 Form MSF 4159 for the recording of occupational diseases is available to download from http://www.gov.uk.

MSN 1850 (M) Amendment 1; MSF 4159

7.3.5 Send the completed form to the MCA, which will keep the report for statistical purposes. The MCA will then take any necessary action in relation to any occupational disease identified, such as issuing safety alerts or further guidance.

MSN 1888 (M+F) Amendment 3

- Health surveillance needs to follow the principles and practices of occupational medicine.
- Risk assessments are required to identify risks to seafarer health.
- Seafarers should have a copy of their health surveillance record when leaving a company.
- The company must keep health surveillance records for up to 40 years from date of last entry because onset of any illness can take many years to develop.

8 Personal protective equipment

8.1 Introduction

8.1.1 The company must identify and assess all risks to the health and safety of seafarers. It will often not be possible to remove all risks, but the company should consider which control measures will make the working environment and methods as safe as reasonably practicable.

 📖 *SI 1999/2205 Reg 5; MSN 1870 (M+F) Amendment 5*

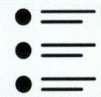

Key points
- Use personal protective equipment (PPE) only when it is not possible to avoid risks or reduce them to an acceptable level through safe working practices. This is because PPE does nothing to reduce the hazard and can only protect the person wearing it, leaving others vulnerable.
- Defective or ineffective protective equipment provides no defence. Therefore you must choose the correct items of equipment and maintain them properly. Keep the manufacturer's instructions safe with the equipment. Refer to them before use and when doing maintenance, as required.

 📖 *Reg 8(4)*

- PPE must be assessed to ensure it effectively reduces the risk to a safe level. If you cannot reduce the risk to a safe level even with relevant PPE then the activity must not be carried out.

8.1.2 Consider all the following controls, ranked in order of effectiveness:

- eliminating the task
- substituting the task with something less hazardous and risky
- enclosing the hazard to eliminate or control the risk
- guarding/segregating workers from the hazard
- introducing a safe system of work that reduces the risk to an acceptable level
- having written procedures that workers know and understand
- reviewing whether there is a good blend of technical and procedural controls
- supervising workers adequately
- identifying workers' training needs
- providing information and instructions (eg signs, hand-outs)
- providing PPE only if you cannot control the hazard by any other means.

8.1.3 Remember that the use of PPE may in itself cause a hazard; for example, by reducing the field of vision or causing loss of dexterity or agility.

Your organisation must
- identify and assess all risks to the health and safety of seafarers
- consider any control measures to make the working environment and methods as safe as possible.

SI 1999/2205; MSN 1870 (M+F) Amendment 5; Reg 5

8.1.4 Provide seafarers with suitable PPE where they need it.

8.1.5 Train all seafarers who need PPE in how to use it. Tell them about its limitations and why they need it. Keep a record of who has had training.

Reg 9; Reg 6(1)

8.1.6 PPE should generally be supplied at no cost to the seafarer, unless it is worn outside the workplace, in which case they may need to contribute to the cost. Alternatively they might want equipment that exceeds the minimum legal standards (eg a more attractive design).

Reg 6(3)

Your organisation should
- assess the equipment required to ensure that it is suitable and effective for the task in question, and meets the appropriate standards of design and manufacture
- ensure that PPE is regularly checked and maintained or serviced, and keep records.

Reg 8

8.1.7 Suitable PPE should:
- be appropriate for the risks involved, and the task being performed, without itself causing increased risk
- fit the seafarer correctly after any necessary adjustment
- take account of ergonomic requirements and the seafarer's state of health
- be compatible with any other equipment that the seafarer is using at the same time, so it continues to be effective.

Defective or ineffective protective equipment provides no defence so proper selection, use and maintenance are essential.

MSN 1870 (M+F) Amendment 5; Reg 6(2)

8.1.8 Details of PPE are listed in a merchant shipping notice (MSN), including the full title of each relevant standard. You must supply the appropriate PPE of the required standard for seafarers doing the tasks listed in the M notice (this is not an exhaustive list). You must supply PPE wherever risk assessment shows there is a risk to health and safety from a work process that you cannot adequately control by other means, but which you can alleviate by providing such clothing or equipment.

Reg 8; Reg 9

8.1.9 Therefore you must choose the correct items of equipment and keep them well maintained. Keep the manufacturer's instructions safe with the equipment. Refer to them before use and when doing maintenance as required. Keep PPE clean and disinfect it when necessary for health reasons.

Reg 8(4)

8.1.10 A **competent person** should inspect each item of protective equipment regularly and before and after every use. Keep records of all inspections. Stow equipment properly in a safe place after use.

Reg 8(4)

8.2 Seafarer duties

8.2.1 Seafarers must wear the protective equipment or clothing supplied when carrying out a task for which it is provided, and follow the instructions for use.

8.2.2 The wearer should check their PPE each time before use. Seafarers should use their PPE as they have been trained and follow the manufacturer's instructions.

Reg 10

8.3 Types of equipment

8.3.1 Overalls, gloves and suitable footwear are the proper working dress for most work about ship. However, these may not protect seafarers against particular hazards in particular jobs. The relevant chapters of this Code make recommendations for the use of special PPE. On other occasions the risk assessment carried out by the officer in charge at the time will identify the need for special protection.

8.3.2 Always choose PPE according to the hazard seafarers are facing and the kind of work they are doing, in line with the findings of the risk assessment.

8.3.3 Box 8.1 shows the main types of PPE.

Box 8.1 **Types of personal protective equipment**

Head protection Safety helmets, bump caps, hair protection

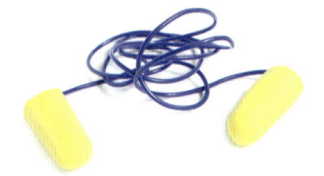

Hearing protection Ear defenders and earplugs

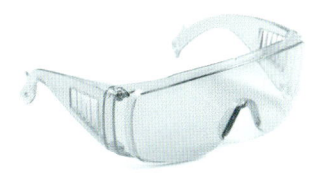

Face and eye protection Goggles and spectacles, facial shields

Respiratory protection Dust masks, respirators, breathing apparatus

Hand and foot protection Gloves, safety boots and shoes

Body protection Safety suits, safety belts, harnesses, aprons, high-visibility clothing

Protection against drowning Lifejackets, buoyancy aids and lifebuoys

Protection against hypothermia Immersion suits and anti-exposure suits

8.4 Head protection

Safety helmets

8.4.1 Safety helmets protect mainly against falling objects, but can also protect against crushing, a sideways blow, and chemical splashes.

8.4.2 Since the hazards may vary, no single type of helmet is the ideal form of protection. Design details are normally decided by the manufacturer who is mainly concerned with compliance with an appropriate standard (see section 8.1.8). The standard selected should reflect the findings of the risk assessment.

8.4.3 The shell of a helmet should be of one-piece seamless construction designed to resist impact. The harness or suspension, when properly adjusted, forms a cradle for supporting the protector on the wearer's head. The crown straps help absorb the force of impact. They are designed to permit a clearance of approximately 25 mm between the shell and the skull of the wearer. The harness or suspension should be properly adjusted before a helmet is worn. Safety equipment should be used in line with the manufacturer's instructions.

Bump caps

8.4.4 A bump cap is an ordinary cap with a hard penetration-resistant shell. Bump caps are useful as protection against bruising and abrasion when working in confined spaces, such as a main engine crankcase or a double-bottom tank. They do not, however, give the same protection as safety helmets and are intended only to protect against minor knocks.

Hairnets and safety caps

8.4.5 Seafarers working on or near to moving machinery should always be aware of the possibility of their hair becoming entangled in the machinery. They should cover long hair with a hairnet or safety cap when working with or near moving machinery.

8.5 Hearing protection

8.5.1 All seafarers exposed to high levels of noise (eg in machinery spaces) should wear ear protection that is suitable for the circumstances. There are three types of protectors: earplugs (disposable or permanent) and ear defenders. See Annex 12.3, the Code of Practice for Controlling Risks due to Noise on Ships, and the Merchant Shipping and Fishing Vessels (Noise at Work) Regulations 2007.

SI 2007/3075; MGN 658 (M+F)

Earplugs

8.5.2 The simplest form of ear protection is the earplug, as shown in Figure 8.1. However, it provides limited noise level reduction.

Figure 8.1   Disposable foam earplugs

8.5.3 Rubber or plastic earplugs also have a limited effect. Extremes of high or low frequency make the plug vibrate in the ear canal, causing a loss in protection. It may be difficult to keep reusable earplugs clean on a ship so disposable ones are recommended. People with ear trouble should use earplugs only after seeking medical advice.

How to fit disposable earplugs

Earplugs only protect against noise if they are fitted correctly, as shown in Box 8.2.

Box 8.2	**How to fit disposable earplugs**

- Make sure that your hands are clean before fitting any earplugs.
- Hold the earplug between your thumb and index finger to then roll and compress the whole earplug.
- Use your other hand to reach over your head and pull up and back on your outer ear.
- This straightens the ear canal and makes way for a tight and snug fit.

Insert an earplug in each ear canal and hold for 20–30 seconds. This enables the earplug to expand and fill your ear canal.

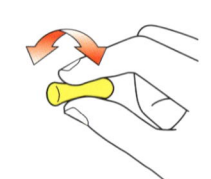

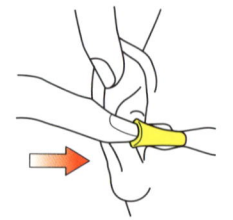

Test the fit of your earplugs

In a noisy place, after putting in your earplugs, cup both your hands over your ears and release them. You should not notice a big difference in the noise level. If the noise level is quieter when your hands are cupped over your ears, your earplugs are probably not correctly fitted. Remove and refit your earplugs.

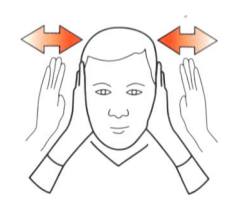

Always remove your earplugs slowly. Twist them to break the seal. Removing your earplugs too quickly could damage your eardrum.

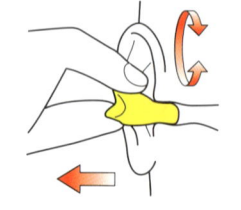

Always read the manufacturer's instructions and get guidance on how to wear earplugs correctly.

Do not reuse disposable earplugs. Do not share your earplugs. Protect your hearing or lose it.

Ear defenders

8.5.4 Ear defenders generally provide better hearing protection than earplugs. They consist of a pair of rigid cups that completely cover the ears, fitted with soft sealing rings to fit closely against the head around the ears. The ear cups are connected by a spring-loaded headband (or neck band), which ensures that the sound seals around the ears are maintained. Different types are available, so provide them according to the circumstances of use and expert advice. Box 8.3 explains how to fit ear defenders.

Box 8.3	**How to fit ear defenders**

Ear defenders protect against noise but only if the cups are fitted and adjusted correctly.

Your ears must be completely enclosed within the ear cups.

Adjust the cups up or down to ensure that the headband fits securely on the crown of your head. They perform best when the cup cushions make a tight seal against your head.

Test the fit of your ear defenders

In a noisy place, place the palms of your hands on both cups, push the cup cushions towards your head then release the cups. You should not notice a big difference in the noise level. If the noise level is quieter when you press the cups, your ear defenders are probably not correctly fitted.

Check the cup cushion regularly for wear and tear. Clean them regularly with a damp hygienic cloth or wipe. If the cup cushions become hard, damaged or deteriorate replace them immediately.

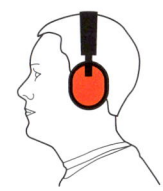

Always read the manufacturer's instructions and get guidance on how to wear ear defenders correctly. Do not share your ear defenders. Protect your hearing or lose it.

8.6 Face and eye protection

8.6.1 The main causes of eye injury are:

- infra-red rays (eg when gas welding)
- ultra-violet rays (eg when electric welding)
- exposure to chemicals
- exposure to flying particles and foreign bodies.

Many types of face and eye protectors are available, designed to international standard specifications, to protect against different types of hazard (see section 8.1.7).

8.6.2 Ordinary prescription (corrective) spectacles do not give any protection unless they are manufactured to a safety standard. Some box-type goggles are designed to be worn over ordinary spectacles.

8.7 Respiratory protective equipment

8.7.1 Respiratory protective equipment is essential when people are working in irritating, dangerous or poisonous dust, fumes or gases. There are two main types of equipment, which perform different functions:

- A respirator filters the air before it is inhaled.
- Breathing apparatus supplies air or oxygen from an uncontaminated source.

8.7.2 The relevant standard gives advice on selection, use and maintenance of the equipment. Make this available to everyone who uses respiratory protective equipment on board ship (see section 8.1.7).

8.7.3 The face-piece of respirators and breathing apparatus must be fitted correctly to avoid leakage. The face seal is affected if the worker wears spectacles, unless they are adequately designed for that purpose, or has a beard. This is particularly important in emergency situations.

Respirators

8.7.4 The respirator must be designed to protect against the hazards being met.

8.7.5 A dust respirator protects against dusts and aerosol sprays but not against gases. Many types of dust respirator are available but they are generally of the ori-nasal type: half-masks covering the nose and mouth.

8.7.6 Many types of light, simple face masks are also available and are useful for protecting against dust nuisance and non-toxic sprays. However, never use them in place of proper protection against harmful dusts or sprays.

8.7.7 Types of respirator include the following:

- **The positive pressure-powered dust respirator** incorporates a battery-powered blower unit, connected by a tube to the face mask to create a positive pressure in the face-piece. This makes breathing easier and reduces face-seal leakage.
- **The cartridge type of respirator** consists of a full face-piece or half-mask connected to a replaceable cartridge containing absorbent or adsorbent material and a particulate filter. It is designed to protect against low concentrations of certain relatively non-toxic gases and vapours.
- **The canister type of respirator** incorporates a full face-piece connected to an absorbent or adsorbent material contained in a replaceable canister carried in a sling on the back or side of the wearer. This type gives more protection than the cartridge type.

8.7.8 The filters, canisters and cartridges in respirators are designed to protect against certain dusts or gases. Different types are available to protect against different hazards; you must choose the right type for the circumstances or conditions. However, remember that respirators have a limited effective life and you must replace or renew them at intervals following the manufacturer's instructions.

> ⚠️ **8.7.9 Warning**
> Respirators do not protect against an oxygen-deficient atmosphere.

Respirators are designed to purify the air of specific contaminants and do not supply any air. Never use them for protection in enclosed spaces such as tanks, cofferdams, double bottoms or other similar spaces against dangerous fumes, gases or vapours. Only breathing apparatus (self-contained or airline) gives protection in such circumstances.

Personal gas monitors

8.7.10 Seafarers should carry personal gas monitors when working in enclosed spaces. A competent person should determine the type of monitor within a safe system of work. It will depend on the circumstances and which contaminants might be present.

8.7.11 Where there is a potential risk of flammable or explosive atmospheres you will need monitors specifically designed to measure for these. All such monitors should be specifically suited for use in potentially flammable or explosive atmospheres.

8.7.12 Monitors should be in good working order. Calibrate and test them in line with either the manufacturer's recommendations or another schedule identified from the findings of the risk assessment.

8.7.13 Seafarers should use personal gas monitors only in line with the procedures set out in Chapter 15 when entering an enclosed space.

Breathing apparatus

8.7.14 Section 15.12 describes the type of breathing apparatus to use when entering a space that is known to be, or suspected of being, deficient in oxygen, or containing toxic gas or vapours.

8.7.15 Do not use breathing apparatus underwater unless it is suitable for that purpose, and then only in an emergency.

Resuscitators

8.7.16 Provide appropriate resuscitators when any person may need to enter an enclosed space (see Chapter 15).

8.8 Hand and foot protection

8.8.1 Gloves

Choose gloves based on the kind of work being done or the substance being handled:

- Use leather gloves when handling rough or sharp objects.
- Use heat-resistant gloves when handling hot objects.
- Use rubber, synthetic or PVC gloves for handling acids, alkalis, various types of oils, solvents and chemicals in general.
- Follow expert advice on the use of gloves.

8.8.2 Footwear

- Foot injuries most often result from the wearing of unsuitable footwear (eg sandals, plimsolls and flip-flops).
- Seafarers should wear safety footwear that is designed to protect against common injuries (impact, penetration through the sole, slipping, heat and crushing) or other hazards identified in the risk assessment.
- Appropriate footwear is manufactured to various standards appropriate to the danger involved (see section 8.1.7).

8.9 Protection from falls

8.9.1 All personnel who are working at height (i.e. in any position from which there is a risk of falling) should wear a safety harness (or belt with shock absorber) attached to a lifeline (Figure 8.2 shows an example). If a vessel is shipping in rough seas, nobody should work on deck unless absolutely necessary. Where this is unavoidable, seafarers on deck should wear a harness secured by lifeline or attached to the ship's structure. This protects them from falls and from being washed overboard, or against the ship's structure (see also Chapter 17).

8.9.2 Inertial clamp devices allow more freedom in movement.

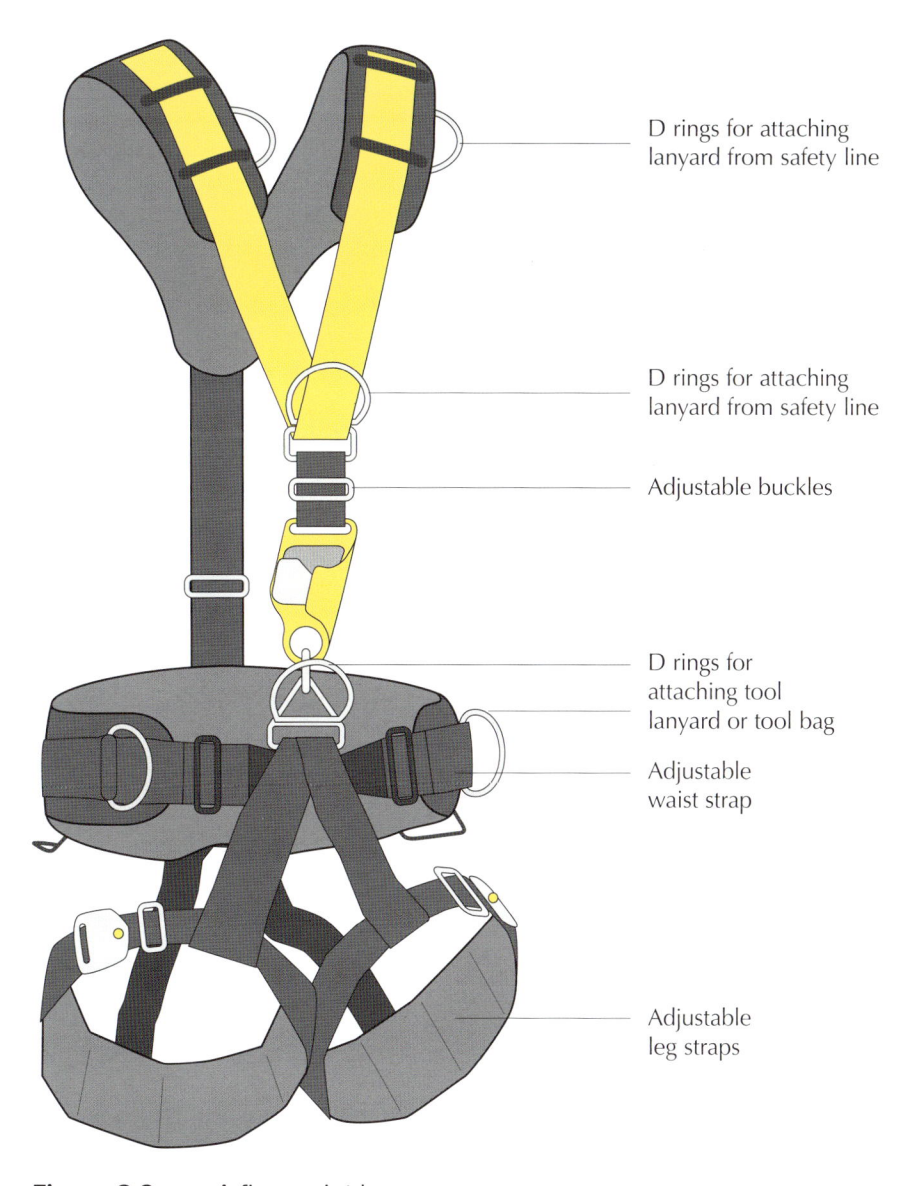

Figure 8.2 A five-point harness

8.10 Body protection

8.10.1 When seafarers are exposed to contaminating or corrosive substances they should wear protective outer clothing. They should wear it only for that purpose and deal with it as directed in the relevant sections of this Code.

8.10.2 Seafarers should wear high-visibility clothing when they need to be seen to be safe (eg during loading and unloading).

8.11 Protection against drowning

8.11.1 Provide a lifebuoy with sufficient line where seafarers are working overside or in an exposed position where there is a risk of falling or being washed overboard, or where they are working in or from a ship's boat. They should also wear a working lifejacket, a personal flotation device or a buoyancy aid, as appropriate (see Figure 8.3). Where necessary, provide thermal protective clothing to reduce the risks of cold shock.

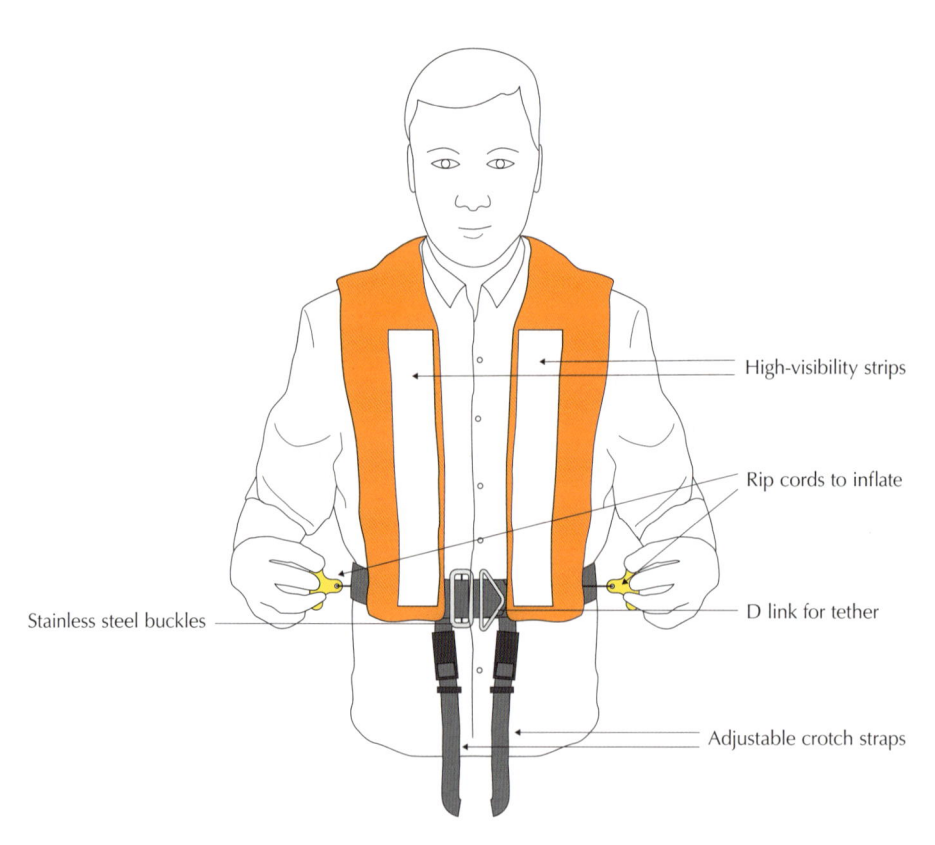

Labels on figure:
- High-visibility strips
- Rip cords to inflate
- D link for tether
- Adjustable crotch straps
- Stainless steel buckles

Figure 8.3 A working lifejacket

- Make sure PPE fits properly, and check it before use.
- Wear PPE every time it is needed.
- Keep the manufacturer's instructions in a safe place.
- Clean PPE after every use.
- If PPE is broken report it immediately.

- The PPE regulations are subject to the general rule that use of PPE is always a last resort. Use PPE only where risks cannot be avoided or reduced to a safe level by means of collective protection, or safe systems of work.
- Do not ask seafarers to pay for their PPE.
- Inappropriate, ill-fitting or defective PPE provides no protection.

9 Safety signs and their use

9.1 Introduction

9.1.1 Any safety signs permanently displayed on board UK-registered ships to give health and safety information or instruction should comply with the applicable regulations and relevant MCA marine notices. Other national and international standards providing an equivalent level of safety will be accepted. The relevant UK notices and International Maritime Organization (IMO) standards relevant to safety signage are listed in Annex 9.1, Parts 1–3.

SI 2001/3444; MGN 556 (M+F) Amendment 1

Key points
- Use safety signs that include hazard warnings to indicate hazards and obstructions or control measures to take where the hazard or obstruction cannot be removed. Pay particular attention on passenger ships to hazards that may be familiar to seafarers but not to passengers.
- Where a language other than English is the working language of the ship, also display any text accompanying a sign in that language.
- Ensure that all seafarers understand the meaning of signs and any colour-coding system in use on their ship and follow the relevant safety procedures.
- Personnel who are aware of any deficiency in their colour vision should tell their supervisor, so that adequate provision can be made where necessary.
- If a seafarer's hearing or sight is impaired (eg by wearing personal protective equipment) take additional measures to ensure that they can see or hear the warning sign or signal; for example, by increasing the brilliance or volume.
- In some cases, more than one type of safety sign may be necessary.
- Avoid placing too many signs together as this may result in confusion and/or overlooking important information.
- If circumstances change, making a particular sign unnecessary (eg if the hazard no longer exists), remove it to avoid displaying misleading information.

9.2 Duty to display signs

Your organisation should
- ensure that safety signs are displayed where appropriate and that the system of signs in use is clearly understood
- take into account the results of the risk assessments and residual risks when determining when and where to use safety signs, to reduce the residual risks even further. See HSE guidance *Safety Signs and Signals*.

9.3 Signs and notices

9.3.1 This section explains the international standards for safety signs. Colours and symbols, when used appropriately, can provide information and warnings of hazards that anyone can understand, regardless of what language they speak. Annex 9.1 shows the types of sign that conform to international systems, where they exist, and European-wide standards.

9.3.2 Symbols relating to life-saving appliances are mandatory and are governed by international standards. Those relating to fire control plans are recommended international standards.

9.3.3 Permanent signs are used to:

- give prohibitions, warnings and mandatory requirements
- mark emergency escape routes
- identify first-aid facilities
- show the location of firefighting equipment.

9.3.4 Red signs mean:

- stop doing something, or do not do it (prohibition)
- stop/shut down or evacuate
- mark the location and type of firefighting equipment.

9.3.5 Prohibition signs are based on a red circular band with a red diagonal bar and white backing. The symbol for the prohibited action is shown in black behind the red diagonal bar; for example, 'No smoking', with a cigarette depicted.

9.3.6 A sign indicating firefighting equipment is a red square or rectangle, with information given in words or by a symbol in white. Alternatively, an IMO sign is a square or rectangle, with information given in words or by a symbol in red.

IMO Resolution A.952(23)

9.3.7 Yellow signs are advisory and mean 'Be careful' or 'Take precautions'. Warning signs are based on a yellow triangle with a black border. The symbol for the hazard is shown in black; for example, 'Poisoning risk' has a black skull and crossbones on a yellow background.

9.3.8 Blue indicates mandatory signs and means 'Take a specific action'. Mandatory signs are based on a blue disc. The symbol for the precaution to take is shown in white; for example, 'Goggles to be worn', with a person's head with goggles depicted. If, exceptionally, no suitable symbol is available, use appropriate wording instead; for example, 'Keep clear'.

9.3.9 Green signs mean emergency escape or a first-aid sign. The sign is a green square or rectangle, with safety information shown by words or a symbol in white. For example, a white arrow on a green background points to an emergency exit.

> *IMO Resolution A.1116(30) (or A.760(18) pre-January 2019); MSN 1676 (M) Amendment 1*

9.3.10 If more information is needed to make clear the meaning of any symbols used in a safety sign or notice, a supplementary sign with text only may appear below the safety sign; for example, 'Not drinking water'.

The supplementary sign should be oblong or square and either:

- white with text in black, or
- the same background colour as the safety colour used on the sign it is supplementing, with the text in the relevant contrasting colour.

9.4 Occasional signs

9.4.1 Illuminated signs, acoustic signals, hand signals and spoken signals may also be used for temporary hazards or circumstances.

9.4.2 Test illuminated signs and acoustic signals regularly to ensure that they are working. Acoustic signals should comply with the IMO Code on Alerts and Indicators, 2009.

9.4.3 Annex 19.3 shows the internationally understood hand signals for the use of lifting appliances.

9.4.4 Spoken signals should comply with the IMO Standard Marine Communication Phrases (SMCP), 2002. This is particularly important when communicating with another ship or with shoreside workers where English is not the official language.

9.5 Electrical wiring

9.5.1 The cores of electrical cables should be readily identifiable throughout their length by colours or numbers. Although various standards (British, other national or international) exist for colour coding of cores, the colours specified in the standards differ. The colours found on any ship will, therefore, depend on the country of building or manufacture of the cables. Take care to positively identify cable duty, and to use colours primarily as a means of conductor tracing.

9.5.2 Be careful when connecting plugs to domestic equipment that has been brought onto a ship, because a wrong connection could prove fatal. UK equipment should be supplied with cable to the EU standard (brown for 'live', blue for 'neutral' and yellow/green for 'earth') but older equipment and that purchased in other countries may have different colours.

9.6 Gas cylinders

9.6.1 There are several standards for the marking of gas cylinders in use globally. Make seafarers aware of the standard in use on board.

BS EN 1089-3:2011

9.6.2 Each cylinder should be clearly marked with the name of the gas and its chemical formula or symbol. Under British standards, the cylinder body should be coloured according to the contents with, where necessary, a secondary colour band painted around the neck of the cylinder to denote the hazards of the gas (eg flammability, toxicity). Figure 9.1 shows an example of colour coding on a gas cylinder label commonly used on board ship. Figure 9.2 shows the colours of gas cylinders.

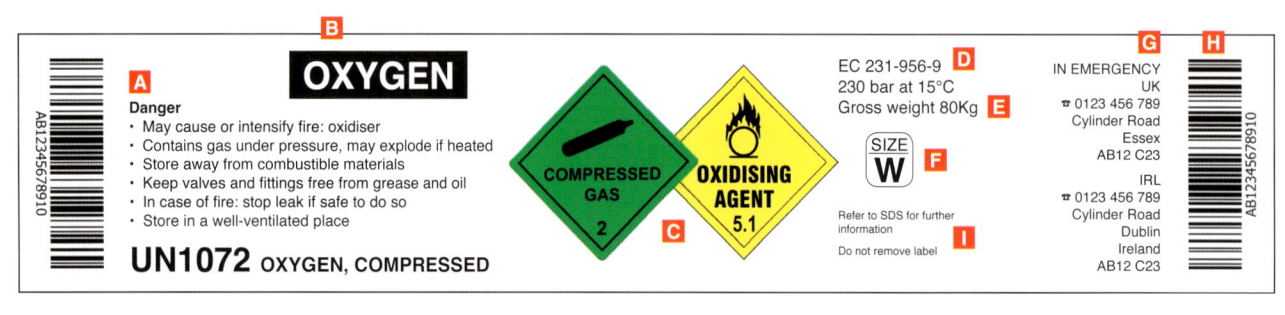

A Hazards and precautions, **B** Name of product, **C** Hazard diamond(s), **D** Filled pressure, **E** Gross weight, **F** Cylinder size, **G** Contact information, **H** Unique cylinder serial number, **I** Additional company information

Figure 9.1 Example of a typical gas cylinder label

Colours and codes are for guidance and illustrative purposes only. **Please always read the label.**

Name of gas	Chemical formula or symbol	Old colour before 2010	New colour of the cylinder after 2010
Oxygen	O_2	Black	Pure white RAL 9010
Carbon dioxide	CO_2	Black	Dusty grey RAL 7037
Compressed air	None – mixed gases	French grey	Grey on cylinder and green on shoulder
Nitrogen	N_2	French grey	Jet black RAL 9005
Acetylene	C_2H_2	Maroon colour BS 541 (Black red RAL 3007)	Maroon colour BS 541 (Black red RAL 3007)
Propane	C_3H_8	Signal red RAL 3001	Signal red RAL 3001
Butane	C_4H_{10}	Not specified	Not specified
Helium	He	Brown	Olive brown RAL 8008

Figure 9.2 Gas cylinder colours

Note: The cylinders of refrigerant gases are not allocated specified ground (cylinder body) or band colours under the British Standard Specification.

9.6.3 Medical gas cylinders carried on board should similarly be marked in line with the relevant British Standard Specification or equivalent (see Appendix 3). The name of the gas or gas mixture in the cylinder should be shown on a label affixed to it. The chemical symbol of the gas should be given on the shoulder of the cylinder. The cylinder should also be colour-coded according to the contents, as shown in Figure 9.3.

The body of the cylinder is to be coloured white (RAL 9010).

Name of gas	Chemical formula or symbol	Old colour before 2010	New colour on the shoulder of the cylinder after 2010
Oxygen	O_2	White on shoulder and black on cylinder body 	Pure white RAL 9010 on both shoulder and cylinder 
Medical oxygen	O_2	White on shoulder and black on cylinder body 	Pure white RAL 9010 on both shoulder and cylinder 
Medical nitrous oxide	N_2O_2	Blue 	Pure white (RAL 9010) on cylinder. Blue RAL 5010 on shoulder 
Compressed air (for breathing apparatus)	None (mixed gases)	French grey BS381 680 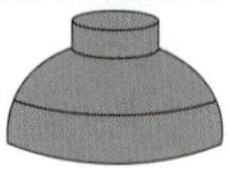	Pure white (RAL 9010) on cylinder. Jet black (RAL 9005) on shoulder 

Figure 9.3 Colour coding of medical gas cylinders

BS EN ISO 407:2004

9.7 Pipelines

9.7.1 Table 9.1 shows the colour-coding system recommended for adoption for the main common pipeline services of UK-registered ships.

BS ISO 14726:2008; BS 4800:2011

Table 9.1 Recommended colour-coding system for pipeline services of UK-registered ships

Pipe contents	Basic identification colour	BS colour reference BS 4800	Colour code band	BS colour reference BS 4800
Water (fresh)	Green	12D 45	Blue	18E 53
Water (salt)	Green	12D 45	None	
Water (fire extinguishing)	Green	12D 45	Safety red	04E 53
Compressed air	Light blue	20E 51	None	
Steam	Silver grey	10A 03	None	
Oil (diesel fuel)	Brown	06C 39	White	
Oil (furnace fuel)	Brown	06C 39	None	
Oil (lubricating)	Brown	06C 39	Emerald green	14E 53

9.7.2 The basic identification colour should be applied on the pipe either along its whole length or as a colour band at regular intervals on the pipe. The colour should similarly be applied at junctions, both sides of valves, service appliances, bulkheads, etc., or at any other place where identification might be necessary. Valves on pipelines used for firefighting should be painted red.

9.7.3 Where applicable, the colour code banding should be in approximately 100 mm widths at regular intervals along the length of the pipe on the basic identification colour or painted between two basic identification colour bands each a width of about 150 mm, as shown in Table 9.2.

Table 9.2 Colour code banding

Pipe contents	Basic colour (150 mm approx)	Colour code (100 mm approx)	Basic colour (150 mm approx)
Water (fresh)	Green	Blue	Green
Water (fire extinguishing)	Green	Safety red	Green
Diesel fuel	Brown	White	Brown

9.7.4 Use the correct colour when replacing or repainting pipes or valves.

9.7.5 When it is necessary to know the direction of the flow of the fluid, show this by placing an arrow near the basic identification colour and painting it white or black to contrast clearly with that colour.

9.7.6 The system recommended in 9.7.5 would be useful, for instance, in tracing a run of pipes, but do not rely on it as a positive identification of the contents of the pipe. Always check before opening the pipe and take precautions in case the content is different from expected.

9.7.7 Other pipeline systems on ships, such as cargo pipelines, may be colour-coded in a similar fashion but no specific recommendations are made here. A comprehensive system to cover the needs of all types of ship would require so many colours that contrasts would be small and easily obscured by fading or dirt.

9.7.8 Colour-coding of pipelines may vary between ships so seafarers moving from one ship to another should check with a competent officer what the colours mean on each particular vessel.

9.8 Portable fire extinguishers

9.8.1 Portable fire extinguishers must comply with the relevant British or European Standard or an equivalent alternative standard.

MSN 1665(M); BS EN 3-10:2009

9.8.2 Fire extinguishers manufactured since May 1997 must comply with BS EN 3. The body of the extinguisher is red, with a zone of colour of between 5% and 10% of the external area to identify the extinguishing agent. Manufacturers have complied with this by printing the operating instructions in the appropriate colour.

9.8.3 It is possible to make extinguishers more visible by highlighting the area around each one with the appropriate colour-coding (as in section 9.8.4). Do not add any colours to the extinguishers because this may invalidate the kitemark.

9.8.4 BS 5423 applied to fire extinguishers manufactured before May 1997. The colour of these extinguishers should not conflict with the recommended systems of colour coding by medium (BS 5306-10:2019) shown in Table 9.3.

BS 5306-10:2019; BS EN 3-7:2004 + A1:2007

Table 9.3 Recommended systems of colour coding by medium

Medium	Colour	Example
Water	Signal red	Water
Foam	Pale cream	Foam
Powder (all types)	French blue	Dry powder
Carbon dioxide	Black	Carbon Dioxide

The colour-coded area should be large enough to be obvious. Where the coding does not cover the whole surface of the extinguisher the remaining area should be either:

- predominantly signal red, or
- self-coloured (natural) metal.

9.8.5 Where there is a mixture of extinguishers of both standards on a ship, try to position them so as to avoid confusion.

- Seafarers must be aware of safety signage symbols, colour coding and their purpose on board ship.
- Safety signage training should form part of an induction process on joining the ship.

- Electrical risks exist with different cabling, connections and colour coding.
- Awareness of illuminated signs, acoustic signals, hand signals and spoken signals is important to maintain accurate communications.

Annex 9.1 Safety signs in accordance with the Merchant Shipping and Fishing Vessels (Safety Signs and Signals) Regulations 2001 and MGN 556 (M+F)

Part 1   Prohibitory signs

Symbol

Meaning

Prohibition – do not do

Examples

No smoking

Smoking and naked flames forbidden

No access for pedestrians

Do not extinguish with water

Not drinkable

No access for unauthorised persons

No access for industrial vehicles

Do not touch

Additional useful prohibition signs

Do not remove electrical earth

Do not operate

Do not use mobile telephones/ Turn off mobile telephones

Unauthorised persons not to use this machine

Do not oil or clean machine whilst in motion

Unauthorised persons may not change grinding wheels

Part 2 Warning signs

Symbol

Meaning

Warning – danger

Examples

Notes

Flammable material or high temperature[1]

Explosive material

Toxic material

Corrosive material

Radioactive material

Overhead load

Industrial vehicles	Danger: electricity	General danger[2]
Laser beam	Oxidant material	Non-ionising radiation
Strong magnetic field	Obstacles	Drop
Biological risk[3]	Low temperature	

1. To be used in the absence of a specific sign for high temperature.

2. This warning sign shall not be used to warn about hazardous chemical substances or mixtures, except for stores containing a number of hazardous substances or mixtures (section 5 of Annex III of MGN 556 (M+F) Amendment 1).

3. Pictogram laid down in Council Directive 90/679/EEC of 26 November 1990 on the protection of workers from the risks related to exposure to biological agents at work (Seventh individual Directive within the meaning of article 16(1) of Directive 89/391/EEC) OJ. No 1,374, 31.12.1990, p. 1.

Additional useful warning signs

Slip or slippery surface	Suffocation (asphyxiation), deficiency in oxygen	Breakthrough hazard

Safety signs and their use

⚠️ Danger of entrapment	 ⚠️ Bump or low deckhead	 ⚠️ **CAUTION** Moving vehicles
 ⚠️ Optical radiation	 ⚠️ Glass hazard, broken glass	 ⚠️ Machinery starts automatically
 ⚠️ Hot surface	 ⚠️ High temperature	  ⚠️ **Danger** Very hot water
  ⚠️ **DANGER** Hot surface	⚠️ **Danger** Explosive atmosphere	 ⚠️ **CAUTION** Moving vehicles
 ⚠️ **Danger** Compressed gas	 ⚠️ **Acetylene**	 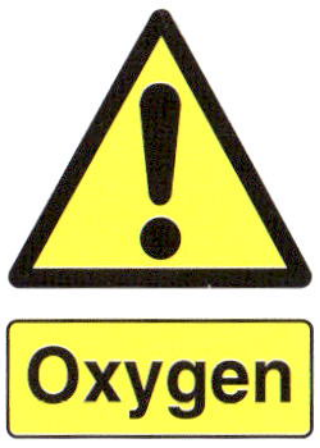 ⚠️ **Oxygen**
 ⚠️ **DANGER** You are entering a CO₂ protected area	 ⚠️ **Caution** Harmful fumes	 ⚠️ **Danger** High noise levels

Part 3   Mandatory signs

Symbol   Meaning

Mandatory – must do

Examples

Eye protection must be worn	Safety helmet must be worn	Ear protection must be worn
Respiratory equipment must be worn	Safety boots must be worn	Safety gloves must be worn
Safety overalls must be worn	Face protection must be worn	Safety harness must be worn
Pedestrians must use this route	General mandatory sign (to be accompanied where necessary by another sign)	

Additional useful mandatory signs

High-visibility clothing must be worn	 Now wash your hands	 Keep clear

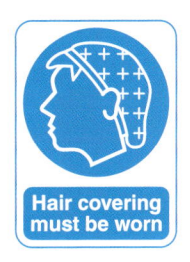

Part 4 Emergency escape, first-aid signs and safe condition

Symbol   Meaning

               Emergency escape, first aid and safe
               condition – the safe way

Examples

Emergency exit/escape route signs

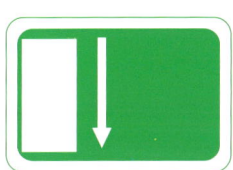

Supplementary information signs

(i) This way

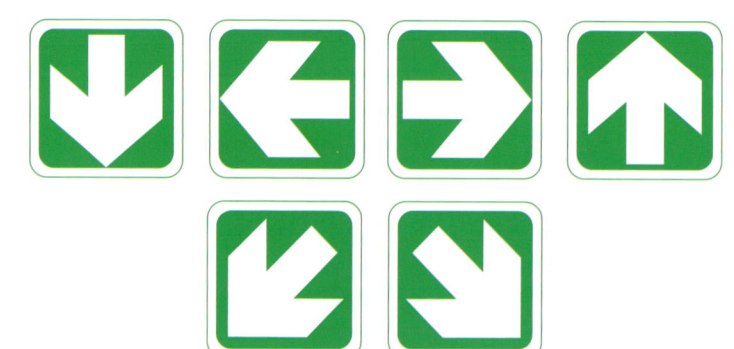

(ii) First-aid signs

First-aid post    Stretcher    Safety shower    Eyewash

(iii) Safe condition signs

Emergency telephone for first aid or escape

Emergency stop for machinery

Additional useful safe condition sign

Drinking water

Part 5 Firefighting symbols

Symbol	Meaning
	Fire equipment – location or use of fire equipment

Examples

| Fire hose | Ladder | Fire extinguisher | Emergency fire telephone |

Supplementary information signs

(i) This way

Fire alarm

Fire extinguisher

Additional useful fire equipment signs

Sprinkler stop valve

Fire flap

Fire extinguishers

Water	Foam	Dry powder
Carbon Dioxide	Water	Foam
Powder	Carbon dioxide	Fire blanket
CO₂ fixed installation	FM 200 fixed installation	NOVEC 1230 fixed installation
Dry powder fixed installation		

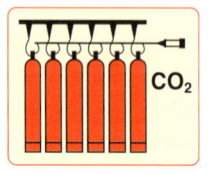

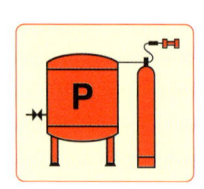

10 Manual handling

10.1 Introduction

10.1.1 This chapter covers manual handling as referenced in the Merchant Shipping and Fishing Vessels (Manual Handling) Regulations 1998. Always use a risk assessment as the basis for appropriate control measures, and put it in place to protect workers who may be affected.

10.1.2 There may be other hazards when handling loads (eg from leakage of a hazardous substance from a package that is being moved). This is covered in Chapter 21 and section 27.11.

For more information on the topics covered in this chapter see MCA's Wellbeing at Sea: A Pocket Guide for Seafarers, section 1.9.

10.1.3 The risk assessment should take full account not only of the characteristics of the load and the physical effort required, but also of the working environment (eg ship movement, confined space, high or low temperature, or physical obstacles such as steps or gangways) and any other relevant factors (eg the age and health of the person, the frequency and duration of the work). Annex 10.1 provides a detailed checklist of factors to consider.

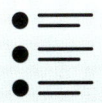

Key points
- The term **manual handling** describes any operation that includes transporting or supporting a load, lifting, putting down, pushing, pulling, carrying or moving by hand or bodily force.
- Move objects by manual handling (as opposed to mechanical means) only when it is safe to do so.
- Take precautions before commencing a lift: assess the load and the area, use the correct technique, and allow lifting only within the person's limitations.

10.1.4 Warning
Accidents, poor organisation or an unsatisfactory working method can cause musculoskeletal injuries.

SI 1998/2857 Reg 5; MGN 90 (M+F) Amendment 1

10.1.5 Limited improvements may be possible on a ship but the company should ensure that, as far as reasonably practicable, it has minimised risks.

Your organisation should
- avoid the need for any hazardous manual-handling operations that may injure seafarers; for example, reorganise, automate or mechanise the work
- train seafarers in appropriate manual-handling techniques. If there is a risk of injury from lifting or carrying by hand, consider whether the job could be done in a different way to reduce the risk
- assess the risk of injury from any hazardous manual-handling activity and reduce it as much as possible
- provide information on the weight of each load and, if appropriate, which side is heaviest.

Reducing risk

10.1.6 Means of reducing the risk of injury may include:
- training seafarers in correct lifting techniques
- reorganising the workplace (so seafarers can keep good posture while lifting or carrying)
- clearly displaying the weight of items, where appropriate, that seafarers may have to move using manual handling
- assessing and removing trip hazards where appropriate
- taking account of an individual's capabilities when allocating tasks.

Personal protective equipment

10.1.7 Seafarers should wear suitable shoes or boots for the job. Protective toecaps help to guard toes from crushing if the load slips.

Instructions

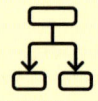

10.1.8 Instruction for personnel should involve experienced and properly trained seafarers demonstrating best practice, especially to new recruits.

Advice to seafarers

Make full and proper use of any system of work provided by the company.

SI 1998/2857 Reg 6

You should:

- use any mechanical aids provided
- follow appropriate systems of work laid down for your health and safety
- take sensible precautions to ensure that you are aware of any risk of injury from a load before picking it up
- cooperate on all health and safety matters
- inform your line manager if you identify hazardous handling activities
- plan the lift – where will you put the load?
- consider whether you need any help. Some loads require two or more people to lift safely. Are appropriate handling aids available? For a long lift, such as deck to shoulder height, consider resting the load midway to change grip (see Figure 10.1)
- assess the load to be lifted, taking account of any information provided (see Annex 10.1)
- look for sharp edges, protruding nails or splinters, surfaces that are greasy or otherwise difficult to grip, and for any other features that may prove awkward or dangerous; for example, sacks of ship's stores may be difficult to get off the deck
- ensure that the deck or area over which the load is to be moved is free from obstructions, especially in narrow accesses, and is not slippery
- check the final stowage location to ensure that it is clear and suitable for the load.

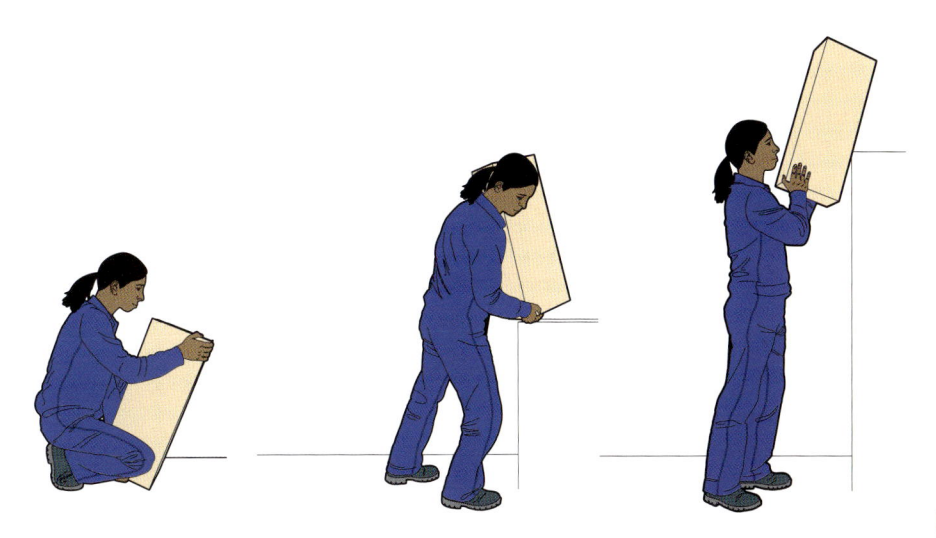

Figure 10.1 Lifting techniques

10.2 Process for good manual handing techniques

10.2.1 Seafarers should wear clothing that does not catch in the load and gives some body protection.

10.2.2 Figure 10.2 shows the stages of the lifting process:

1. Assess the load and the lift before lifting.
2. Take a firm, stable and balanced stance, close to the load. Stand with your feet apart but not too wide, with one leg slightly forward to maintain balance, so that the lift is as straight as possible.
3. At the start of the lift and when lifting from a low level or deck, adopt a crouching position, with knees and hips bent. At the same time maintain the natural curve of the back so your legs do the work. Tuck in your chin while gripping the load, then raise your chin as the lift begins.
4. Grip the load with the whole of your hand, not fingers only. If there is not enough room under a heavy load to do this, put a piece of wood underneath first. A hook grip is less tiring than keeping fingers straight. If you need to vary the grip as the lift proceeds, do this as smoothly as possible.
5. Lift the load by straightening your legs, keeping it close to the body. Keep the heaviest side closest to your trunk. Keep your shoulders level and facing the same direction as your hips. Turning by moving your feet is better than twisting and lifting at the same time. Look ahead, not down at the load, once you are holding it securely.

General lifting

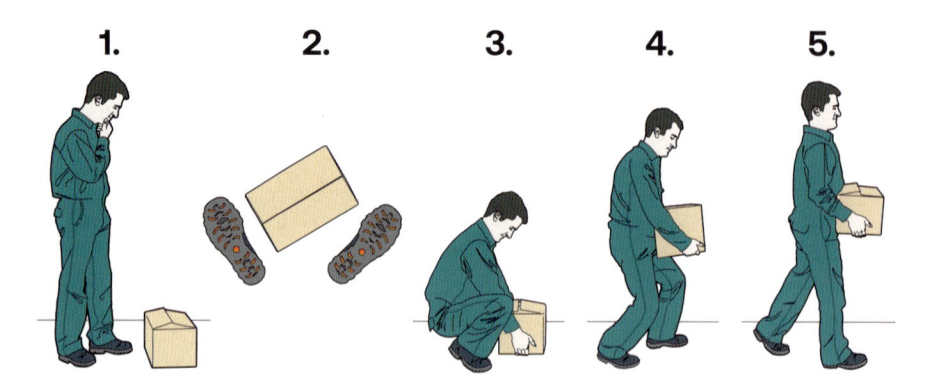

Figure 10.2 Good manual handling techniques

Multi-person lifting

10.2.3 Crews should make sure that there are enough people available to do the task safely.

10.2.4 When two or more people are handling a load, it is better if they are similar height. Lifting, lowering and carrying should, as far as possible, be done in unison to prevent strain and any tendency for either person to overbalance (see Figure 10.3).

10.2.5 Whenever possible organise manual lifting and carrying so that each person has some control over their own rate of work.

Figure 10.3 Multi-person lifting

Carrying a load

10.2.6 Always carry a load in such a way that it does not obscure vision, so you can see any obstructions.

Lowering a load

10.2.7 The procedure for putting a load down is the reverse of that for lifting. Your legs should do the work of lowering with knees bent, back straight and the load close to your body. Take care not to trap fingers. Do not put down the load in an unstable position. If precise positioning is necessary, put the load down first and slide it into position.

It may be possible to slide the load or roll it along. However, uncontrolled sliding or rolling, particularly of large or heavy loads, may introduce fresh risks of injury.

10.2.8 There may be less risk of injury with controlled pushing or pulling instead of lifting.

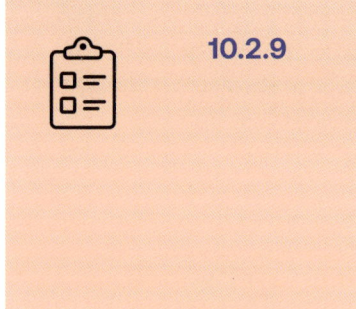

10.2.9 Take particular care if:

- stooping, stretching or twisting is likely
- hands on the load are not between waist and shoulder height
- the deck area is insecure or slippery
- force is applied at an angle to the body
- the load makes sudden or unexpected movements
- the vessel is rolling or pitching.

10.2.10 For pulling and pushing, ensure there is a secure footing. Put your hands on the load between waist and shoulder height wherever possible (see Figure 10.4). The wheels on barrows and trolleys should run smoothly. Tell the supervisor or safety officer if the equipment is not suitable, or is in poor condition.

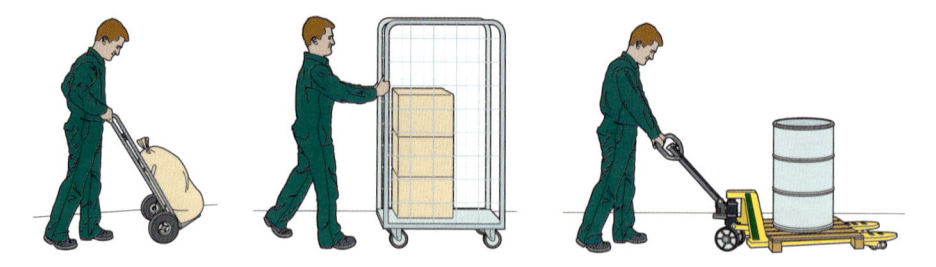

Figure 10.4 Pulling and pushing a load

Where other safety considerations allow, a worker can push with their back against the load, using the strong leg muscles to exert the force (see Figure 10.5).

Figure 10.5 Pushing with back against the load

10.2.11 Even a gentle uphill slope dramatically increases the force needed to push an object, so you may need help when moving a load up a slope or ramp. Take care with unbraked trolleys and sack trucks on a moving/rolling deck, because sudden changes in the angle of deck and direction of the slope may cause loss of control and injury. If a trolley becomes loose, do not try to stop it by standing in its way, but get behind it and try to act as a brake.

10.2.12
- Take care when laying out heavy mooring ropes and wire ropes/hawsers.
- You need a good technique initially when lifting the heavy eye of the rope, then you need a good pulling technique.
- When you are moving a load such as a barrel or drum, it may be safer to roll the load rather than lifting it (see Figure 10.6). You must still take care; consider using a trolley for heavy or large barrels or drums.

Figure 10.6 Moving a barrel or drum

10.2.13 ### Safety for seafarers rigging accommodation and pilot ladders

- Where a work activity involves lifting from deck or overside (eg raising of pilot ladders), follow guidance on body posture and technique to prevent musculoskeletal injury. Where manual handling is unavoidable, lift the ladder from no lower than deck level in stages rather than trying to lift from overside (see section 22.10).
- Get adequate additional manual help and/or appropriate means wherever possible and do a risk assessment.

Do a risk assessment of the dangers associated with this work activity as it involves working overside, which requires a **permit to work** Q (see Chapter 14). It also requires control measures such as a safety line, fall prevention device, safety harness and wearing of lifejackets (see section 17.2.2).

10.2.14 **Warning**
Where the work is very strenuous (eg due to load weight, repetitive effort over a period or environmental factors, such as a confined space or an extreme of temperature), rest at suitable intervals to allow your muscles, heart and lungs to recover. Fatigue makes accidents more likely on this type of work.

- Follow instructions and training on manual handling techniques.
- Everyone involved with manual handling activities needs to be aware of the health and safety responsibilities of themselves and others.
- Communicate clearly so that workers understand instructions; consider any risks and use mechanical means wherever possible.

- Wear appropriate personal protective equipment.
- Notify the supervisor or safety official if mechanical lifting equipment is not working properly.

Annex 10.1 Manual handling factors to consider

Table 10.1 outlines the factors the company should consider when making an assessment of manual-handling operations or when providing instructions for personnel.

The general factors and questions to consider in the risk assessment carried out under the regulations are described in plain text. Some additional factors that may exist on board ship are included for guidance in *italics*.

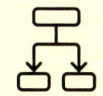

Table 10.1 Manual handling considerations

Factors	Questions
1 The tasks	Does the task involve
	Activity that is too strenuous?
	Holding or manipulating loads at a distance from the trunk?
	Unsatisfactory or unstable bodily movement or posture, especially: • twisting the trunk? • stooping? • reaching upward?
	Excessive movement of loads, especially: • excessive lifting or lowering distances? • excessive carrying distances?
	Risk of sudden movement of loads?
	Frequent or prolonged physical effort, particularly affecting the spine?
	Insufficient rest or recovery periods?
	A rate of work imposed by a process?
	Climbing up or down stairs?
	Handling while seated?
	Use of special equipment?
	Team handling?

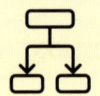

Factors	Questions
2 The loads	Is the load
	Heavy?
	Bulky or unwieldy, or difficult to grasp?
	Unstable or with contents that are likely to shift?
	Likely, because of the contours and/or consistency, to injure workers, particularly if the worker collides with someone or something?
	Wet, slippery, very cold or hot and therefore difficult to hold?
	Sharp?
	Potentially damaging/dangerous if dropped?
3 The working environment	Are there space constraints preventing the handling of loads at a safe height or with good posture?
	Is there an uneven, slippery or unstable deck surface?
	Are there variations in the level of deck surfaces (eg door sills) or work surfaces?
	Are there extremes of temperature or humidity?
	Has account been taken of the sea state, wind speed and unpredictable movement of the vessel?
	Are there steps, stairs, ladders or self-closing doors to be negotiated?
	Is the area adequately lit?
	Is movement or posture hindered by personal protective equipment or clothing?
4 Individual capacity	Is the individual
	Physically unsuited to carrying out the task, either because of the nature of the task or because of a need to protect an individual from a danger that specifically affects them? In other words, does the job require unusual strength, height, etc.?
	Wearing unsuitable clothing, footwear or other personal effects?
	Inadequately experienced or trained?
	Inadequately equipped?
	Is there a hazard to workers who might reasonably be considered unsuited to the task?
	Does the task pose a risk to pregnant women or workers who have a health problem?

Safe weight guidelines

Figure 10.7 shows guidelines on safe weights for manual handling.

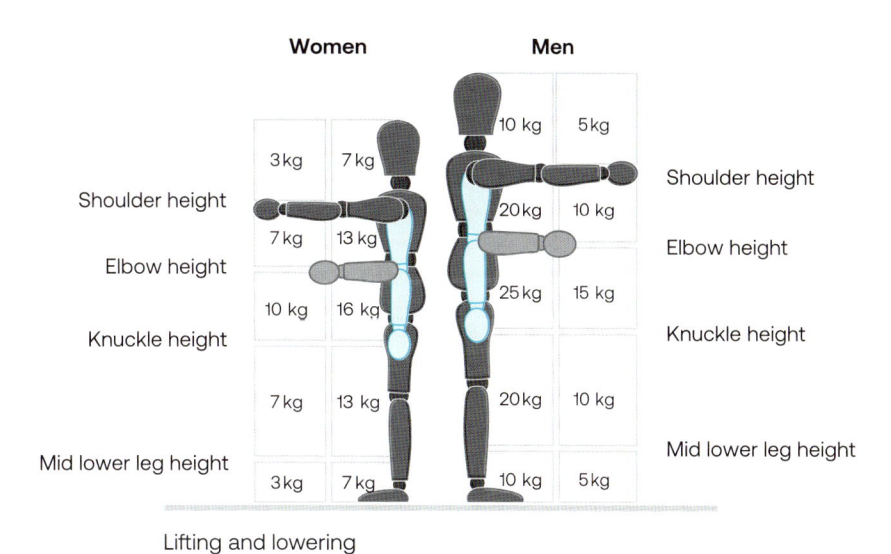

Figure 10.7 Safe weights for manual handling

Safe weights vary depending on the capacity of the individual and the position in which they are holding the weight. Subject to risk assessment, workers may lift lighter weights safely with their arms extended or at high or low levels. Following the guidelines given in Figure 10.7 will reduce the risk of harm. The safe weight is reduced if the seafarer has to twist or carry out the lift repeatedly (say more than 30 times per hour). Also consider the movement of the vessel during the risk assessment for the task. If the load moves through more than one box in the diagram, use the lower weight as the safe weight.

11 Safe movement on board ship

11.1 Introduction

11.1.1 Providing conditions for safe movement on board ship is an integral part of ensuring a safe working environment on board, as required by the Merchant Shipping and Fishing Vessels (Health and Safety at Work) Regulations 1997, regulation 5(2)(e). Following the principles and guidance in this chapter will generally be considered to demonstrate compliance with the duty to ensure a safe working environment on board ship. Where different measures are taken to provide a safe movement, these must provide at least an equivalent level of safety in the operating conditions at the time.

SI 1997/2962; MGN 532 (M) Amendment 2

Key points

This chapter sets out standards to ensure that anyone can move safely to any place on the ship to which they may be expected to go.

These places include accommodation areas, normal places of work and passenger areas. 'People' includes seafarers and other people working on board, passengers, dock workers and other visitors to the ship on business, but excludes people who have no right to be on the ship.

11.1.2 Note particular risks when people work at height, in enclosed spaces or in adverse weather.

11.1.3 **Your organisation should**
- maintain all deck surfaces used for transit about the ship and all passageways, walkways, stairways and fixed ladders
- keep such spaces free from substances liable to cause a person to slip or fall
- provide an adequate level of lighting for areas used for transit, loading or unloading of cargo or for other work processes (see section 11.4 and Annex 11.2)
- ensure that any permanent safety signs providing information for people moving around the ship comply with the regulations and merchant shipping notices
- protect people from falling by use of guards or other appropriate means
- ensure that only competent, authorised people drive the ship's powered vehicles (including mobile lifting equipment) and that the vehicles are properly maintained.

SI 2001/3444; MGN 556 (M+F) Amendment 1

11.2 Drainage

11.2.1 Decks and other places that need washing down frequently, or that can become wet and slippery, should have an effective means of draining water away. Other places include the galley, the ship's laundry and the washing and toilet accommodation.

11.2.2 Inspect drains and scuppers regularly and maintain them properly.

11.2.3 Where drainage is by way of channels in the deck, these should be suitably covered.

11.2.4 Duck boards, where used, should be soundly constructed, designed and maintained. They should be secured to prevent accident.

11.3 Transit areas

11.3.1 Walkways on decks should be clearly marked. Where a normal transit area becomes unsafe to use for any reason, close the area until it can be made safe again. Keep walkways and passageways clear of temporary obstacles.

11.3.2 Transit areas should have slip-resistant surfaces where practicable. Where an area is made slippery by snow, ice or water, spread sand or some other suitable substance over the area. Clean up any spillages of oil, chemicals or grease as soon as possible.

11.3.3 When adverse weather is expected, rig lifelines securely across open decks.

11.3.4	Keep gratings in the deck well maintained and close them when nobody needs access to the space below.
11.3.5	Mark any permanent fittings that may cause hazards to movement (eg pipes, single steps, framing, door arches, and top and bottom rungs of ladders) clearly with contrasting coloured paint, lighting or signage. Temporary obstacles can also be hazardous so mark them appropriately.
11.3.6	When at sea, any gear or equipment that has been stowed to the side of a passageway or walkway should be securely fixed or lashed against the movement of the ship.
11.3.7	Do not leave litter or loose objects, such as tools, lying around. Secure, coil and stow wires and ropes to minimise obstruction.
11.3.8	Look particularly at areas accessed by shore-based workers and passengers, especially on deck, as they will be less familiar with possible hazards.
11.3.9	When lashing and securing deck cargo, including containers, special measures may be needed to ensure safe access to the top of, and across, the cargo.

11.4 Lighting

11.4.1	The level of lighting should be sufficient for all aspects of the work in hand.
11.4.2	Lighting levels should be reasonably constant and minimise glare and dazzle. Avoid deep shadows and sharp contrasts in the levels of lighting between different areas.
11.4.3	Monitor the level of lighting and adjust it, where necessary, to meet the prevailing conditions so that it stays consistent and adequate.
11.4.4	Maintain lighting facilities properly. Report any broken or defective lights to the responsible person and get them repaired as soon as practicable.
11.4.5	Before leaving an illuminated area or space, check that there are no other people there before switching off or removing lights.
11.4.6	Either keep unattended openings in the deck well-lit or close them properly and safely before switching the lights off.
11.4.7	When using portable or temporary lights arrange, secure and cover the light supports and leads to prevent a person tripping, being hit by moving fittings or walking into cables or supports. Keep the leads clear of possible causes of damage (eg running gear, moving parts of machinery, equipment and loads). If leads pass through doorways, secure the doors open. Leads should not pass through doors in watertight bulkheads or fire door openings when the ship is at sea. Never lower or suspend portable lights by their leads.

11.4.8 Where portable or temporary lighting is necessary, fittings and leads should be suitable and safe for the intended usage. To avoid risks of electric shock from mains voltage, in damp or humid conditions use low-voltage (preferably 12-volt) portable lamps, or take other precautions.

11.5 Guarding of openings

11.5.1 Fit any opening, open hatchway or dangerous edge which a person may fall into, through or over with secure guards or fencing of adequate design and construction. Sections 11.5.3 and 11.5.4 give advice on guardrails and safety fencing. However, these requirements do not apply where the opening is a permanent access way or where work is in progress that could not be done with the guards in place.

11.5.2 Close any hatchways that are open for handling cargo or stores, which people may fall through or trip over, as soon as work stops. An exception is during short interruptions or where closing the hatchways would affect safety or mechanical efficiency because of the heel or trim of the ship. In these cases there should be appropriate warning signs next to the open hatchways.

11.5.3 The guardrails or fencing should have no sharp edges and should be properly maintained. Where necessary, provide locking devices and suitable stops or toe-boards. Keep each course of rails substantially horizontal and taut throughout their length.

11.5.4 Guardrails or fencing should consist of an upper rail at a height of 1 metre and an intermediate rail at a height of 0.5 metres. The rails may consist of taut wire or taut chain.

11.5.5 For small motor craft of up to 24 metres length of hull, an equivalent standard can be applied to show compliance with ISO 15085:2003 Small craft – Man-overboard prevention and recovery.

 ISO 15085:2003

11.5.6 Where an opening is a permanent access way, or where work is in progress which could not be done with the guards in place, guards do not have to be fitted during short interruptions in the work (eg for meals). However, post warning signs where the opening is a risk to other people.

11.6 Power-operated watertight doors

11.6.1 Incorrect operation of watertight doors can cause serious injury. All seafarers who may use them should be trained to do so safely and training records kept. Doors should always be used in line with the training.

 MGN 35 (M+F) Amendment 1

Status of watertight door classes A–D during navigation

11.6.2 Class D watertight doors must always be kept closed during navigation.

11.6.3 Class C watertight doors may be opened during navigation to allow passengers or seafarers through. They should close the door immediately afterwards.

11.6.4 Class B watertight doors may be opened during navigation when necessary for work near the door. Close the door as soon as that work is finished.

11.6.5 Class A watertight doors are permitted by the administration to stay open during navigation. Take extra care because if a watertight door is found closed it may automatically close after being opened manually.

11.6.6 Any class of watertight door may be put into bridge operation mode. This means that if it is opened locally it will reclose automatically with enough force to crush anyone in its path as soon as the local control has been released. Always treat doors as if they are in this mode.

11.6.7 The local controls are found on each side of the door and need both hands to operate. A person may open the door using one control, then reach to the other to keep the door open until they are through. No seafarer should try to carry any load through a watertight door unassisted. If they need to do this they should get another person to help.

11.6.8 Notices clearly stating the method of operation of the local controls should be prominently displayed on both sides of each watertight door.

11.6.9 **Warning**
No one should try to pass through a watertight door when it is closing and/or the warning bell is sounding. Always wait until the door is fully open before trying to go through.

11.6.10 Any watertight door found in a closed position must be returned to that position after opening.

11.6.11 Note marine guidance note MGN 35 (M+F) Amendment 1, Accidents when using power-operated watertight doors.

MGN 35 (M+F) Amendment 1

11.7 Stairways, ladders and portable ladders

11.7.1 Stairways on ships are often set at a steep angle so always use handrails. Carry items such as tools in a belt rather than in the hand, to leave hands free.

11.7.2 All ship's ladders should be of good construction and sound material, strong enough for the intended purpose, free from patent defect and properly maintained. Ladders providing access to the hold should comply with the standards in Annex 11.1.

11.7.3 There should be suitable handholds at the top and at any intermediate landing places of all fixed ladders.

11.7.4 Use a portable ladder only where no safer means of access is reasonably practicable (see Figure 11.1).

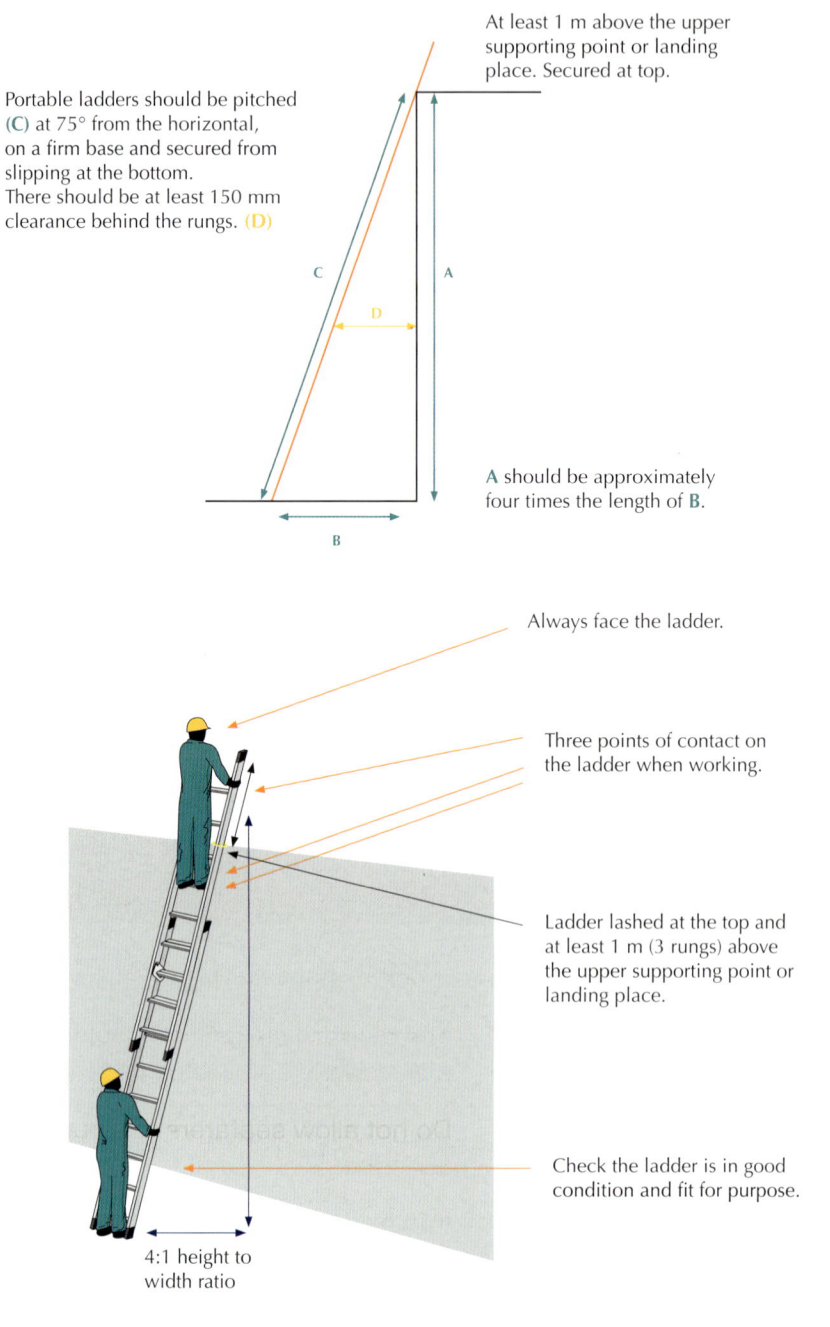

Portable ladders should be pitched (C) at 75° from the horizontal, on a firm base and secured from slipping at the bottom. There should be at least 150 mm clearance behind the rungs. (D)

At least 1 m above the upper supporting point or landing place. Secured at top.

A should be approximately four times the length of B.

Always face the ladder.

Three points of contact on the ladder when working.

Ladder lashed at the top and at least 1 m (3 rungs) above the upper supporting point or landing place.

Check the ladder is in good condition and fit for purpose.

4:1 height to width ratio

Figure 11.1 How to use a portable ladder

> ⚠️ **11.7.5 Warning**
> Use portable ladders only where the motion of the ship does not present a hazard.

11.7.6 Pitch portable ladders at 75° from the horizontal. Secure them properly against slipping or shifting sideways, and to allow a clearance of at least 150 mm behind the rungs. The ladder should extend to at least 1 metre above any upper landing place where practicable, unless there are other suitable handholds.

11.8 Shipboard vehicles

11.8.1 Seafarers who drive the ship's powered vehicles and powered mobile-lifting appliances should have been trained for the category of vehicle or mobile-lifting appliance in use. They should be tested for competence and records maintained.

11.8.2 Either seafarers should be authorised individually in writing to drive vehicles or there should be a list of authorised persons. Make these authorisations available to port authorities for inspection.

11.8.3 Maintain the ship's powered vehicles and mobile-lifting appliances in line with the manufacturer's instructions.

11.8.4 Drivers of the ship's powered vehicles and mobile-lifting appliances should take great care, particularly when reversing.

11.9 Entry into dangerous (enclosed) spaces

11.9.1 A dangerous (enclosed) space is defined in the regulations as 'any enclosed or confined space in which it is foreseeable that the atmosphere may at some stage contain toxic or flammable gases or vapours, or be deficient in oxygen, to the extent that it may endanger the life or health of any person entering that space.' Chapter 15 describes how to identify hazardous spaces and the procedures for entry.

SI 1988/1638

11.10 Working on deck

11.10.1 The responsible officer should ensure that seafarers working on deck are properly instructed in the tasks that they are required to perform.

11.10.2 Do not allow seafarers to sit upon the vessel's bulwark or rail at any time.

11.10.3 Inform deck watchkeeping officers of all work being done on deck or in deck spaces.

11.11 Adverse weather

11.11.1 When expecting adverse weather rig lifelines in appropriate locations on deck.

11.11.2 Nobody should be on deck in adverse weather unless it is necessary for the safety of the ship or life at sea. Where possible delay the work until conditions have improved; for example, until daylight or the next port of call.

11.11.3 Inspect the lashings of all deck cargo. Tighten them, as necessary, when rough weather is expected and check them periodically as conditions allow. Secure the anchors, and fit and seal the hawse and spurling pipe covers, regardless of the expected voyage duration. If ventilation to storerooms has temporarily been stopped during bad weather, seafarers should not enter until the enclosed space entry procedures have been completed (see section 15.1.7).

11.11.4 The master should authorise any work on deck during adverse weather and the bridge watch should be informed. Do a risk assessment, and complete a **permit to work** Q and a company checklist for work on deck in heavy weather.

11.11.5 Any seafarers who need to go on deck during adverse weather should wear a lifejacket suitable for working in, a safety harness (which can be attached to lifelines) and waterproof personal protective equipment (PPE) including full head protection. They should have a water-resistant UHF radio; also consider providing a head-mounted torch.

11.11.6 Seafarers should work in pairs or in teams and be supervised by a **competent person** Q.

11.11.7 Consider using stabilising fins (if fitted) to reduce rolling, and adjusting the vessel's course and speed to mitigate the conditions on deck. If possible, visible communication should be maintained with the bridge; if not, use another continuous means of communication.

 SI 1997/2962; Reg 21

11.11.8 The following list is not exhaustive but identifies points that are often overlooked:

- Watch out for tripping hazards and protrusions such as pipes and framing.
- Be aware that the ship could roll suddenly or heavily at any time.
- It is dangerous to swing on or vault over stair rails, guardrails or pipes.
- Jumping off hatches can cause injuries.

- Keep manholes and other deck accesses closed when not in use; put up guardrails and post warning signs they are open.
- Clean up spillages (eg of oil, chemicals, grease, soapy water) as soon as practicable.
- Treat areas made slippery by snow, ice or water with sand or another suitable substance.
- Place warning signs to indicate where there are temporary obstacles.
- Clear up litter and loose objects such as tools.
- Coil wires and ropes and stow them away securely.
- Rig lifelines securely across open decks in rough weather.
- Stairways and ladders are usually at a steeper angle than is normal ashore.
- Always secure ladders and keep steps in good condition. Take care when using ladders and gangways providing access to or about the vessel, particularly when wearing gloves.
- For guidance on portable ladders see Chapter 22.
- Never obstruct the means of access to firefighting equipment, emergency escape routes and watertight doors.

- Take care while moving about the ship.
- Use appropriate equipment, PPE and clothing.
- Adapt to changes in sea conditions, procedures or equipment that could impact safe movement on the ship.
- Take particular care when working in dangerous areas, such as enclosed spaces or at height.
- Wear suitable footwear. This will protect toes against accidental stubbing and falling loads, give a good hold on deck and provide firm support while using ladders.
- Take extra care when using ladders wearing sea boots.

- Seafarers and other people on board must take care of their own health and safety when moving around the ship.
- Everyone must comply with any measures put in place for their safety.
- Do not operate watertight doors unless you are appropriately trained. Treat watertight doors as if they are in bridge operation mode at all times.

Annex 11.1 Standards for hold access

Hold access: ships built after 31 December 1988

Where the keel of a ship is laid or the ship is at a similar stage of construction after 31 December 1988, the following standards of hold access should be provided:

- The access shall be separate from the hatchway opening, and shall be by a stairway if possible.
- The rungs of a fixed ladder, or a line of fixed rungs, shall have no point where they fill a reverse slope.
- The rungs of a fixed ladder shall be at least 300 mm wide, and so shaped or arranged that a person's foot cannot slip off the ends. Rungs shall be evenly spaced at intervals of not more than 300 mm and there shall be at least 150 mm clear space behind each rung.
- There shall be space outside the stiles of at least 75 mm to allow a person to grip them.
- There shall be a space at least 760 mm wide for the user's body, except that at a hatchway this space may be reduced to a clear space of at least 600 mm by 600 mm.
- Fixed vertical ladders should be provided with a safe intermediate landing platform at intervals of not more than 9 metres.
- Where vertical ladders to lower decks are not in a direct line, a safe intermediate landing shall be provided.
- Intermediate landings shall be of adequate width, afford a secure footing and extend from beneath the foot of the upper ladder to the point of access to the lower ladder. They shall be provided with guardrails.
- Fixed ladders and stairways giving access to holds shall be so placed as to minimise the risk of damage to them from cargo-handling operations.
- Fixed ladders shall, if possible, be so placed or installed as to provide back support for a person using them; but hoops shall be fitted only where they can be protected from damage to them from cargo-handling operations.

Hold access: ships built before 1 January 1989

Where the keel of a ship was laid or the ship was at a similar stage of construction before 1 January 1989, at least the following standards of hold access should be provided:

- Access should be provided by steps or ladder, except:
  - at coamings; and
  - where the provision of a ladder on a bulkhead or in a trunk hatchway is clearly not reasonably practicable.

In such cases ladder cleats or cups may be used.

- All ladders between lower decks should be used in the same line as the ladder from the top deck, unless the position of the lower hatch (or hatches) prevents this.
- Cleats or cups should be at least 250 mm wide and so constructed as to prevent a person's foot slipping off the side.
- Each cleat, cup, step or rung of a ladder shall provide a foothold, including any space behind the ladder, at least 115 mm deep. Cargo should not be stowed as to produce this foothold.
- Ladders which are reached by cleats or cups on a coaming should not be recessed under the deck more than is reasonably necessary to keep the ladder clear of the hatchway.
- Shaft tunnels should be equipped with adequate handholds and footholds on each side.
- All cleats, cups, steps or rungs of ladders should provide adequate handholds.

Annex 11.2 Standards for lighting

- On access routes for people, plant and vehicles, and in lorry parks and similar areas, the minimum level of illumination should not be less than 20 lux (see HSG38).
- For areas used for loading or unloading of cargo or for other work processes, a lighting level of at least 50 lux should be provided.
- In operational areas, where people and vehicles or plant work together, the minimum level of illumination should not be less than 100 lux.
- In adjoining spaces, the maximum contrast ratio for illumination should be no more than 10:1.
- For transit areas, a level of at least 8 lux should be provided (measured at a height of 1 metre above the surface level) unless:
  - a higher level is required by other regulations, e.g. the regulations for seafarer accommodation (see merchant shipping notice MSN 1844(M) and MGN 481(M) Amendment 1); or
  - provision of such levels of lighting would contravene other regulations, e.g. the Convention on the International Regulations for Preventing Collisions at Sea (COLREG), 1972 (as amended), including signals of distress.
- For access equipment and immediate approaches to it, a lighting level of at least 20 lux should be provided (measured at a height of 1 metre above the surface level), unless:
  - a higher level is required by other regulations; or
  - provision of such levels of lighting would contravene other regulations, e.g. the Convention on the International Regulations for Preventing Collisions at Sea (COLREG), 1972 (as amended), including signals of distress; or
  - where the dangers of tripping or falling are greater than usual because of bad weather conditions or where the means of access is obscured, e.g. by the presence of coal dust. In such circumstances, consideration should be given to a higher level, e.g. 30 lux.
- General rules for where these specific regulations do not apply are given in section 11.4 of this chapter.

MSN 1844; MGN 481 (M) Amendment 1; HSG38

12 Noise, vibration and other physical agents

12.1 Introduction

12.1.1 A **physical agent** is an environmental factor such as noise, vibration, optical radiation (eg ultraviolet and infrared light and heat) and electromagnetic fields that may damage the health of those exposed to them.

> *For more information on topics covered in this chapter see section 3.7 of MCA's Wellbeing at Sea: A Guide for Organisations, and section 1.10 of Wellbeing at Sea: A Pocket Guide for Seafarers. Operators of small vessels should also refer to MGN 436 (M+F) Amendment 3 Whole Body Vibration: Guidance on Mitigating Against the Effects of Shocks and Impacts on Small Vessels.*
>
> *Further guidance is available in the MCA's Official Guide to Complying with the Merchant Shipping and Fishing Vessels (Noise at Work) Regulations 2007 and IMO Noise Code 2014, and MIN 588 (M+F) Amendment 2 which outlines the codes of practice for controlling risks due to noise and vibration on ships.*

Key points
- Noise can be a safety hazard at work as it interferes with communication and makes warnings harder to hear.
- The priority should be to prevent risk by removing exposure to a physical agent (**elimination**).
- Failure to control environmental conditions may also make seafarers suffer from increased fatigue.
- Alternative work methods that eliminate or reduce exposure to noise and vibration should be considered.

12.1.2 Where appropriate refer to relevant publications from the Health and Safety Executive (HSE) or other appropriate bodies or get advice from an occupational hygienist or other competent adviser.

Your organisation should

- Identify through risk assessment where personnel are working in the presence of physical agents that are hazardous to health or safety (including fatigue). Evaluate any potential risks from exposure (see Chapter 1). Take appropriate measures to remove, control or minimise the risk (see section 12.2).
- Determine whether surveillance is required. The risk assessment will provide information to establish whether health surveillance is appropriate (see Chapter 7).
- Provide information and relevant training to seafarers so that they know and understand the risks from any physical agents arising from their living and working environment, the precautions to take and the results of any monitoring of exposure.
- Select equipment by assessing where exposure to a physical agent occurs during use. Refer to the company safety procedures and any instructions and operating data supplied by the manufacturer.
- Not charge seafarers for protective equipment such as personal hearing or eye protection, or for hearing examinations or other assessments, as a result of risks at work.
- Provide appropriate personal protective equipment (see Annex 12.3).

12.2 Prevention or control of exposure to a physical agent

12.2.1

When it is not reasonably practicable to prevent or control exposure to a physical agent, consider a combination of the following:

- choosing plant and using processes and systems of work that minimise exposure to the physical agent
- enclosing the equipment either totally or partially
- minimising the number of people who might be exposed to a physical agent and reducing the period of exposure
- designating risk areas that may have hazardous levels of exposure to a physical agent, and displaying suitable and sufficient warning signs
- measuring hazardous levels of exposure to a physical agent, particularly for the early detection of abnormal exposures, resulting from an unforeseeable event or an accident
- taking collective or individual protection measures
- making plans to deal with emergency situations that could result in abnormally high exposure to physical agents.

Your organisation should

- reduce the risk to seafarers as much as reasonably practicable but, where they do not adequately control the risk to health, appropriate PPE must be provided
- take reasonable steps to ensure that all control measures are properly used and maintained. Seafarers should comply fully with the control measures in force
- be aware that for certain physical agents, specific control measures apply (eg noise and vibration). In cases where failure of the control measures could result in risk to health and safety, or where their adequacy or efficiency is in doubt, the exposure of seafarers should be monitored and a record kept for future reference.

12.3 Consultation

12.3.1 **Your organisation should**

Encourage consultation with ship safety representatives and seafarers about proposals to manage risks from exposure to physical agents and health problems arising from such exposure. Consultation should cover the results of the risk assessment, proposals for control, procedures for providing information and training for seafarers, and any health-monitoring system.

12.4 Seafarer information and training

12.4.1 The company should provide seafarers with sufficient information and training to ensure that they are aware of potential risks to their health from exposure to physical agents. Such information should be provided in the working language of the ship. Training should be in a language understood by the seafarer and should include:

- the nature of such risks
- details of the measures taken to eliminate or reduce to a minimum the risks from the physical agent
- any exposure limit values (ELVs) and the exposure action values (EAVs) or action levels (ALs); see section 12.11
- the results of the risk assessment
- safe working practices to minimise exposure to physical agents
- the correct use of PPE where required
- the circumstances in which seafarers are entitled to health surveillance
- how to detect and report signs of injury
- the importance of detecting and reporting signs of injury.

12.5 Noise: introduction

12.5.1 When exposed to harmful noise levels, which might be too loud or for a long duration, sensitive structures in the inner ear can be damaged, causing permanent hearing loss. This section explains how to assess the levels of noise in the workplace and the steps to prevent any associated problems they may cause.

 SI 2007/3075; MGN 658 (M+F)

12.6 Assessing exposure to noise

12.6.1 Noise is measured in decibels (dB), and is given a weighting depending on the exposure to the type of noise.

12.6.2 'A-weighting', sometimes written as dB(A), is used to measure average noise levels, 'C-weighting', or dB(C), to measure peak, impact or explosive noises. Because of the way our ears work, a 3 dB change in noise level is not very noticeable. Yet every 3 dB doubles the sound pressure level received at the ear, so what might seem like small differences in the numbers can be quite significant.

12.6.3 Annex 12.2 gives guidance on daily exposure to different sound levels and the recommended maximum design limits for different areas on board ship under international standards.

 IMO Resolution MSC 337(91)

12.6.4 Table 12.1 describes the lower and upper noise exposure values, noise exposure limits and, where appropriate, action required to reduce that exposure.

12.6.5 For further information on personal hearing protection see Annex 12.3.

12.6.6 When determining noise exposure action levels, do not take account of the effects of using hearing protection. However, do consider the reduction achieved by hearing protection in the case of ELVs.

12.6.7 Although awareness of decibel levels is important to protect hearing, the distance from the source of the sound and duration of exposure to the sound are equally important.

Table 12.1 Upper and lower noise exposure values, limits and action to take

	Daily/weekly exposure (dB(A))	Peak exposure (dB(C))	Required action
Exposure below action value	0–79	0–134	No action required
Lower exposure action values	80	135	Seafarers should have personal hearing protection that complies with the requirements of the Merchant Shipping and Fishing Vessels (Personal Protective Equipment) Regulations 1999.
Upper exposure action values	85	137	Seafarers must use personal hearing protection that complies with the requirements of the Merchant Shipping and Fishing Vessels (Personal Protective Equipment) Regulations 1999. Seafarers are entitled to have their hearing examined by a doctor, or a suitably qualified person under the supervision of a doctor. Companies should establish and implement a programme of measures to reduce the exposure to noise.
Exposure limit values	87	140	Do not exceed this limit.

12.6.8 As a simple guide, there may be a problem if:

- seafarers have to shout to be clearly heard by someone only 2 metres away
- seafarers' ears are still ringing after leaving the workplace
- seafarers are using equipment that causes loud explosive noises, such as cartridge-operated tools or guns
- seafarers are exposed to high-level impact noise from hammering on metal benches, chipping machines or metal endplates on the decks of roll-on/roll-off ferry vessel ramps
- there is machinery such as diesel engines or generators running in a confined space (eg a ship's engine room)
- seafarers not engaged in the provision of entertainment (eg waiters) have to enter or remain in noisy areas such as discos and nightclubs on cruise ships while working.

12.7 Risk assessment: noise

12.7.1 If exposure to noise may be a problem, a **competent person** should do a risk assessment (see Annex 12.2.)

12.7.2 If any seafarer is likely to be exposed to noise exceeding the lower EAVs set out in Table 12.1, the company should arrange for a competent person to assess the actual level of noise exposure.

12.7.3 The company should:

- keep a record of the noise assessment
- regularly review the noise assessment whenever there is a change in the work being undertaken or when new equipment is introduced that may alter noise levels
- use the assessment to develop an action plan for introducing noise control measures.

12.7.4 It is good practice to review the assessment every two years, because noise levels can change over time as, for example, machinery wears out or working practices change.

12.7.5 Safety signs should be displayed in all areas of the ship where seafarers are likely to be exposed to noise. For further information see Chapter 9.

12.8 Health surveillance: noise

12.8.1 Where risk assessment shows that exposure to noise may be causing problems, the employer is required to provide health surveillance of the seafarers at risk in line with Chapter 7.

12.8.2 Health surveillance should include:

- regular hearing checks to measure the sensitivity of hearing over a range of sound frequencies
- informing employees about the results of their hearing checks
- keeping records
- encouraging seafarers to seek further advice from a doctor when hearing damage is suspected.

12.8.3 The organisation responsible for the carrying out of health surveillance should use the most appropriate form of health surveillance in the circumstances.

12.8.4 Companies should arrange regular hearing checks for all seafarers who are regularly exposed to potentially harmful noise levels.

12.9 Noise arising from music and entertainment

12.9.1 The entertainment industry guidance document *HSG260 Sound advice: Control of noise at work in music and entertainment* is available on the HSE website. The guidance outlined is equally relevant to the provision of music and entertainment on ships, including vessels on inland waterways, although the applicable legislation for ships will be the Merchant Shipping and Fishing Vessels (Control of Noise at Work) Regulations 2007 and not HSE's regulations.

 MGN 658 (M+F); HSG260

Vibration

 SI 2007/3077; MGN 353 (M+F) Amendment 2

12.10 Types of vibration and their effects

12.10.1 Hand–arm (or hand-transmitted) vibration comes from the use of hand-held power tools or other vibrating equipment. Regular and frequent exposure to hand–arm vibration can lead to permanent health effects. Occasional exposure is unlikely to cause ill health.

12.10.2 Whole-body vibration occurs through the shaking or jolting of the body through a supporting surface, such as when controlling or riding on a vessel at high speed in choppy seas, using mobile equipment or standing next to a ship's main engines or generators. Whole-body vibration can also be made worse by poor design of the working environment, incorrect seafarer posture, and exposure to shocks and jolts. A primary symptom of whole-body vibration is back pain.

12.11 Exposure limits set by the vibration regulations

Table 12.2 describes the daily EAVs and ELVs for hand–arm and full-body vibration. Note that these are statutory limits; several classification societies publish lower limits/guidance for crew comfort.

Table 12.2 Daily exposure action values and exposure limit values for hand–arm and whole-body vibration

Value	Hand–arm vibration (standardised to eight-hour reference period)	Whole-body vibration (standardised to eight-hour reference period)	Comments
Daily exposure action value	2.5 m/s² A(8)	0.5 m/s² A(8)	Above this limit companies are required to reduce seafarers' exposure to vibration.
Daily exposure limit value	5 m/s² A(8)	1.15 m/s² A(8)	This is the maximum amount of vibration an employee may be exposed to on any single day.

12.12 Determining vibration levels

12.12.1 The company is required to control the risks from hand–arm and whole-body vibration. In most cases, it is simpler to make an initial broad assessment of the risk rather than try to assess exposure in detail.

12.12.2 During the assessment consider:

- which, if any, processes/operations involve regular exposure to vibration, including that emanating from the vessel itself
- whether there are any warnings of vibration risks in equipment handbooks
- any symptoms that might be caused by hand–arm or whole-body vibration and whether the equipment being used, or the vessel itself, produces high levels of vibration or uncomfortable strains on hands and arms, or is causing back pain.

12.12.3 If the broad assessment shows that exposure to vibration is causing problems, a **competent person** who has read and understood the vibration regulations should do a full risk assessment.

12.12.4 Alternatively, the company may either use available vibration data or, if they want to know for certain if the risk is high, medium or low, take measurements to estimate exposures.

12.12.5 The company may be able to get suitable vibration data from the equipment handbook or the equipment supplier. As long as the data is reasonably representative of the way equipment is used on the vessel it should be suitable for estimating seafarers' exposure.

12.12.6 It is also necessary to note how long seafarers are exposed to vibration. Once the relevant vibration data and exposure times have been collected, it will be necessary to calculate each seafarer's daily exposure. This could be by means of an exposure calculator, such as HSE's on vibration at work available from its website or by using Table 12.3, which shows the simple exposure points system.

Table 12.3   Simple exposure points system

Tool vibration (m/s^2)	3	4	5	6	7	10	12	15
Points per hour (approximate)	20	30	50	70	100	200	300	450

12.12.7 Multiply the points assigned to the tool vibration by the number of hours of daily 'trigger time' for the tool(s) and then compare the total with the EAV and ELV points as shown:

- 100 points per day = exposure action value (EAV)
- 400 points per day = exposure limit value (ELV).

12.12.8 In exceptional circumstances, weekly averaging of daily exposure may be used, which allows for occasional daily exposures above the ELV. However, there are stringent conditions for its use. It will often be practical to spread the exposure over more than one day to keep each day's exposure below the ELV. Also, to qualify for weekly averaging, exposures must be reduced to as low as reasonably practicable, taking into account the special circumstances. They must also usually (on most days) be below the EAV. Where weekly averaging is used, the health surveillance of seafarers should be increased. Weekly averaging is most likely to apply in cases of emergency work.

12.12.9 Reduction below the ELV should not be considered a target – exposure must be reduced as low as reasonably practicable, and routine work should be planned and managed to keep below the EAV. Risks are still significant for exposures between the two values and some people will still be at risk if exposed at the action value. This may mean reducing the time for which the seafarer uses the equipment each day.

12.13 Mitigation

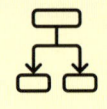

12.13.1 If exposure to vibration is causing problems, the company is required to do all that is practicable to eliminate the risk or minimise it.

12.13.2 The company should group work activities according to whether they are high, medium or low vibration emission risk. It may be useful to identify which tools are high, medium or low vibration emitters by using tags or coloured paint. Action plans should be prioritised for seafarers at greatest risk. As a general guide, follow the controls described in section 12.2.

12.14 Mitigation: hand–arm vibration

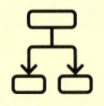

12.14.1 The company should design workstations to minimise the load on seafarers' hands, wrists and arms and, where appropriate, use devices such as jigs and suspension systems, to reduce the need to grip heavy tools tightly.

Selection of efficient equipment

12.14.2 The company should ensure equipment provided for tasks is suitable and can do the work efficiently with the lowest vibration level. The use of high-vibration tools should be avoided wherever possible.

Replacement of equipment

12.14.3 When work equipment needs replacing because it is worn out, the company should choose replacements that are suitable for the work to be carried out efficiently and, wherever possible, cause lower vibration levels. It is strongly recommended that the company has a policy on purchasing suitable equipment, taking account of vibration emissions, efficiency and any specific requirements.

Maintenance of equipment

12.14.4 Maintenance programmes for equipment should be drawn up to prevent avoidable increases in vibration, caused by the use of blunt or damaged equipment or consumable items.

Provide seafarers with training

12.14.5 Seafarers using equipment that can cause vibration should be provided with appropriate training and instruction on its correct use and the signs and symptoms of hand–arm vibration syndrome (HAVS).

Plan to avoid vibration

12.14.6 The company should plan tasks to avoid seafarers being exposed to vibration for long, continuous periods. It is advisable to schedule short periods of exposure with frequent breaks rather than have long uninterrupted exposure.

Provide PPE

12.14.7 Seafarers should be provided with appropriate protective clothing so that they can keep warm and dry. This will help to maintain good blood circulation and reduce the likelihood of vascular symptoms (finger blanching). However, although gloves can be used to keep hands warm, they may not in themselves provide protection from vibration. Anti-vibration work gloves, rather than standard work gloves, should be considered as part of PPE if seafarers are exposed to work involving vibration.

MIN 588 (M+F) Amendment 2

Further guidance is available in the MCA's Official Guide to Complying with the Merchant Shipping and Fishing Vessels (Control of Vibration at Work) Regulations 2007.

12.15 Mitigation: whole-body vibration

12.15.1 Vibration may be reduced by regular maintenance of engines and machinery, and by adjusting the speed of operation or other settings. Seafarers should be provided with information on how to minimise vibration in this way. Severe shocks or jolts should be avoided as far as possible.

12.15.2 When exposure to vibration is unavoidable, reduce the risk of harm by:

- scheduling work to avoid long periods of exposure to vibration in a single day
- planning work so that seafarers do not have to sit in one position for too long
- ensuring that seafarers maintain good posture while working; for example, arranging tasks as far as possible to avoid twisting and stretching
- where possible, adjusting seating to provide good lines of sight, adequate support to the back, buttocks, thighs and feet, and ease of reach for foot and hand controls
- providing adequate rest periods; for example, allow a short break between operations in small fast vessels or mobile machinery and **manual handling**, to give tired muscles time to recover before handling heavy loads
- ensuring that seafarers wear warm, and (if necessary) waterproof clothing in cold and damp conditions; cold exposure may accelerate the onset or worsen the severity of back pain.

When all reasonable steps have been taken to avoid exposure to vibration and to reduce the level of vibration, the final resort for compliance with the ELV is to limit the duration of exposure.

12.15.3 Marine Guidance Note MGN 436 (M+F) Amendment 3 gives guidance on mitigating the risks from whole-body vibration for those working in small, fast craft.

MGN 436 (M+F) Amendment 3

Further guidance is available in the MCA's Official Guide to Complying with the Merchant Shipping and Fishing Vessels (Control of Vibration at Work) Regulations 2007.

12.16 Health surveillance and health monitoring: vibration

12.16.1 If there is a potential risk of harm to seafarers from hand–arm vibration, provide health surveillance for vibration-exposed seafarers in line with Chapter 7. This will apply when seafarers:

- are likely to be regularly exposed above the EAV of 2.5 m/s² A(8)
- are likely to be exposed occasionally above the EAV and where the risk assessment identifies that the frequency and severity of exposure may pose a risk to health
- have a diagnosis of HAVS (even when exposed below the EAV).

12.16.2 Specific guidance on health surveillance for hand–arm vibration risks is available on the HSE website.

https://www.hse.gov.uk/vibration/hav/advicetoemployers/assessrisks.htm

12.16.3 It may be useful to monitor symptoms of back pain to identify health problems and intervene to prevent problems caused or made worse by work activities. Monitoring can also provide information on the effectiveness of the current control methods in place, and help to identify those who are particularly sensitive to whole-body vibration. Older seafarers, people with back problems, young seafarers and pregnant seafarers are at greater risk. Guidance on health monitoring for those at risk from whole-body vibration, including the use of health monitoring questionnaires to monitor seafarers' symptoms, is available on the HSE website.

12.17 Additional guidance

12.17.1 Sources of additional guidance are listed in Marine Guidance Note MGN 353 (M+F) Amendment 2.

📖 *MGN 353 (M+F) Amendment 2*

12.18 Other physical agents

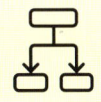

12.18.1 Guidance on protection from artificial optical radiation and electromagnetic fields is listed in the Appendices.

📖 *MGN 428 (M+F) Amendment 2*

- Ensure all seafarers are trained and instructed in using work equipment and have information on how to reduce the effects of noise and vibration.
- Schedule breaks to ensure there are sufficient gaps in work activities, to reduce continuous exposure to noise and vibration.

- Wear PPE as instructed and ensure it is appropriate for the work environment or equipment being used.
- Inspect and maintain the work equipment you use to help reduce the effects of vibration.
- Be aware that back pain is a primary symptom of whole-body vibration.

Annex 12.1 Examples of typical db(A) levels

Noise levels in different locations are given in Table 12.4, to enable personnel to appreciate when and where a potentially harmful noise exposure can exist. These levels are only illustrative and noise levels can vary between similar locations. This is especially true of engine rooms because engine noise can vary considerably with the type of installation.

Table 12.4   Noise levels in different locations

Noise level	Location(s)
120 dB(A)	60 metres from a jet aircraft taking off. Between two running 1800 rpm diesel generators.
110 dB(A)	1 metre from a riveting machine. In a small ship engine room with 900 rpm diesel main engines and a 1550 rpm diesel generator.
105 dB(A)	1 metre from cylinder tops of a slow speed (120 rpm) main diesel engine.
100 dB(A)	Between two running diesel generators (600 rpm).
95 dB(A)	In a slow speed (120 rpm) diesel main engine room at the after end on the floor plate level or in an open side flat.
90 dB(A)	In a noisy factory. In a machine shop with machinery running. In quieter parts of ships' engine rooms.
80 dB(A)	15 metres from a pneumatic drill.
70 dB(A)	Noisy domestic machinery (eg vacuum cleaner at 3 metres).
60 dB(A)	Inside large public building (eg busy supermarket).
50 dB(A)	Inside a house in a suburban area during daytime.
40 dB(A)	In a quiet city area outdoors at night. Whispering in a library at 1 metre.
25–30 dB(A)	In the countryside at night with no wind. In an empty church.
0	Threshold of hearing of young persons of normal hearing.

Annex 12.2 Daily exposure to different sound levels

Seafarers may be exposed to varying levels of noise throughout the day, depending on the length of time they may spend in any particular area of the ship.

Table 12.5 gives a guide to the acceptable maximum daily noise doses for unprotected ears, based on dB(A) sound energy received. The darker area shows that hearing protection is required.

As an alternative illustration and equivalent to the figures in Table 12.5, the maximum daily noise dose for unprotected ears is halved for each increase of 3 dB(A).

Table 12.5 Maximum acceptable daily noise doses for unprotected ears

Less than	Duration	Hearing protection needed?
80 dB(A)	No limit (24 hours)	No
82 dB(A)	16 hours	No
85 dB(A)	8 hours	Yes
90 dB(A)	2 hours	Yes
95 dB(A)	50 minutes	Yes
100 dB(A)	15 minutes	Yes
105 dB(A)	5 minutes	Yes
110 dB(A)	1 minute	Yes

Recommended maximum limits for different areas on board ship

The limits shown in Table 12.6 should be regarded as maximum levels rather than desirable levels and, as appropriate, take account of the attenuation (noise reduction) that can be achieved with ear protectors.

MGN 658 (M+F) Annex 1; IMO Resolution MSC.337(91)

Table 12.6 Recommended maximum limits for different areas on board ship

Designation of rooms and spaces	Ship size 1,600–10,000 GT	>10,000 GT
	Maximum limits (dB(A))	
Work spaces		
Machinery spaces	110	110
Machinery control rooms	75	75
Workshops other than those forming part of machinery spaces	85	85
Non-specified work spaces (other work areas)	85	85
Navigation spaces		
Navigating bridge and chart rooms	65	65
Lookout posts including navigating bridge wings and windows	70	70
Radio rooms (with radio equipment operating but not producing audio signals)	60	60
Radar rooms	65	65
Accommodation spaces		
Cabins and hospitals	60	55
Messrooms	65	60
Recreation rooms	65	60
Open (external) recreation areas	75	75
Offices	65	60
Service spaces		
Galleys, without food processing equipment operating	75	75
Serveries and pantries	75	75
Normally unoccupied spaces		
Cargo holds	90	90
Deck areas	90	90

Annex 12.3 Requirements to provide appropriate personal hearing protection

1. The hearing protection required to be provided by virtue of the Merchant Shipping and Fishing Vessels (Control of Noise at Work) Regulations 2007 is a last resort to control noise exposure.

Key points

Hearing protection should only be used:

- as a short-term measure until other controls to reduce the noise exposure have been introduced
- when all reasonably practicable measures have been taken and a risk to hearing remains.

2. Any hearing protection provided to seafarers is required to comply with the requirements of the Merchant Shipping and Fishing Vessels (Personal Protective Equipment) Regulations 1999. However, not all hearing protectors are the same and different types may be more suitable for different seafarers or indeed the work being undertaken. In this respect the main types of hearing protection are:

    - ear defenders, which completely cover the ear – however, the effectiveness of ear defenders may be reduced if the wearer is also wearing glasses (see section 8.5.4)
    - earplugs, which are inserted in the ear canal (see section 8.5.2)
    - semi-inserts (also called 'canal caps'), which cover the entrance to the ear canal.

    As seafarers frequently have contaminated hands, hearing protection should be of a design to avoid touching the ear canal.

3. In choosing what form of hearing protection to provide, companies should use the results from their noise assessment and information from hearing protection suppliers to make the best choice of hearing protection for the work being undertaken.

Whatever form of protection is chosen, it must:

- reduce employees' noise exposure to below 85 dB(A)
- be suitable for the employees' working environment – consider comfort and hygiene
- be compatible with other protective equipment used by the employee (eg safety helmet, dust mask and eye protection).

HSE advises adding a 4dB(A) real-world correction factor to hearing protection calculations.

Wherever possible, seafarers should be provided with a suitable range of effective hearing protection so they can choose the one that suits them best. Some seafarers may prefer a particular type, or may not be able to use some types of hearing protection because of the risk of ear infections.

Particular consideration should be given to those seafarers who wear spectacles or eye protection similar to spectacles, which have arms that go over the ear. In such cases, ear defenders may not fit securely against the ear because of the presence of the spectacle arms and thus provide inadequate protection against noise. In such circumstances, another form of ear protection may be more suitable.

Companies should ensure that hearing protection works effectively and check that:

- its overall condition is still good and it is clean
- ear defender seals are undamaged
- the tension of the headbands is not reduced
- there are no unofficial modifications
- compressible earplugs are soft, pliable and clean.

5. Companies should ensure that seafarers use hearing protection when required to.

In this context companies may want to

- include the need to wear hearing protection in their safety policy, and put someone in authority in overall charge of issuing it and making sure that replacement hearing protection is readily available
- carry out spot checks to see that the rules are being followed and that hearing protection is being used properly
- consider whether failure to use hearing protectors when required to do so should be included in the company's disciplinary procedures
- ensure that all managers and supervisors set a good example and wear hearing protection at all times when in ear-protection zones.

13 Safety officials

13.1 Introduction

13.1.1 This chapter outlines the regulations, guidance and processes to follow when managing, investigating or reporting safety incidents on board ship.

13.1.2 Some sections apply equally on all ships, whether or not safety officials are appointed or elected by law. Other sections, where indicated, apply only to ships with five or more seafarers, where safety officials are appointed or elected as required by law. The information and guidance here is designed to assist them and to advise companies and masters how to assist them.

Key points

Every person on board is responsible for safety:

- The company is responsible for ensuring the overall safety of the ship and that safety on board is properly organised and coordinated.
- The master has the day-to-day responsibility for the safe operation of the ship and the safety of everyone on board.
- Each employer is responsible for the health and safety of its workers.
- Heads of department are responsible for health and safety in their own department.
- Each officer/manager is responsible for the health and safety of people they supervise and others affected.
- Each individual seafarer or worker is responsible for their own health and safety and that of anyone affected by their acts and omissions.

Your organisation should

- Note that under merchant shipping legislation, specific responsibilities are also given to people with designated roles in ensuring the safety of everyone on the ship. In this chapter, people with a designated safety role on board are called 'safety officials'. This term includes safety officers, safety representatives and other members of safety committees.
- Follow the guidance in Chapter 1. The development of a positive 'safety culture' and high safety standards depend on good organisation and the whole-hearted support of management and all seafarers. People with specific safety responsibilities are more likely to perform well when management is clearly committed to health and safety. It is also important that procedures are in place so that all seafarers can cooperate and participate in establishing and maintaining safe working conditions and practices.

SI 1997/2962

13.2 Employer duties

13.2.1 This section applies to all ships. Every employer is required to appoint one or more competent people to promote health and safety. On board some large ships, where personnel are employed by different employers, each employer must appoint competent people. They do not have to work on the ship themselves, but must be 'competent' for the task. They should know the duties of the workers they are responsible for. They should ensure that any risks encountered as a result of that working environment are dealt with appropriately; for example, by checking that the company has adequate safety procedures for all on board, and by coordinating risk assessments with the company.

Reg 14(1)

13.2.2 The employer may 'appoint' itself where, in a small organisation, there is no one else available to take on this responsibility. Alternatively it may employ someone from outside its own undertaking to advise on health and safety, provided that person is competent.

13.2.3 The employer must provide the competent person(s) with all the information they need to do their job. This includes a copy of the company's safety policy and risk assessments, information about the duties of personnel, and any information provided by other employers about risks and safety procedures in shared workplaces.

Reg 19(1)

13.2.4 The employer must consult workers or their elected representatives on health and safety matters, in particular:

- arrangements to appoint a **competent person** 🔍
- the findings of the risk assessment
- arrangements for health and safety training
- the introduction of new technology.

📑 *Reg 20(1)*

The matters to be discussed might also include selecting work equipment and/or protective clothing and equipment, installation of safety signs, follow-up on accidents and other incidents, and arrangements for health surveillance.

13.2.5 Seafarers and other workers on board, or their elected representatives, must be allowed to make representations to the company or their employer about health and safety matters without disadvantage to themselves. The company should consider such representations, perhaps with the safety committee. Any agreed measures to improve safety should be implemented as soon as possible.

13.2.6 It is also the company's and the employer's responsibility to ensure that workers or their elected representatives have access to relevant information and advice about health and safety matters from inspection agencies and health and safety authorities and, from their own records, about accidents, serious injuries and dangerous occurrences.

📑 *Reg 20(3)*

13.2.7 The company and the employer must give elected representatives adequate time away from their normal duties, without loss of pay, to exercise their rights and carry out their function effectively. Safety representatives must not suffer any disadvantage for doing this.

📑 *Reg 20(4)*

13.3 Company duties

13.3.1 The regulations

13.3.1.1 This section applies only to ships on which five or more seafarers are working. The regulations dealing with safety officials give duties to the company for the appointment of ships' safety officers (see section 13.3.2), the election of safety representatives with specified powers (section 13.3.3) and the appointment of a safety committee (section 13.3.4).

📑 *SI 1997/2962; Regs 15–18*

13.3.1.2 The Secretary of State may grant ad hoc exemptions to specific ships or classes of ships subject to any relevant special conditions. This is to allow different arrangements to be made in cases where the requirements of the regulations would be difficult to apply; for example, on a multi-crew ship with alternate crews working on a regular shift basis. In considering a request for exemption, the MCA needs to be satisfied that effective alternative arrangements exist, and makes it a condition of the exemption that these continue.

13.3.2 Appointment of safety officers

13.3.2.1 On every seagoing ship employing five or more seafarers, the company must appoint a safety officer. The master must record the appointment of a safety officer in the official logbook.

Reg 15

13.3.2.2 The safety officer is the safety adviser aboard the ship. They can assist the company and everyone on board in meeting the statutory responsibilities for health and safety. Some training may be provided on board, but the safety officer should have attended a suitable safety officer's training course. Safety officer training should cover:

- the tasks of the safety committee
- the rights and roles of members of the safety committee
- how to carry out risk assessment and risk management
- how to provide the necessary advice to resolve safety concerns or problems and to encourage adherence to prevention principles
- supervision of safety tasks assigned to crew and other seafarers on board, and passengers where applicable
- accident and incident investigation, analysis and making appropriate corrective and preventive recommendations to prevent their recurrence
- human and organisational factors in safety-critical work
- how to obtain relevant information on a safe and healthy working environment from the competent authority and the company
- effective means of communication with a multinational crew
- the commitment required to promote a safe working environment on board.

In addition, the safety officer should be familiar with the following:

- the occupational safety and health policy and programmes used on board
- the safety tasks assigned to crew and other personnel on board, and passengers where applicable.

- the principles and practice of risk assessment. They should be available to advise people preparing and reviewing risk assessments. It is recognised that, where the safety officer also has other responsibilities (eg chief officer), they may conduct risk assessments themselves. However, the general principle is that the safety officer takes an independent view of safety on behalf of the company.

13.3.2.3 Although not prohibited by the regulations, it is not advisable to appoint the master as the safety officer. This is because the safety officer is required, among other duties, to make representations and recommendations on health and safety to the master.

13.3.2.4 If possible, the company should avoid appointing as safety officer anyone to whom the master has delegated the task of giving medical treatment. This is because one of the duties of the safety officer is to investigate incidents, and they would not be able to give proper attention to this function while providing medical treatment for any casualties.

13.3.3 Election of safety representatives

13.3.3.1 On every ship on which five or more seafarers are working, the company must arrange for the election of safety representatives. The regulations specify that no safety representative may have fewer than two years' consecutive sea service since reaching the age of 18. In the case of a safety representative on board a tanker this must include at least six months' service on such a ship.

Reg 17

13.3.3.2 The company must make rules for the election of safety representatives and cannot disqualify particular people. The company should consult any organisations representing seafarers on the ship when making these rules. The master should organise the election of a safety representative within three days of being requested to do so by any two persons entitled to vote.

13.3.3.3 The number of safety representatives to elect will vary according to the number of seafarers on board and, where appropriate, the number of different departments or working groups. As far as practicable, seafarers at all levels and in all departments should have effective representation.

13.3.3.4 The master must record the election or appointment of every safety representative in writing. This should be either in the official logbook or in the minutes of safety committee meetings (see section 13.3.4).

13.3.3.5		When there is a substantial change in seafarers working on board, the master should remind them of their right to elect new safety representatives.
13.3.3.6		Regulation 17, governing arrangements for the election of safety representatives, does not apply where there are existing agreed arrangements under land-based legislation (The Safety Representatives and Safety Committee Regulations 1977 or The Offshore Installations (Safety Representatives and Safety Committees) Regulations 1989 or the Health and Safety (Consultation with Employees) Regulations 1996).

13.3.4 Appointment of a safety committee

13.3.4.1 The company must appoint a safety committee on every ship with five or more seafarers. The committee must be chaired by the master, and members will include, as a minimum, the safety officer and any elected safety representatives. If practical, in addition to the company's competent person, any competent person appointed by other employers should be invited to attend.

 Reg 17(4)

13.3.4.2 The master must record the appointment of a safety committee in writing. This should normally be in the official logbook or minutes of the committee's meetings.

 Reg 17(5)

13.3.4.3 The composition of a safety committee recommended above does not preclude the appointment of other temporary members. However, the committee should be kept small enough to maintain the interest of members and to function efficiently. Where possible, the relevant shore managers responsible for safety on board may attend safety committee meetings on board ship and should in any event see the committee's minutes. On short-haul ferries where different crews work a shift system, a scheme of alternate committee members may be adopted to secure proper representation.

13.3.4.4 Where large numbers of seafarers work in separate departments (eg passenger ship galleys and restaurants), departmental sub-committees may be formed on lines similar to those of the main committee. These must be chaired by a senior member of the department who should serve as a member of the main safety committee in order to report the views of the sub-committee.

13.3.4.5 It is preferable to appoint as secretary someone other than a safety official, as officials need to concentrate on the discussion rather than on recording it.

13.3.5 Termination of appointments

13.3.5.1 A safety officer's appointment ends when the officer ceases to work in the particular ship or their appointment is otherwise terminated.

13.3.5.2 The appointment of a safety representative cannot be terminated by the company, employer or master. The representative can resign or seafarers can elect a replacement. Otherwise they remain a safety representative for as long as they serve on the ship.

13.3.5.3 A safety committee may be disbanded only when there are fewer than five seafarers working on board the ship. A safety committee can operate whether or not there is an elected safety representative.

13.3.5.4 For ships with fewer than five seafarers on board, the master should ensure that information sharing, training and consultations on health and safety issues are carried out on board.

13.3.6 Support for safety officials

13.3.6.1 The company and master have a duty to facilitate the work of any safety official, providing them with access to a copy of this Code and any relevant legislation, marine notices and other information, including:

- findings of the risk assessment and measures for protection in place
- any other factors affecting the health and safety of people working on the ship
- details of firefighting, first aid and other emergency procedures
- statistical information taken into account when conducting risk assessments.

Reg 19(1)

13.3.6.2 Relevant information might include that concerning dangerous cargoes, maintenance work, the hazards of machinery, plant and equipment, processes and substances in use, and appropriate precautions. All employers should cooperate to obtain information about the findings of their risk assessment.

13.3.6.3 The company and master, cooperating with other employers, must also ensure that safety officials have the necessary resources and means. This will include providing any necessary accommodation and office supplies. They should also allow them sufficient time off from their duties without loss of pay so they can fulfil their functions or undertake any necessary health and safety training.

13.3.7 Company recording of accidents

13.3.7.1 On a ship where no safety officer is appointed under the regulations, the company must ensure that a record is kept of all incidents resulting in death, or serious injury as defined in the Merchant Shipping (Accident Reporting and Investigation) Regulations 2012. This record must be available on request to any elected representative, and any person duly authorised by the Secretary of State.

Reg 19(2); SI 2012/1743 as amended

13.3.8 Receiving representations about health and safety

13.3.8.1 The company and employers must enable seafarers and others working on board or their elected representatives to make representations about health and safety, and should also accept representations or recommendations from the safety officer. The company and master will also receive representations from competent people appointed under Regulation 15, safety officers and safety committees. These should be carefully considered and any agreed measures should be implemented as soon as reasonably practicable.

13.3.8.2 The reaction to such representations will be seen as a measure of commitment to health and safety on board. All representations received, from any source, should be considered carefully. If there is likely to be a delay in giving an answer, tell whoever has made the representations as soon as possible. Implement any safety suggestions, when it is feasible and reasonable to do so, as soon as possible. Give reasons in writing if suggestions for health and safety measures are rejected. Acknowledge all suggestions put forward, whether or not a written response is needed.

13.3.8.3 The master must take a close interest in the work of the safety officials on board. They should check that the safety officer is doing their duties effectively, but should also give encouragement and support. The master is in the best position to ensure that the committee works successfully, by encouraging participation and cooperation from all members.

13.3.8.4 The accident reporting regulations govern when to report an incident to the Marine Accident Investigation Branch (MAIB) of the Department for Transport. It may sometimes be appropriate for the company to inform other ships in the fleet of an incident, and give appropriate recommendations on action to take, in line with the company's safety management system.

SI 2012/1743 as amended

13.4 Duties of safety officers

SI 1997/2962; Reg 16

13.4.1 General advice to safety officers

13.4.1.1 The safety officer must maintain a good working relationship with the safety representatives by, for example, inviting them to join in regular inspections of each part of the ship or, while carrying out an investigation, consulting them on safety matters and arrangements, and in particular on any follow-up action proposed.

13.4.1.2 The safety officer's relationship with the safety committee is different. The safety officer is both a member of the committee and to some extent subject to its direction. A committee has the right to inspect any of the records that a safety officer is required by law to keep, and has the power to require the safety officer to carry out any health or safety inspections considered necessary.

13.4.2 Advice on compliance with safety requirements

13.4.2.1 The safety officer is required by the regulations to try to ensure compliance with the provisions of this Code and any health and safety guidance and instructions for the ship.

Reg 16

13.4.2.2 The safety officer's role should be a positive one, seeking to initiate or develop safety measures before an incident occurs rather than afterwards. They should:

- look out for any potential hazards and the means of preventing incidents
- try to develop and sustain a high level of safety consciousness among seafarers so that individuals work and react instinctively in a safe manner and are aware of the safety of themselves and others. The objective is to become the ship's adviser on safety, who the master, officers and all seafarers will naturally turn to for advice or help on safe working procedures
- where unsafe practice is observed, approach the individual or responsible officer concerned to suggest improvements in working methods or use the safety committee to discuss examples of dangerous or unsafe practices in a particular area. If this brings no improvement, the safety officer should approach the head of department or, as a last resort, the master to use their influence
- ensure that each worker joining the ship is instructed in all relevant health and safety arrangements, and of their importance, before starting work. A suggested outline for this induction is given in Chapter 2

- where possible, ensure that arrangements are made for each new entrant to work with a safety-conscious seafarer
- remind experienced seafarers joining the ship for the first time of the importance of a high level of safety consciousness and of setting a good example to less experienced seafarers.

13.4.2.3 The safety officer should also promote safety on board, subject to the agreement of the master, by:

- arranging the distribution of booklets, leaflets and other advisory material on safety matters
- supervising the display of posters and notices, replacing and renewing them regularly
- arranging for the showing of films on safety publicity and, where appropriate, organising discussions on the subjects covered
- encouraging seafarers to send in ideas and suggestions for improving safety and getting their support for any proposed safety measures which may affect them (the person making a suggestion should always be told about decisions reached and any action taken)
- effectively communicating new requirements or advice in relevant shipping legislation, marine notices and company and ship's rules and instructions relating to safety at work on the ship.

13.4.3 Investigation of accidents and dangerous occurrences

13.4.3.1 The safety officer must investigate notifiable accidents or dangerous occurrences affecting people on board ship or during access, as well as potential hazards to health and safety and any reasonable complaints made by any personnel. They must make recommendations to the master. They should record and investigate, as appropriate, all incidents reported by personnel or passengers.

13.4.3.2 The safety committee may commission additional health or safety investigations or inspections.

13.4.4 Safety inspections

13.4.4.1 The safety officer must ensure that each accessible part of the ship has a health and safety inspection at least once every three months, or more frequently if there have been substantial changes in the conditions of work.

13.4.4.2 'Accessible' means all parts of the ship to which any seafarer has access without prior authority.

13.4.4.3 Deciding whether 'substantial changes in the conditions of work' have taken place is a matter of judgement. Changes are not limited to physical matters, such as new machinery, but can include changes in working practices or possible new hazards. Keep a record of all inspections.

13.4.4.4 It is not necessary to inspect the whole ship at one time, as long as each accessible part is inspected every three months. It may be easier to get quick and effective action on recommendations arising out of an inspection dealing with one section at a time. The safety officer should ensure that the inspections are carried out when necessary. Before beginning any inspection, read previous reports of inspections of the section, together with the recommendations made and the subsequent action taken. Read the control measures identified in any relevant risk assessment, and check compliance with them during the inspection. Note any recurring problems and, in particular, recommendations for actions that have not been done. However, do not allow the findings of previous inspections to prejudice any new recommendations.

13.4.4.5 Safe access, the environment and working conditions are major items to consider, but it is not possible to give a definitive checklist of everything to look for. Suggestions for consideration on these issues are given in Annex 13.1.

13.4.4.6 The safety officer must make representations and, where appropriate, recommendations to the master, and through the master to the company, about any deficiency in the ship in respect of statutory requirements relating to health and safety, relevant merchant shipping notices and the provisions of this Code.

13.4.4.7 To fulfil this function properly, the safety officer needs to be familiar with the appropriate regulations. The introduction of new regulations or of amendments to existing regulations will be announced in marine notices issued by the MCA.

13.4.5 Record of accidents and dangerous occurrences

13.4.5.1 The safety officer must maintain a record of all accidents and dangerous occurrences in line with procedures in the ship's safety management system. On a ship where no safety officer is appointed, this duty falls to the company. These records must be made available on request to any safety representative, the master or to any person duly authorised by the Secretary of State.

13.4.6 Duty to stop dangerous work

13.4.6.1 The safety officer must stop any work in progress which they reasonably believe may cause an accident and immediately inform the master (or a nominated deputy) who is responsible for deciding when work can safely continue.

Reg 16(1)9f

13.4.6.2 This does not apply to an emergency action to safeguard life, even though the action itself may involve a risk to life. These regulations do not require the safety officer to carry out their duties to inspect, keep records or make recommendations at a time when emergency action to safeguard life or the ship is being taken.

13.4.6.3 The safety officer should also encourage other seafarers to stop any work that the seafarer reasonably believes could cause an accident.

13.5 Powers of safety representatives

13.5.1 Unlike the safety officer, the safety representative has powers, not duties, although membership of the safety committee imposes certain obligations.

13.5.2 Safety representatives may, with the agreement of the safety officer, participate in investigations and inspections carried out by the safety officer or, after notifying the master or a nominated deputy, may carry out their own investigation or inspection.

13.5.3 They may also make representations to the company or the relevant employer on potential hazards and dangerous occurrences, and to the master, company or employer on general health and safety matters, such as the findings of the risk assessment, health and safety training, and the introduction of new technology.

13.5.4 They may request, through the safety committee, that the safety officer investigates and reports back to them, and may inspect any of the records the safety officer is required to keep under the regulations. They should ensure that they see all incident reports submitted to the MAIB under the accident reporting regulations (see section 13.3.8.4).

13.6 Advice to safety representatives

13.6.1 Safety representatives should be familiar with the relevant safety regulations and guidance for UK-registered ships, regulations, marine notices and guidance issued by the MCA.

13.6.2 The effectiveness of safety representatives will depend mainly on good cooperation between them, the company, other employers, the master, heads of department and safety officer.

13.6.3 Safety representatives should:

- put forward their views and recommendations in a firm but reasonable and helpful manner
- be sure of the facts
- be aware of the legal position
- be conscious of what is reasonably practicable.

13.6.4 Having made recommendations, they should ask to be kept informed of any follow-up actions taken, or the reasons why such action was not possible.

13.6.5 If a safety representative finds that their efforts are being obstructed, or they are denied facilities, they should bring the matter to the attention of the safety officer or of the master through the safety committee. They should aim to settle any difficulties on board ship or through the relevant employer or the company. If this proves impossible, the problem should be referred to the trade union or to the MCA.

13.7 Advice to safety committees

13.7.1 The safety committee is a forum for consultation between the master, safety officials and others of matters relating to health and safety. It may be used by individual employers for consultation with the company and seafarers. Its effectiveness will depend on the commitment of its members, in particular that of the master. Because of its broad membership, and because the master chairs it, the committee can take effective action in all matters it discusses other than those requiring the authorisation of the company and individual employers. Do not use safety committee meetings for instruction or training.

13.7.2 The frequency of meetings will be determined by circumstances, but the committee should meet regularly, considering the pattern of operation of the ship and the arrangement for manning, and frequently enough to ensure continuous improvement in safety. In particular, a meeting should also take place after any serious incident or accident on the ship, if the normal meeting is not due within a week.

13.7.3 Circulate an agenda (together with any associated documents and papers, and the minutes of the previous meeting) to all committee members, giving them enough time to read the contents and prepare for the meeting.

13.7.4 If there is a long agenda it may be better to hold two meetings close together rather than one long one. If two meetings are held, discuss the most urgent matters at the first meeting.

13.7.5 The first item on the agenda should be the minutes of the previous meeting. This allows any correction to the minutes to be recorded and gives the opportunity to report any follow-up action taken.

13.7.6 The last but one item should be 'any other business'. This enables last-minute items to be introduced, and prevents the written agenda being a stop on discussion. Limit 'any other business' to important issues that have arisen since the agenda was prepared. Include all other items in the agenda of the next meeting.

13.7.7 The last item on the agenda should be the date, time and place of the next meeting.

13.7.8 In the minutes of each meeting record concisely the business discussed and conclusions reached. Give a copy to each committee member. Agree the minutes as soon after the meeting as possible, or amend them if necessary, and then agree them under the first agenda item of the following meeting (see section 13.7.5).

13.7.9 Keep a minutes file or book together with a summary of recommendations recording the conclusions reached. This provides a permanent source of reference and ensures continuity if there are changes in personnel serving on the committee.

13.7.10 Keep all seafarers informed on matters of interest that have been discussed; for example, by posting summaries or extracts from the minutes on the ship's noticeboards. Invite suggestions by similarly posting the agenda before meetings.

13.7.11 Send the relevant extracts from the agreed minutes via the master to the company and, where appropriate, to individual employers, even when the matters referred to have already been taken up with them. Keep a record of the response or action taken by the company.

13.8 Accident investigation

13.8.1 The investigation of accidents and incidents plays a very important part in safety. The identification and study of accidents, principally through the MAIB's accident reporting system, helps prevent similar events happening in future.

📖 *SI 2012/1743*

📖 *Marine Guidance Note MGN 564 (M+F) Amendment 1 explains how to comply with the statutory requirements.*

13.8.2 The master is responsible for the statutory reporting of accidents and dangerous occurrences covered by the regulations. Where a safety officer is on board, however, it is their statutory duty to investigate every such incident. The master will rely extensively on the results and record of the safety officer's investigation when completing their report. The stages of a typical investigation might be as follows:

13.8.3
- When an incident occurs, prioritise the safety of the injured and those assisting them, and the immediate safety of the area.
- Once sufficient help is available, however, the safety officer should avoid getting involved with the rescue operation and concentrate on establishing the immediate facts about the incident.
- Record the names – and addresses in the case of non-crew personnel – of everyone present near the incident. Not all will be witnesses but this can be ascertained later.
- Note and mark the position of the injured, and the use and condition of any protective clothing, equipment or tools likely to have been used.
- Put any portable items that might be relevant to the investigation into safe storage.
- Sketches and photographs are often useful.
- When the injured people have been removed, the safety officer should do a more detailed examination at the scene of the incident, watching out for any changes that might have occurred since the incident and any remaining hazards.

📖 *SI 1997/2962; Reg 16(1)(b)*

The points concerning an incident to look out for will depend on the circumstances. For example, after an incident during boarding, note the following:

- compliance with control measures identified by the risk assessment
- the type of access equipment in use
- the origin of the access equipment (eg ship's own or provided from shore)
- the condition of the access equipment, noting any damage such as a broken guardrail or rung. Examine the position and extent of any damage so that it may be compared with witnesses' statements. Note whether the damage was present before the incident, during or as a result of the incident. (If it was present before the incident it might have been potentially dangerous but may not have been a factor in the incident.)
- any effect of external factors on the condition of the equipment (eg ice, water or oil on the surface)
- the deployment of the equipment (eg location of the quayside and shipboard ends of the equipment)
- the rigging of the equipment, the method of securing, the approximate angle of inclination
- the use of ancillary equipment (eg safety net, lifebuoy and lifeline, lighting)
- the safety of shipboard and quayside approaches to the equipment (eg adequate guardrails, obstructions and obstacles)
- any indication of how the incident might have happened (but approach subsequent interviews with witnesses with an open mind)
- the weather conditions at the time
- the distances, where helpful or relevant.

For all incidents, consider human and organisational factors that may have contributed to the accident. Marine Guidance Note MGN 520 (M) includes common factors in accidents.

The following are examples of issues that may be relevant:

Individual factors

- honest errors and mistakes versus violations and recklessness.

Job factors

- physiological and psychological problems caused by the ship environment (eg noise affecting communication, vibration, changes in workload/duties)
- working pattern and likelihood of fatigue and stress
- any recent technical changes and the associated training or instruction
- job and equipment design
- effectiveness of procedures
- performance-influencing factors
- teamwork and communication.

Organisational factors

- policies, organisation, culture
- recruitment and competence assurance
- bullying and harassment.

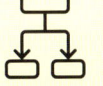

13.8.4 Interview the witnesses as soon as possible after the incident while memories are still fresh. There may be people who were not actually witnesses but who may make valuable contributions, such as a seafarer who was present when an order was given. Do not overlook these people. If it is not possible for some reason to interview a particular person, ask them to send the safety officer their own account of the incident.

13.8.5 Do the interview in an informal atmosphere to put the witness at their ease. The safety officer should first explain the purpose of the interview and take some details of the witness's background. Keep any personal bias out of the interview. Ask the witness to talk about the event in their own way with as few interruptions as possible. Test the accuracy of what they say. There may, for example, be discrepancies between the accounts of different witnesses, between different parts of a statement, or with the safety officer's own observations, which the safety officer may want to query. Avoid asking leading questions which imply a particular answer. Also avoid asking simple questions requiring only a yes/no answer as these stop the witness from thinking about what they are saying. Finally, the safety officer should go over the statement with the witness to ensure that it has been accurately recorded.

13.8.6	Prepare statements for signature by the witness as quickly as possible. However, if the witness changes their mind about signing a statement, the safety officer should annotate it to say it was prepared on the basis of an interview with the witness who subsequently refused to sign it or comment further. Where the witness asks for extensive alterations to the original statement a fresh statement may have to be prepared, but the original statement should be annotated by the safety officer and retained.
13.8.7	It is helpful to adopt a standard format for statements by incident witnesses. A suggested format is shown in Annex 13.2.
13.8.8	It is important to distinguish between facts and opinions. Facts can normally be supported by evidence whereas opinions are personal beliefs. An investigation must depend on the facts gathered, but opinions can be helpful in pursuing a particular line of enquiry so do not disregard them.
13.8.9	Any record of incidents and dangerous occurrences (see section 13.4.5.1) should contain at least the following: • details of incidents/dangerous occurrences/investigations/complaints/inspections • date • people involved • nature of injuries suffered • all statements made by witnesses • any recommendations/representations • any action taken.
13.8.10	Additionally, it is suggested that it should contain: • a list of witnesses; their addresses, positions and occupations • the whereabouts of original signed statements made by the witnesses • the date when the accident/dangerous occurrence reports were sent to the MAIB (if applicable) • a list of items collected; why and where they are stored • an index.
13.8.11	Keep the record with the ship because it must be made available on request to the safety representative and safety committee, if any. It is also a necessary item of reference for safety officers on board the ship. If the ship is sold and remains on the UK register, the record should be transferred with the ship. Where the ship becomes a foreign ship the original owners should keep the record.

- The safety officer should encourage other seafarers to stop any work that a seafarer reasonably believes could cause an accident.
- Keep seafarers informed of matters of interest from safety committee meetings and details that could help prevent future incidents.

- The company must appoint a safety committee on every ship with five or more seafarers.
- The master is responsible for the statutory reporting of accidents and dangerous occurrences covered by the regulations.

Annex 13.1 Checklist for safety officer's inspection

The following are examples of questions that the safety officer should ask. This is not an exhaustive list; it will vary according to the design or conditions on a particular ship.

Means of access/safe movement

- Are the means of access, if any, to the area being inspected (particularly ladders and stairs), in a safe condition, well lit and unobstructed?
- If any means of access is in a dangerous condition (eg when a ladder has been removed) is the danger suitably blocked off and have warning notices been posted?
- Is access through the area of inspection both for transit and working purposes clearly marked, well lit, unobstructed and safe?
- Are fixtures and fittings over which seafarers might trip or which project (particularly overhead, thereby causing potential hazards), suitably painted, cushioned or marked?
- Is any gear that has to be stowed within the area suitably secured?
- Are all guardrails in place, secure and in good condition?
- Are all openings through which a person could fall suitably fenced?
- If portable ladders are used, are they properly secured and at a safe angle?

Working environment

- Is the area safe to enter?
- Are lighting levels adequate?
- Is the area clear of rubbish, combustible material, spilled oil, etc?
- Is ventilation adequate?
- Are seafarers adequately protected from exposure to noise where necessary?
- Are dangerous goods and substances left unnecessarily in the area or stored in a dangerous manner?
- Are loose tools, stores and similar items left lying around unnecessarily?

Working conditions

- Is machinery adequately guarded where necessary?
- Are any necessary safe operating instructions clearly displayed?
- Are any necessary safety signs clearly displayed?
- Are permits to work used when necessary?
- Are seafarers working in the area wearing any necessary protective clothing and equipment?
- Is protective clothing and equipment in good condition and being used correctly?
- Is there any evidence of defective plant or equipment, and if so, what is being done about it?
- Is the level of supervision adequate, particularly for inexperienced seafarers?
- What practicable safety improvements could be made?

General

- Are all statutory regulations and company safety procedures being complied with?
- Is the safety advice in publications such as this Code, merchant shipping notices, etc. being followed where possible?
- Can the seafarers in the area make any safety suggestions?
- Have any faults identified in previous inspections been rectified?

Annex 13.2 Voluntary statement

Relating to an accident on board/name of ship/official number .... on/date of accident/at/time of accident.

Particulars of witness

Name ....

Rank and occupation ....

Home address of crew members

....
....
....
....
....

Address of employment of others

....
....
....
....
....

Statement of witness

I make this statement voluntarily, having read it before signing it and believing the same to be true.

Signature of witness ....

Date .... Time ....

Particulars of interviewer

Name ....

Rank ....

14 Permit to work systems

14.1 Introduction

14.1.1 There are many types of operation on board ship when the routine actions of one person may inadvertently endanger another, or when a series of steps needs taking to protect the people doing specific work.

14.1.2 In this chapter and its annexes:

- **'competent person'** means someone designated and authorised for the task covered by a permit to work under the safety management system (SMS)
- an 'authorised officer' means someone designated and authorised to issue and close permits to work under the SMS.

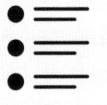

Key points

- Always use a permit to work where the ship's SMS requires it. It is based on the company's risk assessment procedures.
- Using a permit to work ensures that you will follow the necessary control measures to do the work safely.
- Do only the work specified on the permit to work.
- Ensure that the permit to work is completed and followed until the authorised officer has closed it.

Your organisation should

- identify the hazards and ensure they are eliminated or effectively controlled. The company is responsible for seeing that this happens
- put in place appropriate control measures to protect people who may be affected, based on the hazards and findings identified in the risk assessment. A permit to work is a formal record to confirm that control measures are in place when particular work is being done
- be aware that the competent person doing the specified work should not be the same person as the authorised officer for the same task.

14.2 Permit to work systems

14.2.1 A **permit to work** provides an organised and predefined safety procedure. It does not in itself make the job safe but ensures that you follow measures for safe working.

14.2.2 Your ship's SMS will tell you when to use a permit to work, and will provide the form for it. Each permit to work should be relevant and as accurate as possible for the task.

A permit to work should include the:

- location of the work to be done
- details of the work to be done
- nature of any preparatory tests undertaken, and the results
- measures undertaken to make the job safe
- safeguards to take during the operation
- period of validity of the permit to work (should not exceed 24 hours)
- time limits applicable to the work that it authorises.

- Before signing the permit, the authorised officer should ensure that all safety precautions and measures specified as necessary have been taken, or that procedures are in place.
- The authorised officer is responsible for the work until they have either closed the permit or formally transferred it to another authorised officer and fully briefed them on the situation.
- Anyone who takes over from the authorised officer, either routinely or in an emergency, should sign the permit to indicate transfer of full responsibility.
- The competent person responsible for carrying out the specified work should countersign the permit to indicate that they understand the safety precautions and measures to be observed.
- When the work is complete, the competent person should notify the authorised officer and ensure that the authorised officer has closed the permit.

14.2.3 Annex 14.1 gives examples of permits to work for various types of activity, showing different approaches. Annexes 14.1.1 and 14.1.2 are permits to work which record that safety measures have been put in place before the work begins (Annex 14.1.1 is for enclosed spaces and Annex 14.1.2 is for work at height/over the side). Annex 14.1.3 is a general permit to work for handing the site over to the competent person doing the work, and sets out the safety measures to put in place. Use Annex 14.1.3 only when alternative safety procedures are in place to ensure that measures have been carried out before work begins. You may adapt these examples to the circumstances of the individual ship or the job to be carried out, in the light of the risk assessment.

14.3 Sanction to test systems

14.3.1 A sanction to test may be required when additional controls are needed for the testing of high-risk systems, such as **high-voltage** systems. You should issue this following the same procedure as a permit to work.

14.3.2 Only one permit to work or sanction to test should be in force at any one time for the same apparatus/equipment.

14.3.3 Issue a sanction to test when testing operations require the removal of the **circuit main earth**. Note: maintenance and repair cannot be carried out under a sanction to test.

14.3.4 Annex 14.2 is an example of a sanction to test for testing work carried out on electrical high-voltage systems over 1,000 volts. It shows the headings and requirements for each section. You can adapt these to the circumstances of the individual ship, the ship's electrical high-voltage system, or the job to be carried out, in light of the risk assessment.

Annex 14.1 Permits to work

Permits to work are normally required for the following categories of work:

- entry into an enclosed space
- any work requiring use of gas testing/equipment
- any work requiring isolation of machinery or power system
- hot work
- working at height/over the side
- general electrical (under 1000 volts)
- electrical high voltage (over 1000 volts)
- working on deck during adverse weather
- lifts, lift trunks and machinery.

This list is not exhaustive; permits to work, following a similar format, may be required and developed by the company for other categories of work.

Annex 14.1.1

PERMIT TO WORK: ENTRY INTO ENCLOSED SPACES

Note (i): The authorised officer should insert the appropriate details when the sections for other work or additional precautions are used.

Note (ii): The competent person should tick each applicable box as they make their check.

Note (iii): This permit to work contains five sections.

SECTION A – Scope of work

Location (name of space) ....

Plant apparatus/identification (designation of machinery/equipment)

....

Work to be done (reason for entry)

....

Permit issued to (name of competent person carrying out work or in charge of the work party)

....

This permit is valid: from .... hours   Date ....

                                         to .... hours

SECTION B – Checklists

Has a risk assessment been carried out for the proposed work? Y/N

Has a toolbox talk been carried out? Y/N

No conflict with any other permit to work in force? Y/N

B1 – Pre-entry preparation

To be completed by the authorised officer

Checklist	Checked
1. Space thoroughly ventilated. If not, ensure that section B3 is also completed.	
2. Atmosphere tested and found safe. If safe atmosphere not tested, ensure that section B3 is also completed.	
3. Space secured for entry (verified all isolations, lock outs and tag outs are in place for safe entry and work).	
4. Testing equipment available for regular checks while space is occupied and after breaks.	
5. Arrangements for ventilation for duration of permit to work. If ventilation not possible, ensure section B3 is also completed.	
6. Adequate access and lighting.	
7. Rescue and resuscitation equipment available at entrance.	
8. Competent person in attendance at entrance.	
9. Relevant officer of the watch advised of planned entry.	
10. Appropriate communication arrangements agreed between attendant and those entering, including emergency signals.	
11. Emergency and evacuation procedures agreed and in place.	
12. All equipment to be used of appropriate type.	
13. Personal protective equipment to be used: safety helmet, safety harness as necessary.	

B2 – Pre-entry checklist

To be completed by each person entering the space

	Names of persons entering the space
I have received instructions and authorisation from the authorised officer to enter the dangerous space.	
Section 1 of this permit has been completed by the authorised person.	
I have agreed and understand the communication procedures.	
I have agreed upon a reporting interval of …….. minutes.	
Emergency and evacuation procedures have been agreed and are understood.	
I have witnessed the testing of the atmosphere within the space and am satisfied it is safe to enter.	
I am aware that the space must be vacated immediately in the event of ventilation failure or if the atmosphere test shows a change from agreed safe criteria.	
Signatures of persons entering the space	

B3 – Breathing apparatus and other equipment

To be completed by the competent person

Breathing apparatus should only be used to enter unsafe spaces in cases of emergency or for the purpose of atmosphere testing.

Checklist	Checked
1. Those entering the space familiar with any breathing apparatus to be used	
2. Breathing apparatus tested and found satisfactory	
3. Means of communication tested and found satisfactory	
4. Those entering wearing rescue harnesses and lifelines where practicable	
5. Although the enclosed space is potentially unsafe, the adequate breathing apparatus and relevant procedures will be used to mitigate risk, and the reason for entry into the unsafe space is due to an emergency, or for essential atmosphere testing.	

B4 – Other work/additional precautions

To be completed by the authorised officer

Checklist	Checked

SECTION C – Certificate of checks

I confirm that I am satisfied that all precautions have been taken and that safety arrangements will be maintained for the duration of the work.

Competent person

Name .... Signature ....

Time .... Date ....

I am satisfied that all precautions set out in B1 to B4 have been taken and that safety arrangements will be maintained for the duration of the work.

Authorised officer

Name .... Signature ....

Time .... Date ....

Note: After signing the receipt, this permit to work should be retained by the competent person in charge at the place where the work is being carried out until the work is complete and the clearance section signed.

SECTION D – Personnel entry

Names	Time in	Time out

SECTION E – Cancellation of certificate

The work has been completed*/cancelled* and all persons under my supervision, materials and equipment have been withdrawn.

Competent person

Name .... Signature ....

Time .... Date ....

* Delete words not applicable and where appropriate state: The work is complete*/incomplete* as follows: [description]

....

....

....

....

The worksite has been inspected; I accept that all persons, material and equipment have been withdrawn, and the site is secured against entry*/ safe for entry*.

Authorised officer

Name .... Signature ....

Time .... Date ....

*Delete words not applicable.

Annex 14.1.2

PERMIT TO WORK: WORKING AT HEIGHT/OVER THE SIDE

Note (i): The authorised officer should indicate the sections applicable by ticks in the left-hand boxes next to headings, deleting any subheading not applicable.

Note (ii): The authorised officer should insert the appropriate details when the sections for other work or additional precautions are used.

Note (iii): The competent person should tick each applicable right-hand box as they make their check.

Note (iv): This permit to work contains four sections.

SECTION A – Scope of work

Location (name of space)

....

Plant apparatus/identification (designation of machinery/equipment)

....

Work to be done [description]

....

Permit issued to (name of competent person carrying out work or in charge of the work party)

....

This permit is valid: from .... hours   Date ....

to .... hours

NOTE: The validity of this permit to work should not exceed 24 hours.

SECTION B – Checklists

To be completed by the authorised officer (AO) and competent person (CP)

B1 – Preliminary checklist

Checklist	Checked (AO)	Checked (CP)
Has a risk assessment been carried out for the proposed work?		
Has a toolbox talk been carried out?		
No conflict with any other permit to work in force?		

B2 – Preparation checklist

Checklist	Checked (AO)	Checked (CP)
1. Duty officer informed		
2. Warning notices posted		
3. On-deck supervisor identified		
4. Equipment in good order		
5. Work on funnel: • advise duty engineer; • isolate whistle.		
6. Work near radar scanners/radio aerials: • isolate radar and scanner/radio room notified; • notices placed to stop the use of radar/radio.		
7. Work over the side: • advise duty officer/engineer; • lifebuoy and lifeline ready.		

8. Personal protective equipment to be used: • safety helmet; • safety harness and line attached to a strong point; • lifejacket; • other (please list).		
9. All tools to be raised and lowered secured on a lanyard/belt or in a bag.		
10. Has a plan been agreed and necessary equipment been put in place to achieve an effective rescue?		

SECTION C – Certificate of checks

I confirm that I am satisfied that all precautions have been taken and that safety arrangements will be maintained for the duration of the work and no attempt will be made by me or people under my charge to work on any other apparatus or in any other area.

Competent person

Name ………………………………… Signature …………………………………………………

Time ………………………………… Date ………………………………………………………

I am satisfied that all precautions have been taken and that safety arrangements will be maintained for the duration of the work and no attempt will be made by me or people under my charge to work on any other apparatus or in any other area.

Authorised officer

Name ………………………………… Signature …………………………………………………

Time ………………………………… Date ………………………………………………………

Note: After signing the receipt, this permit to work should be retained by the competent person in charge at the place where the work is being carried out until work is complete and the clearance section signed.

SECTION D – Cancellation of certificate

The work has been completed*/cancelled* and all persons under my supervision, materials and equipment have been withdrawn and warned that it is no longer safe to work on the apparatus detailed in this permit to work.

Competent person

Name .... Signature ....

Time .... Date ....

* Delete words not applicable and where appropriate state:

The work is complete*/incomplete* as follows: [description] (*Delete words not applicable)

....

....

....

....

The worksite has been inspected; I accept that all persons, material and equipment have been withdrawn and all persons warned that it is no longer safe to work on the apparatus detailed in this permit to work.

Authorised officer

Name .... Signature ....

Time .... Date ....

Annex 14.1.3

PERMIT TO WORK: GENERAL

VESSEL:		
AREA OF VESSEL:	PTW No: _____/_____	

1. AUTHORISED OFFICER (AO):	TIME:
1.1 WORK DESCRIPTION (Please use BLOCK CAPITALS at all times except for signatures)	
Category of work (delete as required): **HOT WORK/DANGEROUS SPACE WORK/WORK AT HEIGHT/ OVERSIDE WORK/GAS TESTING/EQUIPMENT**	
Equipment to be worked on:	

Proposed work description:	COMPETENT PERSONS (CPs)	4
	1	5
	2	6
	3	7

Requested by (AO):	Signature:	Company:

HAS RISK ASSESSMENT OF THIS TASK BEEN CONDUCTED? YES/NO – If no, **STOP THE JOB** until one is provided.
HAS TOOLBOX TALK BEEN CONDUCTED? YES/NO   HAS LOLER LIFTING PLAN BEEN CONDUCTED? YES/NO

2. PRECAUTIONS: Delete as required. Leave applicable item clear.

2.1 GENERAL
(a) Inform other personnel who may be affected	YES/NO	(d) Rope off area	YES/NO
(b) Provide additional access, lighting, ventilation	YES/NO	(e) Post warning signs	YES/NO
(c) Visit to work site required	YES/NO	(f) Provide radio communications	YES/NO

2.2 PERSONAL PROTECTIVE EQUIPMENT REQUIRED
Safety helmet	YES/NO	Coveralls	YES/NO	Safety goggles	YES/NO	Other (list)
Safety boots	YES/NO	Gloves	YES/NO	Safety harness	YES/NO	

2.3 HOT WORK
(a) Fire watch required	YES/NO
(b) Safety watch during breaks	YES/NO
(c) Portable extinguisher to be at site	YES/NO
(d) Shielding to prevent spread of sparks	YES/NO

2.4 ENTRY INTO DANGEROUS (ENCLOSED) SPACES
(a) Isolate from systems	YES/NO
(b) Risk of oxygen deficiency	YES/NO
(c) Need to wear breathing apparatus/mask	YES/NO
(d) Provide additional ventilation	YES/NO
(e) Provide additional lighting	YES/NO
(f) Rescue plan provided *required*	YES/NO
(g) Portable monitor in use by entry team *required*	YES/NO
(h) Appoint crew standby person	YES/NO

CONTACT MASTER

If third-party action, crew standby person must be additional to contractor standby person.

2.5 ISOLATION/ENERGY RELEASE PROTECTION
If isolation is required, lock down to be confirmed by chief engineer.
Certificate to be provided and attached to permit to work.
Certificate number: _____

2.6 WORKING AT HEIGHT/OVER THE SIDE
Have alternative means for task been explored?	YES/NO	Rescue plan provided *required*	YES/NO	Suitable bridge equipment to be locked out
Work crew at least two people *required*	YES/NO	Continuous comms est. with bridge	YES/NO	
Life ring with line readily available	YES/NO	Warning notices posted	YES/NO	

2.7 CONTRACTOR CONTROL OF WORK SITE
I hereby sign that I agree to take complete control of [insert work site] _____ at the time of issuance of this permit to work and have satisfied myself that all required isolations are in place to secure this site. – **NO CREW IS NOW TO ENTER THIS SITE.**
AO TO SIGN TO TAKE CONTROL OF SITE   Signed (AO): _____

2.8 GAS TESTING/EQUIPMENT
Vessels: Serial no. _____ Calib. date: _____ Gas testing result: _____
If outside contractor, certificate must be attached.

2.9 ENSURE NO CONFLICT WITH ANY OTHER PERMIT   Completed by (AO signature):

3. AUTHORISATION BY AO (as named in Part 1)

Authority is hereby given for the work detailed in Part 1.1 to be carried out, provided the precautions in Part 2 are strictly observed.	
AUTHORISED BY (AO):	SIGNATURE:
TIME:	DATE:

PERMIT IS VALID FOR A PERIOD OF 12 HOURS FROM THE TIME AUTHORISED, UNLESS EXTENDED UNDER SECTION 5

4. ACCEPTANCE

We accept the conditions of the permit as stated above and will inform all competent persons involved in the work of the precautions to be taken. We will display a copy of the permit at the work site at all times during the task.	
ACCEPTED BY CP:	SIGNATURE:
ACCEPTED BY AO:	SIGNATURE:
TIME:	DATE:

5. EXTENSION (Site should be inspected prior to extension being granted)

Permit must not exceed 24 hrs	Signature
First extension granted at: 6hrs on:	
Second extension granted at: 6hrs on:	

6. COMPLETION (Delete as required for 6.1)

6.1 Work is complete	YES/NO	All tools/apparatus have been removed and secured	YES/NO
Site is in a safe condition	YES/NO	Normal operations may be resumed	YES/NO
Name (AO):	Signature:	Date:	Time:

6.2 The worksite has been inspected; I accept that all equipment is operational, the site is safe, and that I take back full control of this site.			
CLOSED BY (AO):	Signature:	Date:	Time:

Annex 14.2 Sanction to test: electrical high voltage (over 1000 volts)

Note (i): The authorised officer should indicate the sections applicable by ticks in the left-hand boxes, deleting any subheading not applicable.

Note (ii): The authorised officer should insert the appropriate details when the sections for other work or additional precautions are used.

Note (iii): The competent person should tick each applicable right-hand box as they make their check.

Note (iv): This sanction to test contains six sections.

SECTION A – Scope of work

Location (name of space)

....

Plant apparatus/identification (designation of machinery/equipment)

....

Work to be done [description]

....

Sanction to test issued to (name of competent person carrying out work or in charge of the work party)

....

This sanction to: from .... hours   Date ....

test is valid: to .... hours

NOTE: The validity of this sanction to test should not exceed 24 hours.

SECTION B – Checklists

To be completed by the authorised officer (AO) and competent person (CP)

B1 – Preliminary checklist

Checklist	Checked (AO)	Checked (CP)
Has a risk assessment been carried out for the proposed work?		
Has a toolbox talk been carried out?		
No conflict with any other permit to work in force?		

B2 – Isolation data checklist

Checklist	Checked (AO)	Checked (CP)
1. The above apparatus is dead and has been isolated from the system at the following points: [description]		
2. Circuit main earths have been applied to the equipment at the following points. These earths may be removed and replaced to your instructions. [description]		
3. Safety locks [detail location fitted and identify lock set]		
4. Additional precautions to avoid danger have been taken by: [description]		

5. Caution/danger notices have been applied at all points of isolation as listed above and safety signs positioned as follows:		
6. Presence of other hazards including list of other relevant permits to work/ sanctions to test:		
7. Treat all other apparatus and areas as dangerous. Further precautions:		

SECTION C – Certificate of checks

I accept responsibility for carrying out the work on the apparatus detailed in this sanction to test and no attempt will be made by me or by persons under my charge to work on any other apparatus or in any other area.

I confirm that the above equipment is dead and isolated from all live conductors and connected to earth.

I am satisfied that all precautions set out in Section B have been taken and that safety arrangements will be maintained for the duration of the work.

Safety key no.. Received/applied*

(*Delete word not applicable)

Competent person

Name .... Signature ....

Time .... Date ....

I confirm that the above equipment is dead and isolated from all live conductors and connected to earth. I am satisfied that all precautions set out in Section B have been taken and that safety arrangements will be maintained for the duration of the work.

Safety key no.. Received/applied*

(*Delete word not applicable)

Authorised officer

Name .... Signature ....

Time .... Date ....

Note: After signing the confirmation, this sanction to test should be retained by the competent person in charge at the place where the work is being carried out until the work is complete and the clearance section signed.

SECTION D – Clearance of sanction to test

The work for which this sanction to test has been completed*/ cancelled* and all persons under my supervision have been withdrawn and warned that it is no longer safe to work on the apparatus detailed in this sanction to test.

All work materials and equipment have been removed.

Competent person

Name .... Signature ....

Time .... Date ....

Safety key no.. Received/applied*

The worksite has been inspected; I accept that all persons, material and equipment have been withdrawn, and the site is secured against entry*/ safe for entry.*

(*Delete words not applicable)

SECTION E – Cancellation of sanction to test

This sanction to test is cancelled.

Authorised officer

Name .... Signature ....

Time .... Date ....

Safety key no.. Received/applied*

* Delete words not applicable and where appropriate state:

The work is complete*/incomplete* as follows: [description]

....

....

....

....

15 Entering enclosed spaces

15.1 Introduction

15.1.1 An enclosed space is one that is not designed for continuous worker occupancy and has either or both of the following characteristics:

- limited openings for entry and exit
- inadequate ventilation.

  SI 2022/96

15.1.2 Any enclosed space deprived of regular and constant ventilation may become dangerous. The atmosphere in any enclosed space may at some stage contain toxic or flammable gases or vapours, or be deficient in oxygen, to the extent that it may endanger the life or health of any person entering that space. Gases are often invisible to the human eye.

15.1.3 Some spaces may be a dangerous enclosed space only intermittently, perhaps due to the type of cargo carried or work to be undertaken (eg a compartment during spray painting).

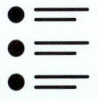

Key points
- Enclosed spaces pose a particular risk to seafarers. Risk assessment must be undertaken and appropriate procedures followed.
- A register should be made of any enclosed spaces that seafarers may enter.
- If in doubt, treat a space as 'enclosed' until the atmosphere has been tested.
- If the atmosphere is considered to be suspect or unsafe to enter, then the space should only be entered if it is essential for the safety of life or of the ship or for testing purposes in such a scenario and breathing apparatus should always be worn.

> **Your organisation should**
> - comply with the 2022 enclosed spaces regulations (if applicable) as well as best practice, to mitigate the risk of enclosed space entry to seafarers
> - provide all guidance, training and equipment necessary to ensure safety when entering enclosed spaces.

15.1.4 Any enclosed space is potentially life threatening and every precaution should be taken both prior to entry and while inside. The dangers may not be readily apparent and, following testing, atmospheres can change (eg hot work and use of certain chemicals). In addition, isolated areas with very low oxygen content or small concentrations of toxic gases may exist.

Note: A single inhalation with a 5% oxygen content may result in instantaneous loss of consciousness and subsequent death. Similarly, small concentrations of a toxic substance may result in loss of consciousness and subsequent death. Therefore, it is essential that all necessary precautions are taken, including a risk assessment, the completion of a **permit to work** and frequent testing of the atmosphere when the space is entered, according to the schedule set out in the risk assessment.

15.1.5 Awareness of any risks is necessary for all spaces on board ship. If in doubt, any such space should be regarded as enclosed and appropriate action taken based on the findings of the risk assessment. Appropriate control measures should be put in place to protect anyone who may enter an enclosed space. Procedures (such as systems of work, permits to work and emergency procedures) should be part of a ship's safety management system.

15.1.6 In carrying out their assessment, the **competent person** must take into account any cargo previously carried in the space, ventilation, the coating of the space, the degree of corrosion and any other relevant factors. The factors affecting adjacent and connected spaces may be different from those affecting the space to be entered, but may affect the atmosphere in the space to be entered.

15.1.7 An enclosed space may not necessarily be enclosed on all sides; for example, ships' holds may have open tops but the nature of the cargo makes the atmosphere in the lower hold dangerous. Such places are not usually considered to be enclosed spaces but the atmosphere may become dangerous because of a change in the condition inside or in the degree of enclosure or confinement, which may occur intermittently (eg in diving bells or saturation chambers). Personnel need to exercise caution before entering any space on board a ship that has not been opened for some time. Examples of such spaces include:

- cargo spaces
- double bottoms
- fuel tanks
- ballast tanks
- cargo pump rooms
- cargo compressor rooms
- cofferdams
- chain lockers
- void spaces
- duct keels
- inter-barrier spaces
- sewage tanks
- boilers
- engine crankcases
- engine scavenge air receivers
- carbon dioxide (CO_2) rooms
- battery lockers
- certain hold ladders enclosed in trunks
- any spaces adjacent or connected to an enclosed space (eg cargo space access ways).

This is not an exhaustive list, and awareness of potential risks is necessary for all spaces on board ship. If in any doubt, any such space should be regarded as potentially enclosed and appropriate action taken.

15.1.8 Enclosed spaces can change over time. Risk assessments should consider how works undertaken may change the atmosphere. Spaces connected in any way to an enclosed space can become dangerous due to the migration of gases from the enclosed space. Migration of gases from another space can also cause the enclosed space to become dangerous. Precautions must be taken where this is a possibility.

15.1.9 The possibility of leaks from one space to another should be considered in risk assessments of any space.

15.1.10 Incidents involving fumigants are likely to occur outside of the enclosed space due to gas migration.

15.1.11 Any enclosed spaces on the ship should be identified using risk assessment (carried out in accordance with Annex 1.2). A register should be made of any enclosed spaces that seafarers may enter. Throughout the assessment process, there should be an assumption that the space to be entered is hazardous until proved positively to be safe for entry. Spaces where the risk of a change in atmosphere is significant should be included on this register.

15.1.12 The register should record:

- the characteristics of the space, including physical layout of the space and access and egress points
- any potential hazards
- measures to prevent entry, including locking and signage arrangements
- procedures to follow when entering, including details such as estimated time needed to ventilate the area
- information related to ventilation, including equipment and where the equipment is stored
- lighting and requirements for temporary lighting
- requirements for atmospheric testing.

Any difficulties inherent in rescue from the space should also be considered, and solutions identified, so that in the event of an emergency, the crew is in the best position to respond quickly.

15.1.13 The register should be reviewed regularly and should be available to any seafarer that may need to enter the space. This is equally important when a space is not entered regularly. The register should be updated where new hazards and features of the space are identified. The register should also be available to those ashore – for reference when considering time required for entry and repairs.

15.1.14 In addition:

- if there is any unexpected reduction in or loss of the means of ventilating spaces that are usually continuously or adequately ventilated, such spaces should also be dealt with as enclosed spaces
- when it is suspected that there could be a deficiency of oxygen in any space, or that toxic gases, vapours or fumes could be present, then such a space should be treated as enclosed.

15.1.15 Entrances to all unattended enclosed spaces on a ship should be kept locked or secured against entry. Any hatches to readily accessible enclosed spaces should be marked at the entrance to a dangerous space. When the space is open for work to be carried out, an attendant should be posted or a barrier and warning sign put in place. As far as possible, work should be arranged in such a way that no one has to enter the space.

15.1.16 Some hold ladders are enclosed in trunks that are open at the top and bottom. These are connected to the hold space and will share any hazardous atmosphere within the parent space. Hazardous atmosphere can be trapped in these spaces, which may not be adequately ventilated when hatches are opened. Such spaces should remain secured until atmosphere testing of the space has taken place.

15.1.17 All crew should be given on-board training and familiarisation with the risks of entry into dangerous spaces on board. Training should include as a minimum:

- the vessel's register of enclosed spaces
- identification of the hazards likely to be faced during entry into enclosed spaces
- any steps taken to reduce those hazards to an acceptable level
- knowledge of the procedures for assessment of the space
- knowledge of how to test the atmosphere of an enclosed space (where responsibilities require this)
- knowledge of the procedures for safe entry
- knowledge of the duties of those assigned to stand by during an entry
- means of communication to be used in the event of an emergency
- recognition of the signs of adverse health effects caused by exposure to hazards during entry.

15.2 Duties and responsibilities of a competent person and an authorised officer

15.2.1 A competent person means a person with sufficient theoretical knowledge and practical experience to make an informed assessment of the likelihood of a dangerous atmosphere being present or subsequently arising in the space, including taking measurements of the atmosphere.

IMO Resolution A.1050 (27)

15.2.2 An authorised officer means a person authorised to permit entry into an enclosed space and with sufficient knowledge of control and **elimination** of hazards, and of the procedures to be established and complied with on board, to be able to ensure that the space is safe for entry. (For further details on the role of an authorised officer see Chapter 14, in which International Maritime Organization (IMO) Resolution A.1050(27) refers to this person as the 'responsible person'.)

15.2.3 On the basis of their risk assessment, the authorised officer should decide the procedures to be followed for entry into a potentially dangerous space. These will depend on whether the assessment shows that:

- there is minimal risk to the life or health of a person entering the space then or at any future time
- there is no immediate risk to health and life but a risk could arise during the course of work in the space
- the risk to life or health is immediate.

15.2.4 Where the assessment shows that there is no immediate risk to health or life but that a risk could arise during the course of the work in the space, the precautions described in sections 15.3 to 15.9 should be taken as appropriate.

15.2.5 Where the risk to health or life is immediate, then the additional requirements specified in section 15.10 are necessary.

15.2.6 On vessels where seafarers are never expected to enter an enclosed space, or no accessible enclosed spaces exist, some of the recommendations of this chapter may not apply; for example, the requirement to have atmosphere-testing equipment on board the ship at all times, and the requirement for entry drills. However, all seafarers should have on-board training to help them recognise the risks from enclosed spaces and to familiarise them with any applicable procedures. When the competent person and authorised officer are shore-based personnel, no entry into a potentially dangerous space should be permitted until such suitably competent persons are present.

15.2.7 When non-ship staff (i.e. shoreside personnel) are contracted to enter an enclosed space on a ship in a UK port, the requirements of the Confined Spaces Regulations 1997 apply. While the master retains overall authority for any activity on board their ship, the employer of the shoreside team is responsible to ensure compliance with the Confined Spaces Regulations 1997. The company should satisfy themselves that the shoreside personnel have sufficient training, equipment and arrangements for rescue in accordance with the Confined Spaces Regulations 1997, Approved Code of Practice, or suitable equivalent arrangements. Once a permit to work has been issued, the shoreside personnel should take responsibility for the operation and for rescue arrangements. It is not sufficient to rely on emergency services for rescue arrangements, although they should still be notified in the event of any such emergency; nor should the ship's crew be designated to provide back-up support.

15.3 Precautions before entering an enclosed space

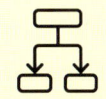

15.3.1 The following precautions should be taken, before any potentially enclosed space is entered, so as to make the space safe for entry without breathing apparatus and to ensure it remains safe whilst persons are within the space:

- A competent person should review the ship's register of enclosed spaces. If the register needs to be amended, the designated person ashore (DPA) should be informed.
- A competent person should make an assessment of the space and an authorised officer to take charge of the operation should be appointed – see section 15.2.
- The potential hazards should be identified – see section 15.4.
- The space should be prepared, vented and secured for entry – see section 15.4.
- The atmosphere of the space should be tested – see section 15.5.
- A permit to work system that includes a record of the gas measurements should be used.
- Procedures for preparation and entry should be agreed – see sections 15.8 and 15.9.
- Emergency procedures should be in place.

15.3.2 When the procedures listed in the previous paragraph have been followed and it has been established that the atmosphere in the space is or could be unsafe, then the additional requirements, as specified in section 15.10, should be followed.

15.3.3 In addition to pre-entry testing of the atmosphere, it is recommended that any person entering a potentially enclosed space should wear a personal atmosphere-monitoring device capable of detecting oxygen deficiency, carbon monoxide, toxic gases and explosive atmospheres. The specification of the monitoring equipment should be relevant to the hazardous atmosphere which may be found in the space. It is important to recognise that carrying a personal atmosphere monitor is no substitute for pre-entry testing.

15.4 Preparing and securing the space for entry

15.4.1 Prior to opening the entrance to an enclosed space, the space should be depressurised. Precautions should be taken in case pressurised or unpressurised vapour or gases are released from it. The space should be thoroughly ventilated, preferably by mechanical means, or by natural means, and then tested (see section 15.5) to ensure that all harmful gases have been removed and no pockets of oxygen-deficient atmosphere remain. Any vented gases should be discharged away from the area, thereby not contaminating the immediate area of the entry point to the space or other spaces.

15.4.2 The space should be isolated and secured against the ingress of dangerous substances by blanking off pipelines or other openings and by closing and securing valves, in accordance with the risk assessment and on-board procedures. The valves should be secured in the closed position, or some other means used to indicate that they are not to be opened. Remote-operated valves should, where practicable, have their remote actuators inhibited with notices placed locally and on the relevant controls. The officer on watch should be informed.

15.4.3 Where necessary, any sludge, scale or other deposit liable to give off fumes should be cleaned out. This may in itself lead to the release of gases, and precautions should be taken (see section 15.10).

15.4.4 Compressed oxygen should not be used to ventilate any space.

15.4.5 When appropriate, pumping operations or cargo movements should be suspended when entry is being made into an enclosed space.

15.5 Testing the atmosphere of the space

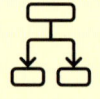

15.5.1 From May 2022, ships to which The Merchant Shipping and Fishing Vessels (Entry into Enclosed Spaces) Regulations 2022 apply are required to carry atmosphere-testing equipment. This must be capable of measuring concentrations of oxygen, flammable gases or vapours, hydrogen sulphide and carbon monoxide before any person enters the space.

SI 2022/96

15.5.2 Atmosphere testing equipment should have clearly defined calibration procedures included within the manufacturer's instructions. Where the operation of the ship permits access more frequently than the manufacturer's recommended calibration renewal period (eg ferry services), the calibration equipment may be kept ashore, and arrangements for calibration should be clear from the ship's SMS. Otherwise, calibration equipment should be carried on board.

MSC.1/Circ.1477

15.5.3 Testing should be carried out by remote means before entry and at regular intervals thereafter. Testing of a space should be carried out using properly calibrated and maintained equipment, and only by competent persons trained in the use of the equipment. Testing of any space should be carried out at different height levels.

15.5.4 If testing by remote means is not possible (eg where remote double-bottom tanks have to be entered), it should be assumed that the atmosphere is hazardous until proven otherwise. The procedures in section 15.12 should be followed, and see A.1050(27)(9).

A.1050(27)(9)

15.5.5 Personal monitoring equipment is designed for personal use only and to provide a warning against oxygen deficiency, toxic gases and explosive atmospheres whilst the wearer is in the space. This equipment should not be used as a means of determining whether an enclosed space is safe prior to entry, unless the specific equipment has the necessary certified/approved additional capability to conduct remote readings (i.e. pumped capability).

15.5.6 Personal monitoring equipment that has been designed and has the certified/approved capability to remote sense, by the addition of a length of sample hose and provision of either an electronic external or internal air pump to draw a sample of the atmosphere over the sensors, is acceptable as long as it is used with the additional equipment for remote sensing. The risk assessment should take account of the size and shape of the space, the reach of the remote sensor and the ability to thoroughly test the atmosphere throughout the space.

Testing for oxygen deficiency

15.5.7 The normal level of oxygen in the atmosphere is 20.8%. Any variation from that may indicate a problem and should be investigated further. For example, when the oxygen reading is 20%, consideration should be given to further testing for toxic gases, where appropriate, because toxic gases may have displaced some oxygen (see section 15.5.12). Once other risks are discounted, a steady reading of at least 20% oxygen by volume should be obtained before entry is permitted. Minor changes in temperature, pressure, condition of the cargo used and other conditions can influence the speed of oxygen depletion, so the atmosphere should be tested regularly when enclosed spaces are opened.

15.5.8 A combustible gas detector cannot be used to detect oxygen deficiency.

Testing for flammable gases and vapours

15.5.9 The combustible gas element of the detector detects the amount of flammable gas or vapour in the air. An instrument capable of providing an accurate reading at low concentrations should be used to judge whether the atmosphere is safe for entry.

15.5.10 The combustible gas element of the detector is calibrated on a standard gas. When testing for other gases and vapours, reference should be made to the calibration curves supplied with the instrument. Particular care is required should accumulations of hydrogen and methane be suspected, because high levels may affect the accuracy of the device.

15.5.11 In deciding whether the atmosphere is safe to work in, a 'nil' reading on an appropriate combustible gas detector is desirable but, where the readings have been steady for some time, up to 1% of the lower flammable limit may be accepted (eg for hydrocarbons in conjunction with an oxygen reading of at least 20% by volume).

Testing for toxic gases

15.5.12 The presence of certain gases and vapours requires specialised equipment and trained personnel to undertake accurate and reliable testing. If this equipment is not available for use do not enter, and the period of gas freeing should be considerably extended. Where measurement can be carried out, the readings obtained by this equipment should be compared with the workplace exposure limit (WEL) for the contaminant given in the latest edition of the Health and Safety Executive (HSE) Guidance Note EH40, which can be found on the HSE website. (Workplace exposure limits are sometimes known as occupational exposure limits (OELs) and are given in international industry safety guides.) Workplace exposure limits provide guidance on the level of exposure to toxic substances. Entry should not be authorised if the atmosphere measures over 50% of the WEL. However, it is necessary to know for which chemical a test is being made in order to use the equipment correctly, and it is important to note that not all chemicals may be tested by these means. Tests for specific toxic contaminants, such as benzene or hydrogen sulphide, should be undertaken depending on the nature of the previous contents of the space. The safety data sheets for previous cargoes or fuel carried should be referred to.

IMO Resolution A.1050(27)

Summary

15.5.13 In all spaces, the following conditions should be tested as a minimum. Entry should not be permitted, and personnel should not remain in an enclosed space (unless in an emergency; see section 15.10) until the following readings are obtained.

The four conditions shown in Table 15.1 should be tested as a minimum in all spaces.

15.5.14 When a toxic chemical is encountered for which there is no means of testing, then the additional requirements specified in section 15.10 should also be followed.

Table 15.1 Minimum content for enclosed spaces

Oxygen (O_2)	At least 20% by volume (see section 15.5.7)
Flammable gas	Nil Note: where readings have been steady for some time, up to 1% of the lower flammable limit (LFL) may be acceptable in conjunction with a 20% oxygen level
Carbon monoxide (CO)	Content is less than: **100 ppm** short-term exposure limit (STEL): maximum exposure is 15 minutes* **20 ppm** time weighted: maximum exposure is 8 hours*
Hydrogen sulphide	Content is less than: **10 ppm** STEL: maximum exposure is 15 minutes* **5 ppm** time weighted: maximum exposure is 8 hours*
Toxic gases	Less than 50% of the WEL*

* Current limits within EH40

15.5.15 If a separate combustible gas detector is used, this is not suitable for measuring levels of gas at or around its workplace exposure limit, where there is solely a toxic, rather than a flammable, risk. This level will be much lower than the flammable limit, and the detector may not be sufficiently sensitive to give accurate readings.

15.6 Use of control systems

15.6.1 Entry into an enclosed space should be planned in advance and use should be made of a permit to work system. Details of the arrangements to be followed in a permit to work system are described in Chapter 14 (a sample permit to work can be found in Annex 14.1.11).

15.6.2 For situations for which a well-established safe system of work exists, a checklist may exceptionally be accepted as an alternative to a full permit to work, provided that the principles of the permit to work system are covered and the risks arising in the enclosed space are low.

15.6.3 No person should enter an enclosed space unless authorised to do so by an authorised officer. Only the minimum number of trained personnel required to do the work should be authorised to enter. Those entering must be wearing appropriate clothing and PPE. All equipment used must be in good working condition and inspected before use.

15.7 Safety precautions before entry

15.7.1 The space and its access areas should be adequately illuminated.

15.7.2 No source of ignition should be taken or put into the space unless the authorised officer is satisfied that it is safe to do so.

15.7.3 A rescue plan should be in place (see section 15.13). In all cases, rescue and resuscitation equipment should be positioned ready for use at the entrance to the space. A risk assessment should identify what rescue equipment may be required for the particular circumstances but, as a minimum, this should include:

- appropriate breathing apparatus, with fully charged spare cylinders of air
- lifelines and rescue harnesses
- torches or a lamp (approved for use in a flammable atmosphere, if appropriate)
- a means of hoisting an incapacitated person from the confined space, if appropriate.

15.7.4	Breathing equipment may be bulky and limit movement in the space. Before entry is permitted, it should be established that entry with breathing apparatus is possible. Any difficulty of movement within any part of the space, or any problems if any incapacitated person had to be removed from the space (as a result of breathing apparatus or lifelines or rescue harnesses being used), should be considered. Risks should be minimised or entry prohibited.
15.7.5	Lifelines should be long enough for the purpose and capable of being firmly attached to the harness, but the wearer should be able to detach them easily should they become tangled. They should not be relied upon as the sole means of recovering a casualty from a space.
15.7.6	When necessary, a rescue harness should be worn to make it easier to recover a casualty in the event of an accident.
15.7.7	In addition to rescue harnesses, wherever practicable, hoisting equipment should be used. Hoisting equipment should be attended by personnel stationed at the entrance who have been trained in how to pull an unconscious person from an enclosed space.
15.7.8	At least one competent person, with appropriate equipment, should be posted to remain as an attendant at the entrance to the space whilst it is occupied.
15.7.9	An agreed and tested system of communication should be established: <ul><li>between any person entering the space and the attendant at the entrance</li><li>between the attendant at the entrance to the space and the officer on watch.</li></ul>
15.7.10	Communication systems should be appropriate for the operation, taking into consideration whether persons outside the enclosed space have line of sight and, in the event of loss of consciousness, how much time it would take to reach the casualty, taking appropriate safety measures.

15.8 Procedures and arrangements during entry

15.8.1	Ventilation should continue during the period that the space is occupied and during any temporary breaks. In the event of a failure of the ventilation system, all personnel in the space should leave immediately.
15.8.2	The atmosphere should be tested periodically and the results recorded whilst the space is occupied and personnel should be instructed to leave the space should there be any deterioration of the conditions. Testing should be carried out more frequently if there is any possibility of change in the conditions in the space. Should a personal gas detector give an alarm, everybody should leave the space immediately.

15.8.3 If unforeseen difficulties or hazards develop, the work in the space should be stopped and everybody should leave the space so that the situation can be reassessed. Permits should be withdrawn and only reissued, with any appropriate revisions, after the situation has been reassessed.

15.8.4 If any personnel in a space feel in any way adversely affected, they should give the prearranged signal to the attendant standing by the entrance and immediately leave the space.

15.8.5 Should an emergency occur, the general (or crew) alarm should be sounded so that back-up is immediately available to the rescue team. Under no circumstances should the attendant enter the space. See section 15.13.

15.9 Procedures on completion

15.9.1 On expiry of the permit to work, everyone should leave the space and the entrance to the space should be closed or otherwise secured against entry. See Annex 14.1.1.

15.10 Additional requirements for entry into a space where the atmosphere is suspect or known to be unsafe

15.10.1 If the atmosphere is considered to be suspect or unsafe to enter, no one should enter the space unless it is essential for the safety of life or of the ship or for testing prior to entry in such a scenario, when no practical alternative exists, such as if remote testing is not possible. Breathing apparatus must always be worn (see sections 15.12 on equipment and 15.13 on emergency rescue). The number of competent persons entering the space should be the minimum compatible with the task to be performed.

15.10.2 When appropriate, portable lights and other electrical equipment should be of a type approved for use in a flammable atmosphere.

15.10.3 Should there be a risk of chemicals, whether in liquid, gaseous or vapour form, coming into contact with the skin and/or eyes, then appropriate PPE should be worn.

15.11 Training, instruction and information

15.11.1 The company should provide any necessary training, instruction and information to seafarers in order to ensure that the requirements of the Entry into Enclosed Spaces Regulations are complied with. This should include:

- recognition of the circumstances and activities likely to lead to the presence of a dangerous atmosphere
- the hazards associated with entry into enclosed spaces, and the precautions to be taken
- the use and maintenance of equipment and clothing required for entry into enclosed spaces
- instruction and drills in rescue from enclosed spaces.

15.11.2 It is recommended that all seafarers whose duties may involve entry into enclosed spaces attend a dedicated course for entry into enclosed spaces.

   SI 1988/1638

15.12 Breathing apparatus

15.12.1 Entry into a space that has or is suspected to have a dangerous atmosphere should only be permitted when it is essential for rescue or other safety-critical purposes.

15.12.2 No one should enter a space where the atmosphere is unsafe or suspect without wearing breathing apparatus that is of an approved type; see section 15.12.4. This also applies to entering an enclosed space to rescue another person. Prior to entry instruction in the use of breathing apparatus must be given by a competent person.

15.12.3 When using self-contained breathing apparatus, it should not be necessary to remove any part of the breathing apparatus or any protective clothing to change over to the self-contained supply. Air-purifying respirators are not suitable because they cannot supply clean air from an independent source.

15.12.4 Equipment for use with two air supplies may consist of:

- conventional self-contained breathing apparatus of the open circuit, compressed air type that is approved to EN 137:2006 and has been additionally tested for use with an airline connection; or
- compressed airline breathing apparatus incorporating an emergency self-contained supply. The compressed airline breathing apparatus should be of the demand-valve type and approved to BS EN 14593-1:2018 or BS EN 14594:2018 or, for self-rescue purposes, to BS 1146:2005 (or equivalent standard). The emergency self-contained supply should comply with the relevant parts of the appropriate standard.

   EN 137:2006; BS EN 14593-1:2018; BS EN 14594:2018; BS 1146:2005

The capacity of the self-contained supply should be sufficient for the wearer to escape to a safe atmosphere. When determining this capacity, it should be recognised that, under stress or in difficult conditions, the wearer's breathing rate may be in excess of the nominal breathing rate of 40 litres per minute.

15.12.5 The authorised officer should make sure that the supply of air to breathing apparatus from outside the space is continuous and available only to those working in the space. A competent person must be assigned responsibility to monitor the air supply and ensure it is not interrupted. Place signage at positions where airlines are in use to warn other personnel not engaged in the enclosed entry. When a mechanical pump is being used, it should frequently be checked carefully to ensure that it continues to operate properly. Any air pumped directly into a pipeline or put into reserve bottles must be filtered and should be as fresh as possible. Pipelines or hoses used to supply air should be thoroughly blown through to remove moisture and freshen the air before connection to breathing apparatus and face masks. It is essential that where the air supply is from a compressor sited in a machinery space, the engineer of the watch is informed so that the compressor is not shut down until the work is completed.

15.12.6 The authorised officer and the person about to enter the space should undertake the full pre-wearing check and donning procedures recommended in the manufacturer's instructions for the breathing apparatus.

15.12.7 In particular, they should check that:

- there will be sufficient clean air at the correct pressure
- low-pressure alarms are working properly
- the face mask fits correctly against the user's face so that, combined with pressure of the air coming into the mask, there will be no ingress of oxygen-deficient air or toxic vapours when the user inhales (it should be noted that facial hair or spectacles may prevent the formation of an air-tight seal between a person's face and the face mask)
- the wearer of the breathing apparatus understands whether or not their air supply may be shared with another person and, if so, is also aware that such procedures should only be used in an extreme emergency
- when work is being undertaken in the space, the wearer should keep the self-contained supply for use if there is a failure of the continuous supply from outside the space.

15.12.8 When in an enclosed space, breathing apparatus should not be removed unless it is an emergency. This includes to share an air supply; see section 15.13.

Maintenance of equipment for entry into enclosed spaces

15.12.9 All breathing apparatus, rescue harnesses, lifelines, resuscitation equipment and any other equipment provided for use in, or in connection with, entry into enclosed spaces, or for use in emergencies, should be properly maintained, inspected periodically and checked for correct operation by a competent person, and a record of the inspections and checks should be kept. All items of breathing apparatus should be inspected for correct operation before and after use.

15.12.10 Equipment for testing the atmosphere of enclosed spaces, including oxygen meters, should be kept in good working order and, where applicable, regularly serviced and calibrated. Manufacturers' recommendations, which should always be stored with the equipment, should be complied with at all times.

15.13 Preparation for an emergency

15.13.1 Safety drills for entry into and rescue from an enclosed space should be carried out every two months (see section 4.8 on enclosed space entry drills). Figure 15.1 shows a flowchart to follow. Drills should, as a minimum, include:

- checking and use of personal protective equipment required for entry
- checking and use of communication equipment and procedures
- checking and use of instruments for measuring the atmosphere in enclosed spaces
- checking and use of rescue equipment and procedures
- instructions in first aid and resuscitation techniques.

15.13.2 For every entry to an enclosed space, a rescue plan should be in place (see section 4.9 on action in event of an enclosed space emergency). Suitable rescue equipment should be available at the entry to the space, and roles allocated in the event of an emergency arising. Selection of such equipment should take into account the depth and volume of the space, the size of the access way, the potential distance of the casualty from the point of entry and the resources available to assist in the rescue.

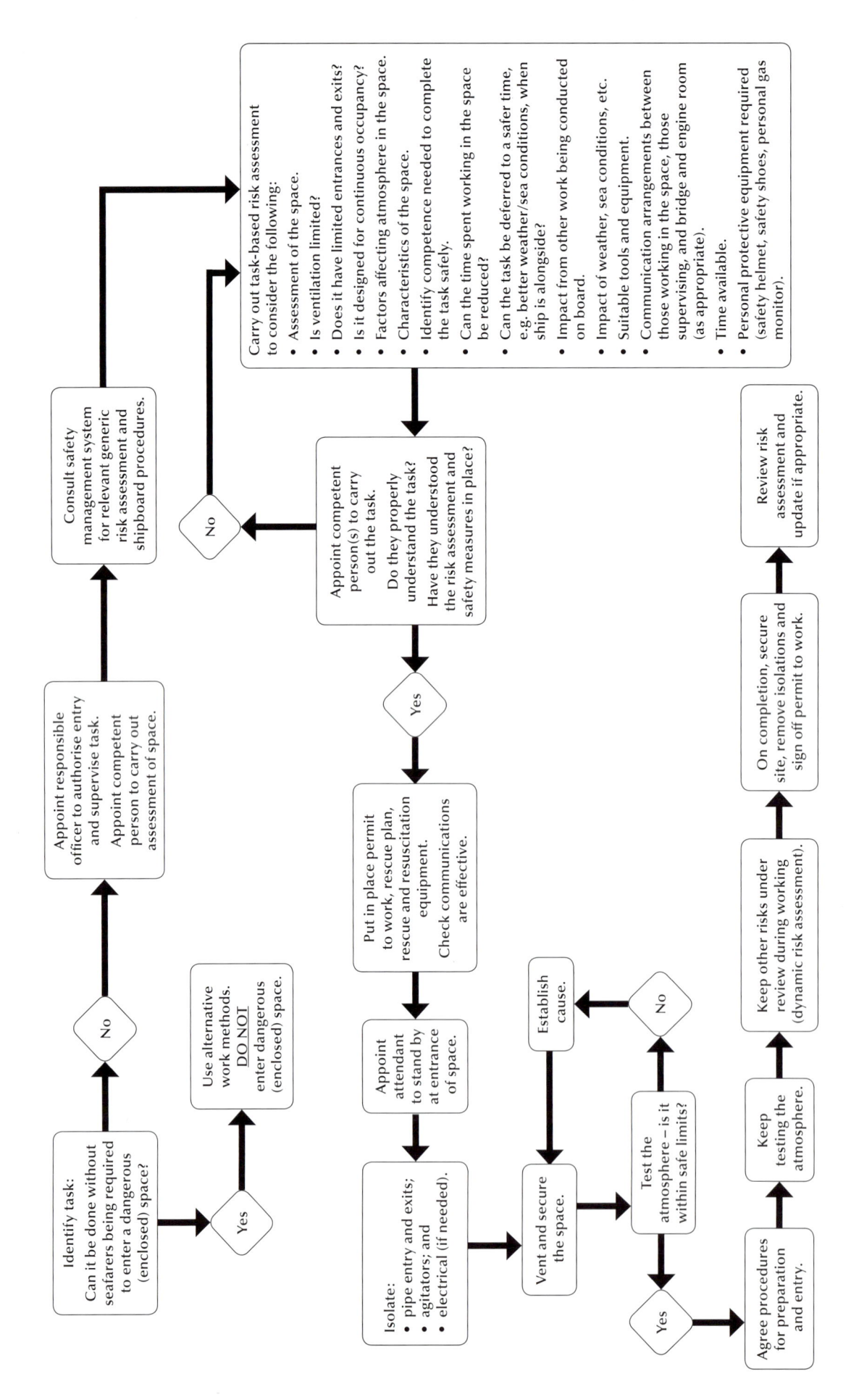

Figure 15.1   Procedure for entering an enclosed space

Emergency rescue arrangements

15.13.3 No one should enter any enclosed space to attempt a rescue without taking suitable precautions for their own safety. Failure to do so will put the would-be rescuer's life at risk and almost certainly prevent the person they intended to rescue being brought out alive. Many multiple fatalities have occurred as a result of individuals attempting a rescue without taking adequate precautions.

15.13.4 Should an emergency occur, the general (or crew) alarm should be sounded so that back-up is immediately available to the rescue team. Under no circumstances should the attendant enter the space.

15.13.5 Once help has arrived, the situation should be evaluated, considering what rescue equipment is needed, and the rescue plan should be put into effect. An attendant should remain outside the space at all times to ensure the safety of those entering the space to undertake the rescue.

15.13.6 Once the casualty is reached, the checking of the air supply must be the first priority, to assess if the casualty is breathing and whether resuscitation outside of the space is required. Unless they are gravely injured, they should be removed from the enclosed space as quickly as possible. Where rescue of a casualty is required, or injuries need to be treated/stabilised, this should commence as soon as possible (including before rescue activities commence) if safe to do so.

15.13.7 Self-contained breathing apparatus that is specifically suited for such applications must be worn. If it is found that it is not possible to enter a tank wearing a self-contained breathing apparatus, the bottle harness may be removed and passed through the access but the face mask must always be worn. Care should be taken to ensure that the harness does not drop onto or pull on the supply tube and dislodge the face mask.

15.13.8 An emergency escape breathing device (EEBD) is not suitable for use by a rescuer. This is a supplied air or oxygen device, with a limited supply, designed only to be used for escape from a compartment that has a hazardous atmosphere. It should not be worn by a rescuer entering a space to attempt a rescue.

15.13.9 When entering a space to carry out a rescue, it is important to ensure that the area adjacent to the space of entry is free from any hazard and cordoned off accordingly.

15.14 Potential hazards associated with enclosed spaces

Oxygen deficiency

15.14.1 If an empty tank or other confined space has been closed for a time, the oxygen content may have been reduced for a number of reasons.

15.14.2 The following are examples only:

- Rusting may have occurred due to oxygen combining with steel.
- Oxygen-absorbing chemicals may have been present.
- Oxygen-absorbing cargoes may have been carried, including:
  - grain, grain products and residues from grain processing (such as bran, crushed grain, crushed malt or meal) hops, malt husks and spent malt
  - oilseeds, products from oil seeds (such as seed expellers, seed cake, oil cake and meal)
  - copra
  - wood in such forms as packaged timber, round wood logs, pulpwood, props (pit props and other prop-wood), woodchips, wood-shavings, wood pellets and sawdust
  - jute, hemp, flax, sisal, kapok, cotton and other vegetable fibres, empty bags, cotton waste, animal fibres, animal and vegetable fabric, wool waste and rags
  - fish, fishmeal and fish-scrap
  - guano
  - sulphidic ores and ore concentrates
  - charcoal, coal lignite and coal products
  - direct reduced iron
  - dry ice
  - metal wastes and chops, iron swarf, steel and other turnings, borings, drillings, shavings, filings and cuttings
  - scrap metal.
- Gases from volatile cargoes may have displaced the oxygen in tanks.
- Hydrogen may have been produced in a cathodically protected cargo tank used for ballast.
- Oxygen may have been displaced by the use of carbon dioxide or other fire-extinguishing or preventing media, or inert gas in the tanks or inter-barrier spaces of tankers or gas carriers.
- Nitrogen or another inert gas may have been used to inert, purge or top-up the tanks.

Oxygen-enriched atmosphere

15.14.3 This may arise from:

- leaks from damaged or poorly maintained hoses, pipes and valves
- leaks from poor connections
- opening valves deliberately or accidentally
- not closing valves properly after use
- using an excess of oxygen in welding, flame cutting or a similar process
- poor ventilation where oxygen is being used.

Because oxygen is odourless, colourless and tasteless, an oxygen-enhanced atmosphere cannot be easily detected by human senses. However, because oxygen aids combustion, even a small increase in the concentration of oxygen in the air produces an increased risk of fire – including spontaneous combustion – or explosion.

Toxicity of oil

15.14.4　Hydrocarbon gases are flammable as well as toxic and may be present in fuel or cargo tanks that have contained crude oil or its products.

15.14.5　Hydrocarbon gases or vapours may also be present in pump rooms and cofferdams, duct keels or other spaces adjacent to cargo tanks due to the leakage of cargo.

15.14.6　The components in the vapour of some bunker oils and oil cargoes, such as benzene and hydrogen sulphide, are very toxic.

Toxicity of other substances

15.14.7　As appropriate, seafarers should understand the properties and hazards of relevant gases and the mitigation measures required.

15.14.8　Some of the cargoes carried in bulk, liquid, gas or packaged form may be toxic or liable to emit toxic gas; appropriate testing for toxic gas should be carried out as per section 15.5.

15.14.9　There is the possibility of leakage from drums of chemicals or other packages of dangerous goods where there has been mishandling, incorrect stowage, or damage due to heavy weather.

15.14.10　Inert gas does not support life. In addition, trace components that are often present in the inert gas, such as carbon monoxide, sulphur dioxide, nitric oxide and nitrogen dioxide, are very toxic.

15.14.11　The interaction of vegetable or animal oils, sewage or slops from drilling operations with sea water may lead to the release of hydrogen sulphide, which is very toxic.

15.14.12　Hydrogen sulphide or other toxic gases may be generated where the residue of grain or similar cargoes permeates into or chokes bilge-pumping systems.

15.14.13　The chemical cleaning, painting or repair of tank coatings may involve the release of solvent vapours.

15.14.14　Fumigants may have been used on cargoes in the space (see section 21.8).

Flammability

15.14.15 Flammable vapours may still be present in any tanks that have contained oil products or chemical or gas cargoes.

15.14.16 Cofferdams, fuel treatment and processing rooms, and other spaces that are adjacent to tanks, may contain flammable vapours, should there have been leakage into the space.

Other hazards

15.14.17 Although the inhalation of contaminated air is the most likely route through which harmful substances enter the body, some chemicals can be absorbed through the skin.

15.14.18 Some of the cargoes in bulk, liquid, gas or packaged form are irritants or corrosive if permitted to come into contact with skin. Appropriate testing should be carried out as per section 15.5.

15.14.19 Disturbance of rust, scale or sludge residues of cargoes of animal, vegetable or mineral origin, or disturbance of water that could be covering such substances, may lead to the release of toxic or flammable gases.

- Entry into enclosed spaces is a dangerous activity – never rush in without following procedures, even if it is an emergency.
- Even spaces which are not typically enclosed can pose a risk due to gas migration.
- Toxic gases (and oxygen deficiency) can be invisible to the human eye.
- Some spaces may be a dangerous enclosed space only intermittently, perhaps due to the type of cargo carried or work to be undertaken (eg a compartment during spray painting).
- Ships to which The Merchant Shipping and Fishing Vessels (Entry into Enclosed Spaces) Regulations 2022 apply are required to carry atmosphere-testing equipment.

16 Hatch covers and access lids

16.1 Introduction

16.1.1 Any hatch covering used on a ship should be of sound construction and material, strong enough for its purpose, free from patent defect and properly maintained.

16.1.2 Do not use or operate a hatch covering unless it can be removed and replaced without endangering any person.

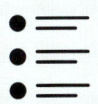

Key points
- Everyone working on board should know the health and safety risks in handling hatch covers and access lids.
- Everyone working on board should take precautions, including use of appropriate personal protective equipment (PPE).

Your organisation should
- plan and implement appropriate control measures to protect workers whose health and safety may be put at risk by the operation of hatch covers and access lids, based on the findings of the risk assessment
- warn seafarers about the risks of handling hatch covers and access lids
- provide appropriate PPE
- ensure that all information regarding correct replacement positions is clearly marked
- ensure hatch cover replacements conform to the ship's plan.

16.2 Precautions

16.2.1 Before a vessel departs, secure all weather deck hatch covers in the correct closed position, and positively confirm this (eg by report or logbook entry). While the vessel is at sea inspect hatch covers regularly to ensure that their integrity is maintained.

16.2.2 When using lifting appliances, attach them to hatch covers from a safe position and without exposing personnel to any danger.

Use and maintenance of hatch and access lids

16.2.3 Do not use a hatch unless its covering has been completely removed or, if not completely removed, made properly secure.

16.2.4 Keep all hatch covers properly maintained. Replace or repair any defective or damaged ones as soon as possible.

16.2.5 **Warning**

All personnel involved with the handling and/or operation of hatch covers should be properly instructed in their handling and operation. Document this in induction (familiarisation) or other training records on board; see section 2.1.2. A responsible person should supervise all stages of opening or closing hatches.

16.2.6 Follow this safety guidance on the use of hatch covers:

- Use hatch covers only in line with the manufacturer's instructions.
- Do not work on, or place loads over, any section of a hatch cover unless it is known that the cover is properly secured and can safely support the load.
- Use covers and beams only if they are a good fit and their end supports overlap by exactly the right amount (enough but not too much).
- When hatches are open, the area around the opening and in the hatchways should be appropriately lit and have guardrails where necessary. Guardrails should be tight, with stanchions secured in position, and properly maintained.
- Never temporarily cover partly opened, unguarded hatches with anything that could hide the opening. This may cause a serious hazard for any person walking across the hatch.
- Unless hatches are fitted with coamings to a height of at least 760 mm (30 inches), cover them securely or fence them to a height of 1 metre (39 inches) when they are not being used for the passage of cargo.

16.3 Mechanical hatch covers

16.3.1 When using mechanical hatch covers:

- Follow the manufacturer's instructions for the safe operation, inspection, maintenance and repair of hatch covers.
- Check the trim of the vessel. Do not remove the hatch locking pins or preventers of rolling hatches until a check wire is fast to prevent premature rolling when the tracking is not horizontal.
- Keep hatch wheels greased and free from dirt and the coaming runways and keep drainage channels clean.
- Secure rubber sealing joints properly and keep them in good condition to provide a proper weathertight seal.

16.3.2 Remember the following:

- Secure all locking and tightening devices in place on a closed hatch at all times when at sea.
- Keep securing cleats greased. Check cleats, top-wedges and other tightening devices regularly while at sea.
- Secure hatch covers properly immediately after closing or opening them. Secure them in the open position with chain preventers or by other suitable means. No one should climb onto any hatch cover unless it is properly secured.
- Keep the area clear of all items that might foul the covers or the handling equipment.

16.3.3 **Warning**
Except in the event of an emergency endangering health or safety, do not operate a power-operated hatch covering unless authorised to do so by a responsible ship's officer.

16.4 Non-mechanical hatch covers

16.4.1 Each non-mechanical hatchway should have an appropriate number of properly fitting hatch covers, pontoons or slab hatches. These should be adequately marked to show the correct replacement position.

16.4.2 Handle pontoon hatches and hatch slabs with care. Stow them properly, stacked so as not to endanger or impede the normal running of the vessel.

16.4.3	When using a crane or derrick to handle pontoons or slab hatches, position it directly over them to lessen the risk of violent swinging once the weight has been taken.
16.4.4	Provide appropriate gear, strong enough to lift pontoons and slab hatches. A **competent person** 🔍 should operate the crane or winch under the direction of a ship's officer or other experienced person.
16.4.5	Do not remove or replace hatch covers until somebody has checked that all people are out of the hold or clear of the hatchway.

16.5 Non-mechanical manually handled hatch covers

16.5.1	Each non-mechanical manually handled hatchway should have an appropriate number of properly fitting beams and hatch covers, pontoons or slab hatches, which should be adequately marked to show the correct replacement position. For wooden hatch covers, provide enough properly fitting tarpaulins, batten bars, side wedges and locking bars so that the hatch will remain secure and weathertight for all weather conditions.
16.5.2	Manually handled hatch covers should be capable of being easily lifted by two people. Such hatch covers should be thick and strong enough and have handgrips. Wooden hatch boards should be strengthened by steel bands at each end. One person should not try to handle hatch covers unaided unless the covers are designed for single-handed operation.
16.5.3	Hatch boards, hatch beams and tarpaulins should be handled with care and properly stowed, stacked and secured so as not to endanger or impede the normal running of the vessel. Remove hatch boards working from the centre towards the sides, and replace them working from the sides towards the centre. Where tarpaulins are used, personnel should walk forwards and not backwards so they can see where they are walking.
16.5.4	When using a crane or derrick to handle a beam, position it directly over the beam to lessen the risk of violent swinging once the weight has been taken.
16.5.5	Appropriate, strong enough gear should be provided specifically for the lifting of beams. Slings should be long enough, secured against accidental dislodgement while in use and fitted with control lanyards. To avoid undue stress the angle between arms of slings at the lifting point should not exceed 120°. A competent person should operate the crane or winch under the direction of a ship's officer or other experienced person.
16.5.6	Beams and hatch covers remaining in position in a partly opened hatchway should be securely pinned, lashed, bolted or otherwise properly secured against accidental dislodgement.

16.5.7	Hatch covers and beams should not be removed or replaced until somebody has checked that all people are out of the hold or clear of the hatchway. Immediately before removing beams check that the pins or other locking devices have been freed.
16.5.8	No one should walk out on a beam for any purpose.
16.5.9	Do not use hatch covers in the construction of deck or cargo stages or place loads on them that are liable to damage them. Do not place loads on hatch coverings without the authority of a ship's officer.

16.6 Steel-hinged inspection/access lids

16.6.1	Inspection/access hatch lids should be constructed of steel or similar material and hinged so they can be easily and safely opened or closed. Lids on weather decks should be seated on watertight rubber gaskets and secured weathertight by adequate dogs, side cleats or equivalent tightening devices.
16.6.2	When not secured, inspection/access hatch lids should be capable of being easily and safely opened from above and from below.
16.6.3	Adequate handgrips should be provided in accessible positions to lift inspection/access hatch lids by hand without straining or endangering personnel.
16.6.4	Heavy or inaccessible hatch lids should be fitted with counterweights so that they can easily be opened by one or two people. Where a counterweight cannot be fitted because a hatch lid is inaccessible, provide the lid with a purchase or pulley with eye-plates or ringbolts fitted in appropriate positions. In this way the hatch can be opened and closed without straining or endangering personnel.
16.6.5	Where hatch lids are fitted with a security device, they should be capable of being opened easily from below in the event of an emergency.
16.6.6	When the hatch lids are open they should be easily and safely secured against movement or accidental closing. Provide adequate locking pins, steel hooks or other means.

16.7 Access to holds/cargo/other spaces

16.7.1	Seafarers should enter holds/cargo/other spaces only on the authority of a responsible ship's officer. Before granting authority the ship's officer should ensure that the space has been adequately ventilated and, where appropriate, tested for noxious gases/oxygen content. All other appropriate pre-entry precautions should be taken (see Chapter 15).
16.7.2	Where possible the permanent means of access should be used for entry. When this is not possible, portable ladders may be used (see section 17.3). Where necessary, lifelines and safety harnesses should be used.

Hatches and access lids must

- be secured when not in use
- have guardrails in place when in use
- be secured and periodically checked when at sea.

Holds and other spaces must be thoroughly checked to ensure that no one is inside before hatches or access lids are fully secured.

17 Work at height

17.1 Introduction

17.1.1 Anyone working where there is a risk of falling may be regarded as working at height. In addition to work on ladders, staging and scaffolding, this includes work inside a tank, near an opening such as a hatch, or on a fixed stairway. Further guidance is available in Marine Guidance Note MGN 410 (M+F) Amendment 2.

SI 2010/332; MGN 410 (M+F) Amendment 2

17.1.2 Working at height (also known as working aloft) should be risk assessed and suitable control measures should be taken to protect people who may be at risk. Depending on the severity of the risk, a **permit to work** may be required.

Key points
- Work at height only if there is no reasonably practicable alternative. Only competent people should do any activity relating to work at height, or use equipment for work at height. This includes organising, planning and supervising such activities. Where a seafarer is being trained to do such work, they must be supervised by another seafarer who is competent both to supervise and undertake that activity. In this context, 'competent people' have sufficient skills, knowledge and experience to perform the task safely.
- Only following company risk assessment and under supervision by a **competent person** should anyone under 18 years of age be working at height; see Merchant Shipping Notice MSN 1838 (M) Amendment 1 on minimum age requirements. Personnel under 18, or with less than 12 months' experience at sea, should not work at height unless it forms part of their planned training, and unless they are supervised by a competent person.
- Choose work equipment that is fit for purpose and meets the requirements of Chapter 18. Use it in line with safe procedures and good practice.

MSN 1838 (M) Amendment 1

17.2 General

Your organisation should
- ensure that work at height, where it is necessary, is properly planned, appropriately supervised and done as safely as is reasonably practicable
- include a risk assessment as part of the planning
- consider the potential risks from falling objects or fragile surfaces and plan for emergency situations.

Annex 17.1 gives guidance on planning for emergency situations while working at height.

> SI 2006/2183; MGN 331 (M+F) Amendment 2; MGN 532 (M) Amendment 2; MGN 533 (M) Amendment 2

17.2.1 Personnel working at height may not be able to give their full attention to the job and guard themselves against falling. Take proper precautions to ensure personal safety when work has to be done at height. Remember that the movement of a ship in a seaway and in poor weather conditions, even when alongside, will add to the hazards involved in work of this type. Use a stage, ladder, scaffolding, aerial work platform, bosun's chair or secured scaffold tower when working beyond normal reach. Inspect any equipment before use to ensure it is in a good state of repair.

17.2.2 Personnel working at height where there is a risk of falling should wear a safety harness with a lifeline or other arresting device at all times (see Figure 8.2). A safety net should be rigged where necessary and appropriate. Additionally, where work is done overside, they should wear a working lifejacket (personal flotation device) or buoyancy garments (see Figure 8.3). Keep a lifebuoy with sufficient line attached ready for immediate use. Personnel should be observed by a person on deck.

> For guidance on the use of equipment for overside working, see MGN 578 (M) Amendment 1 Use of overside working systems on commercial yachts, small commercial vessels and loadline vessels.

Other than in emergency situations, personnel should not work overside while the vessel is under way. If such work is necessary, rescue boats should be ready for immediate use. A responsible person should closely monitor any such work.

17.2.3 Before working near the ship's whistle, the officer responsible should ensure that it is isolated and that warning notices are posted on the bridge and in the machinery spaces.

17.2.4 Before working on the funnel, the officer responsible should inform the duty engineer to ensure that the emission of steam, harmful gases and fumes is reduced as much as possible.

17.2.5 Before working near radio frequency emitting devices, the officer responsible should inform the radio room or person in charge of radio equipment that no transmissions should be made while there is a risk to personnel. To prevent use place a warning notice on the communications equipment or isolate it.

17.2.6 Where work is to be done near the radar scanner, the officer responsible should inform the officer on watch so that the radar and scanner are isolated. Place a warning notice on the radar equipment until the necessary work has been completed.

17.2.7 On completion of the work of the type described in 17.2.6, the person responsible should, where necessary, inform the appropriate person that the precautions taken are no longer required and that isolations and warning notices can be removed.

17.2.8 Do not work at height near cargo working unless it is essential. Take care to avoid risks to anyone working or moving below. Erect suitable barriers and display warning notices.

17.2.9 Tools and stores should be sent up and lowered by line in suitable containers. Secure the containers in place for the stowage of tools or materials not immediately in use. Secure tools with a tool lanyard of the appropriate length and strength rating, connected to a suitable anchor point. Secure heavy tools to an adequately strong structure; do not attach them to the user. Have a clearly signposted exclusion zone at the base of the work area so tools cannot accidentally fall on someone below.

17.2.10 Handle tools carefully where cold or greasy hands can affect grip. Consider wearing appropriate gloves in these situations.

17.3 Portable ladders

17.3.1 Avoid working from ladders as far as possible. Use them only for low-risk, short-duration work where personnel can maintain three points of contact (both feet and a handhold) on the ladder at all times (see Figure 17.1). Where necessary, personnel must use a safety harness with a lifeline secured above the work position, where practicable.

  MGN 410 (M+F) Amendment 2

17.3.2 Use a portable ladder only where no safer means of access is reasonably practicable. The ladder should be suitable for the work and in a safe condition.

Portable ladders should be pitched **(C)** at 75° from the horizontal, on a firm base and secured from slipping at the bottom. There should be at least 150 mm clearance behind the rungs. **(D)**

At least 1 m above the upper supporting point or landing place. Secured at top.

A should be approximately four times the length of **B**.

Always face the ladder.

Three points of contact on the ladder when working.

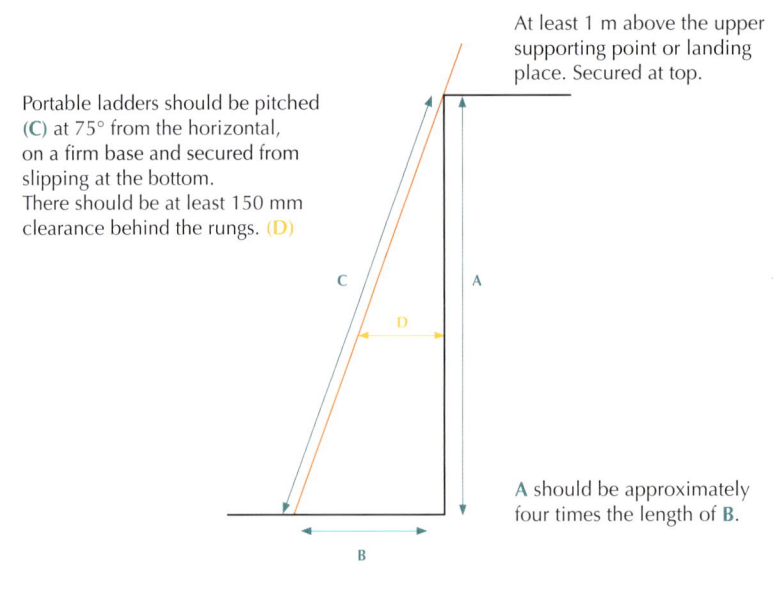

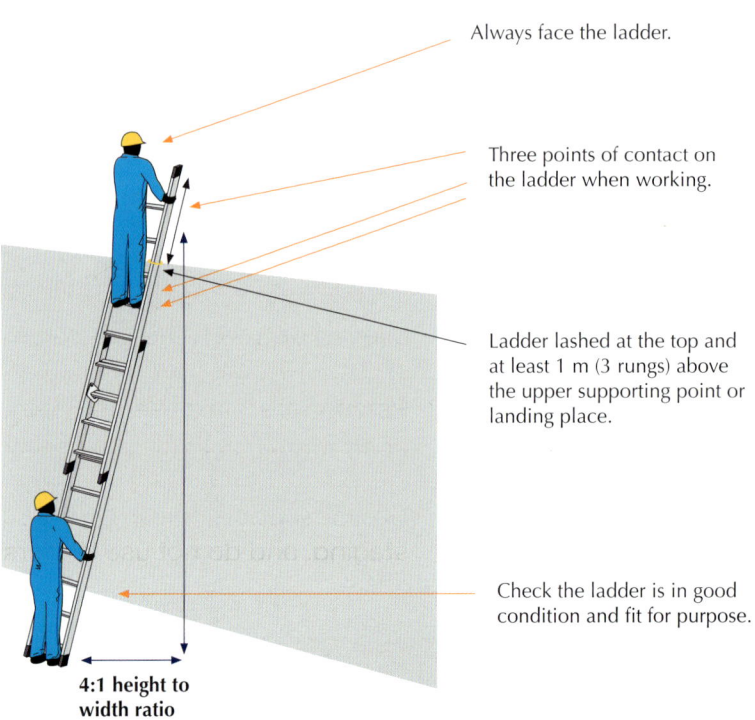

Ladder lashed at the top and at least 1 m (3 rungs) above the upper supporting point or landing place.

Check the ladder is in good condition and fit for purpose.

4:1 height to width ratio

Figure 17.1 Portable ladder

17.3.3 Only use ladders that:

- have no visible defects
- have an up-to-date record of inspection by a competent person; inspect them on a schedule according to the manufacturer's instructions
- are suitable for their intended use; i.e. strong and robust enough
- have been stored and maintained according to the manufacturer's instructions.

17.3.4 A competent person should inspect the condition of the ladder regularly in accordance with the manufacturer's instructions, as well as before use, and keep a record. Look out for:

- damaged or worn ladder feet
- twisted, bent or dented stiles
- cracked, worn, bent or loose rungs
- missing or damaged tie rods
- cracked or damaged welded joints, loose rivets or damaged stays.

HSE LA455

17.3.5 Do not paint or treat wooden ladders to hide defects and cracks. When not in use, stow them safely in a dry, ventilated space away from any heat source.

17.3.6 Pitch portable ladders at 75° from the horizontal and on a firm base, as shown in Figure 17.1. Secure them firmly against slipping or shifting sideways and allow a clearance of at least 150 mm behind the rungs. Where practicable, the ladder should extend to at least 1 metre above any upper landing place, unless there are other suitable handholds.

17.3.7 Annex 17.2 (reproduced from MGN 410 (M+F) Amendment 2) gives further guidance on the requirements for ladders.

17.3.8 When using portable extending ladders there should be sufficient overlap between the extensions.

17.3.9 Personnel should use both hands when ascending or descending a ladder and not try to carry tools or equipment in their hands.

17.3.10 Do not support planks on the rungs of ladders to be used as staging, and do not use ladders horizontally for such purposes.

17.4 Cradles and stages

17.4.1 Cradles should be at least 430 mm (17 inches) wide and fitted with guardrails or stanchions with taut ropes to a height of 1 metre (39 inches) from the floor. Intermediate rails or taut ropes and toe-boards should be fitted. Annex 17.3 (reproduced from MGN 410 (M+F) Amendment 2) gives further guidance.

📖 *MGN 410 (M+F) Amendment 2*

17.4.2 Examine planks and materials used for the construction of ordinary plank stages carefully to ensure they are strong enough and free from defects.

17.4.3 Stow wooden components of staging in a dry, cool, ventilated space.

17.4.4 Examine ancillary equipment such as lizards, blocks and gantlines thoroughly before use.

17.4.5 When rigging a stage overside, the two gantlines used in its rigging should be at least long enough to trail into the water to provide additional lifelines if a person should fall. Keep a lifebuoy and line ready nearby.

17.4.6 Do not use gantlines for working at height for any other purpose and keep them clear of sharp edges when in use.

17.4.7 The anchoring points for lines, blocks and lizards must be strong enough and, where practicable, permanently fixed to the ship's structure. Test integral lugs with a hammer. Do not use portable rails or stanchions as anchoring points. Treat any anchoring points as lifting points and inspect/test them as described in Chapter 19.

17.4.8 Always secure stages and staging that are not suspended against movement. Restrict hanging stages against movement as far as practicable.

17.4.9 In machinery spaces, keep staging and its supports clear of contact with hot surfaces and moving parts of machinery. In the engine room, do not use a crane gantry directly as a platform for cleaning or painting. It can, however, be used as the base for a stable platform if suitable precautions are taken.

17.4.10 Where personnel working from a stage have to raise or lower themselves, they must keep movements of the stage small and closely controlled.

17.4.11 Guidance for rail and trolley systems for overside working is available in MGN 578 (M) Amendment 1.

📖 *MGN 578 (M) Amendment 1*

17.5 Bosun's chair

17.5.1 When using a bosun's chair with a gantline, secure it to the gantline with a double-sheet bend and seize the end to the standing part with adequate tail (see Figure 17.2). Annex 17.3 (reproduced from MGN 410 (M+F) Amendment 2) gives further guidance.

📖 *MGN 410 (M+F) Amendment 2*

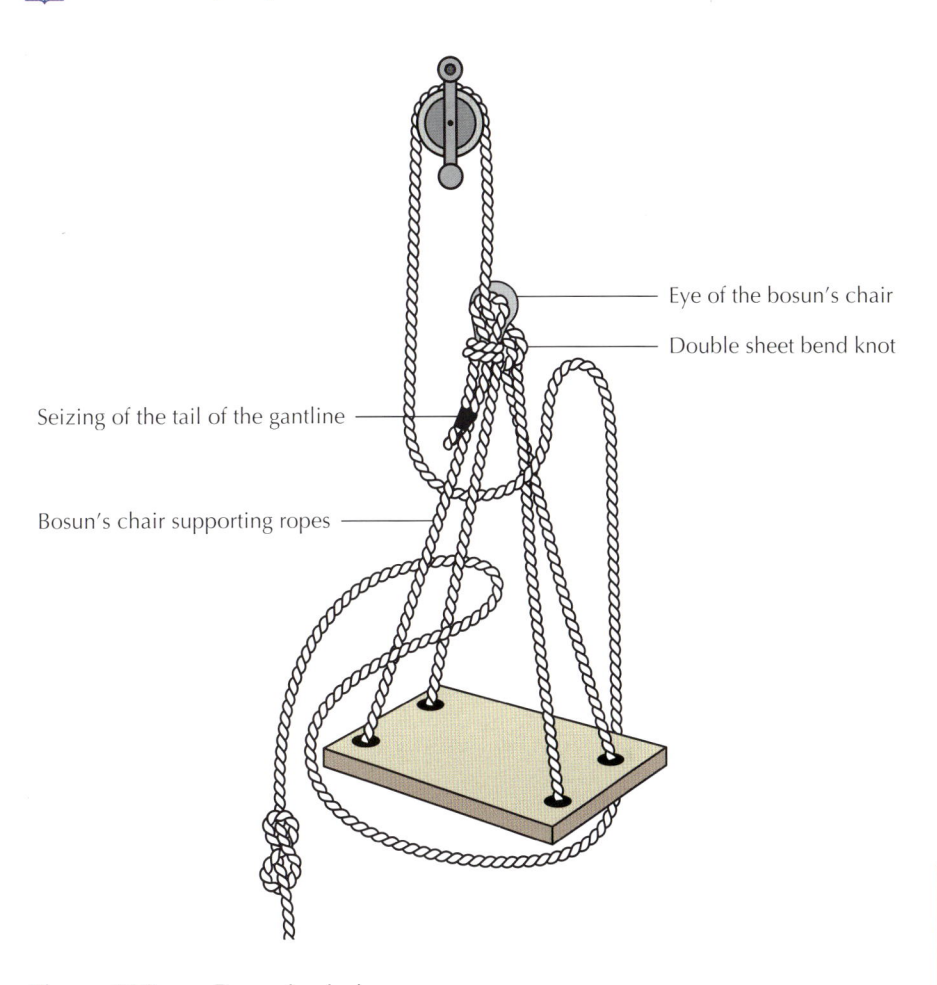

Eye of the bosun's chair
Double sheet bend knot
Seizing of the tail of the gantline
Bosun's chair supporting ropes

Figure 17.2 Bosun's chair

17.5.2 Do not use hooks to secure bosun's chairs unless they are of the type that, because of their special construction, cannot be accidentally dislodged, and unless they have a marked safe working load that is adequate for the purpose.

17.5.3 When rigging a bosun's chair for use examine the chair, gantlines and lizards thoroughly and renew them if there is any sign of damage. Load test them to at least four times the load they will be required to lift before hoisting a person.

17.5.4 When using a chair for riding topping lifts or stays, it is essential that the bow of the shackle, and not the pin, rides on the wire. In any case seize the pin.

	17.5.5	When it is necessary to haul a person aloft in a bosun's chair, it is generally done by hand rather than by using a winch. For access to masts on large sailing yachts a competent person may use a winch providing a risk assessment has been done and effective safety measures are in place to control the risks identified.
	17.5.6	If a seafarer is required to lower themselves while using a bosun's chair, they should first frap both parts of the gantline together with a suitable piece of line to secure the chair before making the lowering hitch. It is dangerous to hold on with one hand while making the lowering hitch with the other. Another person should stand by to tend the lines.

17.6 Working from a floating work platform

	17.6.1	Floating work platforms should be stable and have suitable fencing. Do not use unsecured trestles or planks to give additional height. Use safety lines and working lifejackets.
	17.6.2	Before using a work platform the person in charge should consider the strength of tides and other hazards, such as wash from passing vessels.
	17.6.3	When work is to be done at or near the stern or near bow/stern thrusters, the person in charge should inform the duty deck and engineering officers so they can isolate the equipment and place warning notices in the engine room, on the bridge and at any local controls.
	17.6.4	The person in charge should also inform the duty engineer and deck officers when personnel are working near ship's side discharges so that they are not used until the work is completed. Isolate equipment and attach notices to this effect to the relevant control valves. Do not remove the notices until the work is completed.

17.7 Scaffolding, including scaffolding towers

17.7.1 Only trained, competent and certified workers should erect, adjust or dismantle scaffolding, including towers. Annex 17.4 (reproduced from MGN 410 (M+F) Amendment 2) gives further guidance.

📖 *MGN 410 (M+F) Amendment 2*

17.7.2 Use only scaffolding of an approved design and rig it in conformity with a generally recognised configuration, as shown in Figure 17.3. If necessary, a competent person should calculate its strength and record this information. Follow appropriate procedures for testing. When replacing components ensure they are compatible with the existing scaffolding. The structure may need to be retested.

Figure 17.3 Scaffolding tower

17.7.3 Take care when assembling and dismantling the scaffold (see Annex 17.4).

17.7.4 Ensure that the structure is stable and access to it is safe. If it is mobile fix it securely so it cannot inadvertently move while in use, considering the anticipated weather conditions. Never move a scaffolding tower while people or materials are on the structure.

17.7.5 Anyone rigging or dismantling scaffolding should have received adequate training.

17.7.6 To prevent people or objects falling off, incorporate measures such as adequate safety rails and toe-boards.

17.7.7 Do not exceed the safe working load of the structure.

- Inspect all equipment and consider whether it is safe before use.
- Supervise personnel working at height and ensure they follow safety procedures.
- Ensure that personnel working overside wear personal flotation devices.

- Do risk assessments before anyone works at height.
- When working at height on ladders personnel must use a safety harness (see Figure 8.2) with a lifeline secured above the work position.

Annex 17.1 Emergency planning for work at height

1. Planning should take into account the possibility of emergencies occurring that mean workers need rescuing from where they are working at height. Sources of available guidance are listed in Annex A of MGN 410 (M+F) Amendment 2; in particular, the Work at Height Safety Association guidance regarding rescue. Together with the risk assessment, consider any circumstances that might occur when people are working at height and how to rescue them. The following is a non-exhaustive list of questions to ask:

   - What type of emergency could occur requiring the rescue of a worker? For example, is it likely to be a fall from height to the deck or into an open hold? A fall that leaves a worker suspended from a safety harness or from the equipment on which they were working? Or might it even involve a full or partial collapse of that equipment?
   - Is access likely to be readily available to the worker or workers concerned should a rescue situation occur?
   - How difficult would it be to recover a fallen or suspended worker from a hold to deck level or to lower a fallen or suspended worker to deck level?
   - What level of competence would the people involved in the rescue require?
   - Would any specialist equipment be required?
   - Are there any hazards that could be encountered during the rescue? For example, is the worker requiring rescue in an area where oxygen deficiency or other hazardous atmosphere could be a problem for rescuers?
   - Is appropriate protective equipment readily available to rescuers in the situations described in the bullet point above?
   - Are rescuers, or others on board, appropriately trained to provide appropriate medical care to a worker who has fallen or become suspended while working at height?

   MGN 410 (M+F) Amendment 2

2. Seafarers will obviously wish to rescue a colleague in distress as quickly as possible. However, undue haste can result in additional casualties, because rescuers do not take appropriate safety precautions before attempting a rescue. Therefore, while the aim should be to rescue the casualty as quickly as possible, the rescuers should not put their own health and safety at risk.

3. When a ship is in port, there may be a tendency to await the arrival of the local emergency services. However, there may be delays to the emergency service. It is therefore essential that appropriate procedures and measures are in place on board to deal with any emergencies and rescues that could arise, whether at sea or in port. The aim in any situation requiring the rescue of a person suspended whilst working at height should be to rescue the suspended person, whether injured or uninjured, as safely and promptly as possible, having regard to all the circumstances, including the health and safety of the rescuers.

4. A person left suspended at height in a harness for a significant period of time may suffer from symptoms of suspension syncope or suspension intolerance. This includes tingling in arms and legs and feelings of faintness.

5. During and after rescue, follow standard first-aid guidance. If a rescuer is unable to immediately release a conscious casualty from a suspended position, elevation of the legs by the casualty or the rescuer where safely possible may help prolong tolerance of suspension. Up-to-date guidance on the treatment of suspension syncope/intolerance is given in HSE Research Report RR708 'Evidence-based review of the current guidance on first aid measures for suspension trauma'.

HSE RR708

Annex 17.2 Requirements for ladders

1. A ladder shall be positioned so as to ensure its stability during use.

2. A suspended ladder shall be attached in a manner that:

    - makes it secure
    - ensures it cannot be displaced
    - prevents it from swinging.

    The last two bullet points do not apply to a rope ladder.

3. Portable ladders shall rest on footing that is stable, firm, of sufficient strength and of suitable size and composition safely to support the ladder so that its rungs or steps remain horizontal.

    Where, owing to the movement of the ship, it is not reasonably practicable to ensure that the rungs or steps of a portable ladder remain horizontal, all appropriate measures to ensure the stability of the portable ladder shall be taken.

4. The feet of a portable ladder shall be prevented from slipping during use by:

    - securing the stiles at or near their upper or lower ends
    - using an anti-slip device or
    - any other arrangement of equivalent effectiveness.

5. A ladder used for access shall be long enough to protrude sufficiently above the place of landing to which it provides access, unless other measures have been taken to ensure a firm handhold.

6. No interlocking or extension ladder shall be used unless its sections are prevented from moving relative to each other while in use.

7. A mobile ladder shall be prevented from moving before it is stepped on.

8. A ladder shall be used in such a way that:

    - a secure handhold and secure support are always available to the user; and
    - the user can maintain a safe handhold when carrying a load by hand.

Annex 17.3 Requirements for rope access and positioning techniques

All rope access equipment should only be set up and used by trained and competent personnel. The equipment should be inspected before each use, and thoroughly examined at least every three months, in accordance with a specified schedule.

1. A rope access or positioning technique shall only be used if:

    - subject to the next bullet point, it involves a system comprising at least two separately anchored ropes, of which one ('the working rope') is used as a means of access, egress and support and the other is a safety rope
    - the seafarer is provided with and uses a suitable harness and is connected by it to the working rope and the safety rope
    - the working rope is equipped with safe means of ascent and descent and has a self-locking system to prevent the seafarer falling should they lose control of their movements
    - the safety rope is equipped with a mobile fall prevention system that is connected to and travels with the seafarer
    - the working rope and the safety rope take different leads
    - ropes are protected from right angles or sharp edges
    - the tools and other accessories to be used by the seafarer are secured to their harness or seat or by some other suitable means.

2. A rope access or positioning technique may involve a system comprising a single rope where:

    - the risk assessment has demonstrated that the use of a second line would entail higher risk to persons
    - appropriate measures have been taken to ensure safety.

Annex 17.4 Requirements for scaffolding

1. Strength and stability calculations for scaffolding shall be carried out unless:

   - a note of the calculations, covering the structural arrangements contemplated, is available
   - the scaffolding is assembled in conformity with a generally recognised standard configuration.

2. Depending on the complexity of the scaffolding chosen, an assembly, use and dismantling plan shall be drawn up by a competent person. This may be in the form of a standard plan, supplemented by items relating to specific details of the scaffolding in question.

3. A copy of the plan, including any instructions it may contain, shall be made available for the use of the person supervising and the seafarers concerned in the assembly, use, dismantling or alteration of the scaffolding.

4. The bearing components of the scaffolding shall be prevented from slipping by:

   - attachment to the bearing surface
   - provision of an anti-slip device
   - any other arrangement of equivalent effectiveness.

5. The load-bearing surface of the scaffolding shall be of sufficient capacity.

6. The scaffolding shall be positioned to ensure its stability.

7. Wheeled scaffolding shall be prevented by appropriate devices from moving accidentally during work at height.

8. The dimensions, form and layout of scaffolding decks shall:

   - be appropriate to the nature of the work to be performed
   - be suitable for the loads to be carried
   - permit work and passage in safety.

9. Scaffolding decks shall be assembled in such a way that their components are prevented from moving inadvertently during work at height.

10. There shall be no dangerous gaps between the scaffolding deck components and the vertical collective safeguards to prevent falls.

11. When any part of a scaffold is not available for use, including during the assembly, dismantling or alteration of scaffolding, it shall be:

    - marked with general warning signs in accordance with the Merchant Shipping and Fishing Vessels (Safety Signs and Signals) Regulations 2001
    - suitably delineated by physical means preventing access to the danger zone.

12. Scaffolding shall be assembled, dismantled or significantly altered only under the supervision of a competent person and by seafarers who have received appropriate and specific training in the operations envisaged in accordance with Regulation 12 of the Merchant Shipping and Fishing Vessels (Health and Safety at Work) Regulations 1997 No. 2962 and Regulation 11 of the Merchant Shipping and Fishing Vessels (Provision and Use of Work Equipment) Regulations 2006 No. 2183, which shall include:

    - understanding the plan for the assembly, dismantling or alteration of the scaffolding
    - safety during the assembly, dismantling or alteration of the scaffolding
    - measures to prevent the risk of persons or objects falling
    - safety measures in the event of changing weather conditions that could adversely affect the safety of the scaffolding
    - permissible loads
    - any other risks that the assembly, dismantling or alteration of the scaffolding may entail.

    In addition:

    - follow appropriate procedures for conducting testing as necessary
    - avoid combining old and new components where possible, with regular testing where the compatibility of parts may be in question.

13. For the purposes of this annex, 'competent person' means the person possessing the knowledge or experience necessary for the performance of the duties imposed on that person by this annex.

18 Provision, care and use of work equipment

18.1 Introduction

18.1.1 The term 'work equipment' applies to any machinery, appliance, apparatus, tool or installation provided for use at work.

18.1.2 The exception to the above is any safety equipment or apparatus provided in compliance with the International Convention for the Safety of Life at Sea (SOLAS) requirements, which is subject to other merchant shipping regulations. See the Merchant Shipping and Fishing Vessels (Provision and Use of Work Equipment) Regulations 2006 (PUWER) as referenced and *Merchant Shipping (Marine Equipment) UK Conformity Assessment Procedures for Marine Equipment*.

> SI 2183/2006; MGN 331 (M+F) Amendment 2; SI 2016/1025 as amended; MSN 1874 (M+F) Amendment 7

Key points
- Train seafarers who are unfamiliar the equipment on board in its proper use before allowing them to use it.
- Where any seafarer who is likely to use any item of work equipment does not understand the language in which the information and instructions are written, provide information and instructions in the working language of the vessel or another language that the seafarer understands.
- Carry, supply and use isolation equipment and PPE appropriate to the ship's electrical installation as and when required.

Responsibilities of the organisation
- The company is generally responsible for all work equipment on board.
- Where work equipment is provided from ashore, the shore provider is responsible for its condition. However, the company is responsible for maintaining it while on board and using it safely.
- All work equipment should comply with any relevant standards laid down by merchant shipping or general UK regulations. Any equipment not covered by regulations or type approvals should comply with the appropriate British Standard or its nearest International Organization for Standardization (ISO) equivalent. See Annexes 18.1 and 18.3 for more details.

18.2 Duty of seafarers and workers

18.2.1 Seafarers and workers should comply fully with all instructions or training they have been given for the use of any work equipment.

18.2.2 Personnel should use the correct tools or equipment for a task. They should use tools only for their intended purpose. Use of unsuitable tools or equipment may lead to accidents and incidents.

📖 *PUWER Part 4 Reg 34*

18.2.3
- No-one should operate any item of work equipment unless they are competent, and authorised, to do so.
- Never wear loose clothing or jewellery while using machinery because it may become caught in moving parts.
- Tie back long hair and cover it with a hair net or safety cap.
- Personal protective equipment (PPE) should be provided and worn as required by the Merchant Shipping and Fishing Vessels (Personal Protective Equipment) Regulations 1999 and Merchant Shipping Notice MSN 1870 (M+F) Amendment 5.

📖 *SI 1999/2205; MSN 1870 (M+F) Amendment 5*

18.3 Risk assessment and specific risks

18.3.1 Do a risk assessment and put safety measures in place for the safe operation of the equipment and all expected circumstances.

18.3.2 Consider the following risks, as appropriate to the equipment, to protect workers who may be at risk while using it:

- mechanical risks such as crushing, impact, trapping, entanglement, cutting or friction
- non-mechanical risks such as noise, vibration, electrical hazards, temperature and radiation
- any article or substance falling or being ejected from work equipment
- rupture or disintegration of parts of work equipment
- work equipment overheating or catching fire
- the unintended or premature discharge of any article or any gas, dust, liquid, vapour or other substance that is produced, used or stored in the work equipment
- the unintended explosion of the work equipment or any article or substance produced, used or stored in it
- work equipment being struck by lightning while someone is using it.

📖 *PUWER Reg 9*

18.3.3 Where any seafarer using work equipment is, or could be, exposed to one or more of the above risks or hazards, the company should prevent any significant risks to their health and safety by providing appropriate work equipment or protective devices. If that is not practicable they should control them adequately by appropriate means.

18.3.4 Where an item of work equipment causes a specific risk to health or safety, it must be used, repaired, modified or maintained only by seafarers who have been designated to perform the particular task. They must be competent and must have had appropriate training, either as part of their overall training for their current position or provided by other qualified people on board or ashore (including the manufacturer of the equipment).

18.4 Dangerous parts of work equipment

18.4.1 Every dangerous or exposed working part of work equipment must have appropriate guards or protection devices. These must be maintained and/or replaced as necessary and must be kept in position when the relevant parts are in motion.

18.4.2 Stow equipment in a tidy and correct manner when not in use. Protect any cutting edges.

18.4.3 All guards or protection devices provided should:

- be of substantial construction
- not cause any additional hazard
- not be easily removed
- be situated at a sufficient distance from the danger zone (see section 18.4.4)
- not restrict the view of the operator more than necessary
- be constructed or adapted to allow parts to be fitted or replaced and for maintenance work. Allow access only to the area where work is to be carried out and, where possible, without having to dismantle the guard or protection device.

📖 *PUWER Reg 13*

18.4.4 'A sufficient distance' means as close as possible to the danger zone without affecting the use or performance of the work equipment. 'From the danger zone' means the zone within or around work equipment where a seafarer would be exposed to a significant health or safety risk.

18.5 Lighting

18.5.1 Provide adequate and appropriate lighting at any place where work equipment is used.

📖 *PUWER Reg 24*

18.6 Markings

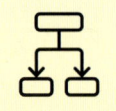

18.6.1 Where any health and/or safety markings are required to comply with the requirements of the Merchant Shipping and Fishing Vessels (Safety Signs and Signals) Regulations 2001, the markings must comply with MGN 556 (M+F) Amendment 1, and Chapter 9 of this Code. Fix them to the equipment so they are clearly visible to any person using the equipment or nearby.

📖 *PUWER Reg 26; MGN 556 (M+F) Amendment 1*

18.7 Warnings

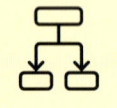

18.7.1 Where any work equipment needs warning signs or warning devices their meaning should be clear and they should be easy to see or hear.

📖 *PUWER Reg 27*

18.8 Stability of work equipment

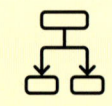

18.8.1 Where work equipment has to be stable to be used safely, stabilise it with clamps or another appropriate method.

18.8.2 When deciding the most appropriate method for stabilising work equipment, consider the potential movement of the ship under all conditions.

📖 *PUWER Reg 23*

18.9 High or very low temperatures

18.9.1 Take appropriate measures to prevent injury where any equipment, parts of equipment or anything produced by, used by, or stored in such equipment could burn, scald or cause any other injury to any seafarer because it is at a high or low temperature.

18.9.2 Appropriate measures may include fitting guards or barriers to the hazardous parts of the equipment, isolating the equipment or providing PPE.

📖 *PUWER Reg 16*

18.10 Maintenance

18.10.1 Keep all work equipment in good repair and efficient working order following the manufacturer's instructions. See section 19.3 for further information.

18.10.2 A **competent person** 🔍 should inspect the equipment regularly. When equipment does not seem to be working properly, or has been treated in a way likely to cause damage, report it and take it out of service until it can be inspected and any necessary repairs or maintenance undertaken.

18.10.3 The company or competent person is responsible for deciding what maintenance work is required, in line with the manufacturer's instructions. The following should normally form part of a maintenance routine:

- Grease bearings thoroughly, following the manufacturer's recommendations. Dry bearings and other moving parts will impose additional loads that can lead to failure.
- Check all ropes, chains and attachment points regularly for wear, damage or corrosion, and replace them as necessary. See Chapter 19 for further guidance.
- Test all controls, emergency stop controls, brakes, safety devices and so on to ensure that they are operating correctly before use. These checks should also be part of a planned maintenance system.

18.10.4 Shut down the equipment before carrying out maintenance work on powered equipment.

18.10.5 Where shutting down is not possible, put in place appropriate protective measures to do the work safely without exposing the worker, or any other person, to any significant risk to their health and safety. Protective measures may include:

- exposing the dangerous part as little as possible
- getting a responsible ship's officer or competent person to authorise the exposure
- permitting only a competent person to carry out the operation
- ensuring that any person working close to the machinery has enough clear space and adequate light while they are working
- ensuring that any person operating close to the machinery has been trained in safe systems of work for that machinery, and the dangers and precautions to take
- displaying a conspicuous warning notice on or close to the machinery
- keeping maintenance logs up to date.

PUWER Regs 7 and 25

18.11 Inspection

When to inspect

18.11.1 Where the safety of work equipment depends on the installation conditions, a competent person should inspect the equipment after initial installation, or after reassembly at a new site or in a new location, and before putting it into service for the first time. This is to ensure that it has been installed correctly, following the manufacturer's instructions, and is safe to use. In this context, inspection means a competent person carrying out visual or more rigorous checks. It may include testing when appropriate.

18.11.2 A competent person should inspect any work equipment that is exposed to conditions causing deterioration as outlined in the safety management system, risk assessment and procedures. When exceptional circumstances occur that jeopardise the safety of the work equipment, further inspection is necessary so any remedial action can be taken to ensure it is still safe to use. Exceptional circumstances include modification work, accidents, exposure to extreme weather, any use that falls outside the equipment's design parameters, and prolonged periods of inactivity.

18.11.3 Any work equipment used for lifting loads, including personnel, is also subject to the provisions of the Merchant Shipping and Fishing Vessels (Lifting Operations and Lifting Equipment) Regulations 2006. These set out specific requirements for the inspection, testing and **thorough examination** 🔍 of such lifting equipment. See Chapter 19 for more details.

📖 *PUWER Reg 8*

How to inspect work equipment

18.11.4 Inspection means a detailed visual and operational check of work equipment to ensure its condition and integrity for ongoing safe use. It should take account of the manufacturer's guidance.

18.11.5 Examine structures frequently for damage such as corrosion, cracks, distortion or wear of bearings, and securing points.
Also check hollow structures, such as gantries or masts, for water trapped inside them. If water is found drain the structure, treat it where practicable, then seal it to prevent further ingress of water.

18.11.6 Record the results of all inspections and keep the records available until the next inspection.

18.11.7 Where any ship's work equipment is used outside the ship, or work equipment from outside the ship is used on the ship, provide physical evidence that the last inspection required under the Merchant Shipping and Fishing Vessels (Provision and Use of Work Equipment) Regulations 2006 has actually been carried out. In this context 'used outside the ship' includes equipment used on the quayside, dock or jetty, the ship's boats, pontoons or on board another ship. It also applies to equipment operated by workers who are not employed by the company.

18.11.8 Inspections should cover factors such as the standard of welding or other fixing and materials used, together with the strength of any part of the ship that supports it and to which it is attached. Take account of any inspection requirements or guidance from the manufacturer. Reinspect work equipment regularly, no less than every five years, or more frequently if the manufacturer recommends, to ensure that its installation has not deteriorated.

18.12 Information and instructions

18.12.1 Incorrect use of tools and equipment can cause accidents and incidents, as well as damage to the equipment.

18.12.2 All seafarers, and any managers or supervisors who use work equipment, should have access to all necessary health and safety information and written instructions relating to its use, including those from manufacturers.

18.12.3 This information should be in an easily understood form. It should include information and, where appropriate, written instructions on the conditions in which the work equipment may be used and its method of use.

18.12.4 It should include foreseeable abnormal situations and what to do if such a situation occurs; and information on any conclusions drawn from previous experience of using that work equipment.

PUWER Reg 10

18.13 Training

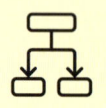

18.13.1 All seafarers who use work equipment, or who supervise its use, should have received adequate training covering the method of use of the equipment, any risks that may arise from its use and any precautions to take.

18.13.2 Similarly, seafarers specifically designated to carry out repairs, modifications, maintenance or servicing to work equipment, or who supervise such work, should have had adequate training for that purpose when the use of that equipment may involve a specific health and safety risk to the person using it; for example, electrical equipment or mechanical cutting equipment.

18.13.3 In accordance with the *International Safety Management Code for Merchant Shipping*, keep records of all such training and indicate when workers become fully competent.

18.13.4 All instruction or information must be in the working language of the vessel.

PUWER Reg 11

18.14 Hand tools

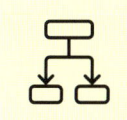

18.14.1 Do not use damaged or worn tools and keep cutting edges sharp and clean. Only a competent person should maintain, repair or dress tools.

18.14.2 The following guidelines apply to the use of tools and equipment:

- When using a tool with a cutting edge, always keep both hands behind the edge/blade.
- Do not use hammers with split, broken or loose shafts and worn or chipped heads. Make sure the heads are properly secured to the shafts.
- Files should have a proper handle. Never use them as levers.
- When using a tool direct it away from the body so that if it slips it does not cause injury (except when using a spanner; see below).
- When using a spanner you will gain more control by pulling towards the body.
- Do not use spanners with splayed jaws. Scrap any that show signs of slipping. Have enough spanners of the right size. Do not improvise, for example, by using pipes as extension handles on spanners.
- Hold a chisel between the thumb and base of the index finger with the thumb and fingers straight, and with the palm of the hand facing towards the hammer blow.
- The cutting edge of chisels should be sharpened to the correct angle. Do not allow the head of a cold chisel to spread to a mushroom shape; grind off the sides regularly.
- Never use screwdrivers as chisels and never use hammers on them. Split handles are dangerous.
- Do not force a saw through the material being cut: push it with a light, even movement.

18.15 Portable power-operated tools and equipment

18.15.1 The following guidelines apply for portable power-operated equipment:

- Portable powered equipment is dangerous unless it is properly maintained, handled and used. Only competent people should operate it.
- The flexible cables of electric tools should comply with the relevant British or international standard.
- Before work begins, ensure that power supply leads and hoses are in good condition, laid safely clear of all potentially damaging obstructions, and do not obstruct safe passage.
- Where cables pass through doorways, secure the doors open.

18.15.2 The risk of electric shock is increased by perspiration and locations that are damp, humid or have large conductive surfaces. In these conditions operate power tools from low voltage supplies; no more than 50 volts AC with a maximum of 30 volts to earth or 50 volts DC.

18.15.3 The risk associated with portable electric tools also applies to portable electric lamps. The supply to these should not exceed 50 volts.

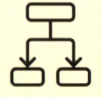

18.15.4 When it is not practicable to use low voltages take other precautions. These could include a local isolating transformer supplying one appliance only, or a high-sensitivity earth leakage circuit breaker (also known as a residual current device), that trips at 30 milliamperes (mA) residual current or less.

18.15.5 Do not use double-insulated tools on ships outside the accommodation because water can provide a contact between live parts and the casing, increasing the risk of a fatal shock. An earth leakage circuit breaker may also fail to operate when used with such tools because there may be no earth wire in the power supply cable fitted to the tool.

18.15.6 Chain linkages (**whip checks** 🔍) or similar devices should be fitted between sections of pneumatic hose to prevent whiplash in the event of breakage. Alternatively you can use safety valves that close off the lines. Accessories and tool pieces (eg drill bits, chisel) should be absolutely secure in the tool. In particular, replace any retaining springs, clamps, locking levers and other built-in safety devices on pneumatic tools after the tool piece has been changed.

18.15.7 Do not change accessories or tool pieces while the tool is connected to a power source.

18.15.8 The correct safety guards should be securely fixed to appliances that require them. Check that they are secure before starting any operation. Remove guards only when the equipment is not operating. If you need to remove guards while operating for maintenance or examination of the equipment, take the following precautions:

- Only a responsible person should authorise the removal of safety guards, or carry out the work or examination.
- There should be adequate clear space and lighting to do the work.
- Tell anyone working close to the machinery what the risks are. Instruct them in a safe system of work and the precautions to take.
- Display a conspicuous warning notice.

18.15.9 During temporary interruptions to work (eg during meal breaks and on completion of a task) **isolate** 🔍 equipment from power sources and leave it safely or stow it away correctly.

18.15.10 When a work operation causes high noise levels, wear hearing protection. When flying particles may be produced protect the face and eyes (see Chapter 8).

18.15.11 The vibration caused by reciprocating tools (eg pneumatic drills, hammers, chisels) or high-speed rotating tools can cause a permanent disablement of the hands known as 'dead' or 'white' fingers. In its initial stages, this appears as a numbness of the fingers and an increasing sensitivity to cold. In more advanced stages the hands become blue and the fingertips swollen.

18.15.12 Seafarers who experience the symptoms above should not use such equipment.

18.15.13 Advise other seafarers not to use reciprocating tools for more than 30 minutes without a break, unless the risk assessment indicates a lesser period of use.

See Chapter 12 for further information.

18.16 Workshop and bench machines (fixed installations)

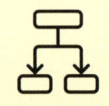

18.16.1 Only competent personnel should operate fixed installations. The operator should check the machine every time before use, and ensure that all safety guards and devices are in position and operative, that all tool pieces (eg drill bits, cutting blades) are in good condition, and that the work area is adequately lit and free from clutter.

18.16.2 Do not use a machine when a guard or safety device is missing, incorrectly adjusted or defective, or when it is itself in any way faulty. If there is any defect isolate the machine from its power source until it has been repaired.

18.16.3 During operations, personnel should ensure that work pieces are correctly secured in position, that machine residues (eg swarf, sandings) do not build up excessively, and that they are disposed of in a correct and safe manner.

18.16.4 Switch off the machinery and isolate the power supply whenever it is left unattended, even if only briefly. Recheck the machinery and any safety guards before resuming work.

18.17 Abrasive wheels

18.17.1 Only competent people should select, mount or use abrasive wheels, and they should follow the manufacturer's instructions. They are relatively fragile so store and handle them with care.

18.17.2 Follow the manufacturer's instructions when choosing the correct type of wheel for the job in hand. Generally, soft wheels are more suitable for hard material and hard wheels for soft material.

18.17.3 Before mounting a wheel, brush it clean and inspect it closely to ensure that it has not been damaged in storage or transit. You can check the soundness of a vitrified wheel further by suspending it vertically and tapping it gently. If the wheel sounds dead, it is probably cracked so do not use it.

18.17.4 Do not mount a wheel on an unsuitable machine. It should fit to the spindle freely but not loosely. If the fit is too tight the wheel may crack as the heat of the operation causes the spindle to expand.

18.17.5 Tighten the clamping nut only enough to hold the wheel firmly. When the flanges are clamped by a series of screws, the screws should be first screwed home with the fingers and diametrically opposite pairs tightened in sequence.

18.17.6		The speed of the spindle should not exceed the stated maximum permissible speed of the wheel.
18.17.7		Provide a strong guard which encloses as much of the wheel as possible. Keep guards in position at every abrasive wheel, both to contain wheel parts in the event of a burst and to prevent an operator touching the wheel (see also section 18.17.8).
18.17.8		Where there is a work rest secure it properly to the machine and adjust it as close as practicable to the wheel. The gap should normally not be more than 1.5 mm.
18.17.9		Do not use the side of a wheel for grinding; this is particularly dangerous when the wheel is badly worn.
18.17.10		Never hold the work piece in a cloth or pliers.
18.17.11		Machines used for dry grinding should have suitable transparent screens. Where an abrasive wheel is being trued or dressed the screen should protect the worker from the exposed part of the wheel. As with all grinding operations, workers should wear properly fitting eye protectors and follow the manufacturer's instructions and procedures.

18.18 Hydraulic/pneumatic/high-pressure jetting equipment

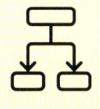

18.18.1	Seafarers using hydraulic/pneumatic/high-pressure systems should be trained and competent in their use. Always follow the manufacturer's guidelines. Do not operate equipment at higher pressures than the manufacturer recommends.
18.18.2	Seafarers using high-pressure jetting equipment should wear the correct protective equipment. These systems may involve a heated supply source so operators should guard against splashing and scalding. Notices should be displayed on approaches to areas where such work is taking place to warn others that a high-pressure system is in the area. Finally, seafarers should ensure that the direction of such jetting is safe.
18.18.3	Before starting work, seafarers should ensure that the equipment and supply systems are in sound condition, and that incorporated safety devices are in place and functioning correctly.
18.18.4	Where equipment is defective or suspect, shut down the systems and isolate and depressurise them to allow effective replacement or repair. Only competent personnel should make these repairs, using approved components.

18.18.5		Before activating a pressure system, and also when closing it down, do the recommended checks to ensure that there are no air pockets or trapped pressure in the system, as these may cause the equipment to work erratically.
18.18.6		When handling hydraulic fluid, personnel should:

- use the correct grade when topping up systems
- clean up spillages immediately
- clean off any splashes of such fluid onto skin areas off immediately (many of these fluids are mineral based)
- keep naked lights away from equipment during service/test periods (hydraulic fluids may give off vapours that may be flammable).

18.18.7 If there is a high-pressure release of oil, air or any other substance that penetrates the skin, get medical advice immediately.

18.18.8 The pressure of compressed air should be no higher than is necessary to do the work satisfactorily.

18.18.9 Do not use compressed air to clean the working space, and never direct it at any part of a person's body.

18.19 Hydraulic jacks

18.19.1 Inspect jacks before use to ensure that they are in a sound condition and that the oil in the reservoir reaches the minimum recommended level.

18.19.2 Before operating a jack check that its lifting capability is adequate for the work and that its foundation is level and strong enough.

18.19.3 Apply jacks only to the recommended or safe jacking points on equipment.

18.19.4 Equipment under which seafarers are working should be properly supported with chocks, wedges or by other safe means – never by jacks alone.

18.19.5 If possible remove jack operating handles when they are not required, so they will be in position for raising or lowering the jack.

18.20 Laundry equipment

18.20.1 All seafarers who work in a laundry, or use any part of the equipment there, must be fully instructed in how to operate the machinery. Seafarers under 18 years of age should not work on industrial washing machines, hydro-extractors, calender presses or garment presses unless they have been fully trained in how to operate the machine and the precautions to take. If appropriate a competent person should supervise them.

📖 *MSN 1838 (M) Amendment 1*

18.20.2 Inspect equipment before use for faults and damage. Pay particular attention to:

- the automatic cut-off or interlocking arrangements on equipment such as washing machines and hydro-extractors
- the guards and emergency stops on presses, calender presses, mangling and wringing machines.

Report any defects or irregularities you find during the inspection immediately, or when the equipment is working. Do not use the machine until any necessary repairs or adjustments have been carried out.

Display a conspicuous notice warning against use on the defective machine.

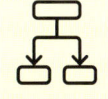

18.20.3 To ensure the necessary standard of maintenance for laundries, the equipment needs frequent and regular inspection, including thorough checking of all electrical equipment and apparatus.

18.20.4 Do not overload machines, and always distribute loads uniformly.

18.20.5 Do not rely entirely on interlocking or cut-off arrangements on the doors of washing machines, hydro-extractors and drying tumblers. Do not open doors until all movement has ceased.

18.21 Controls

18.21.1 Any seafarer operating the controls of any work equipment should be able to ensure from the control position that no other seafarer will be exposed to any significant risk to their health and safety as a result of the starting up or use of that equipment.

18.21.2 Where such an arrangement is not reasonably practicable, introduce appropriate systems of work to ensure that no seafarer is exposed to any significant risk to their health and safety as a result of starting up or using the equipment.

18.21.3 This may include audible, visible or other suitable warning devices, as required by sections 18.6 or 18.7, so that any seafarers likely to be affected are aware that the equipment is about to be started. See Annex 18.3 for more details.

18.21.4 Any seafarers who could be exposed to a risk to their health and safety because work equipment is being started or stopped must be given enough time and a means to get out of the way first.

📖 *PUWER Reg 18*

18.22 Use of mobile work equipment

18.22.1 Where mobile work equipment is used on board a ship:

- ship's powered vehicles or powered mobile lifting appliance must not be driven during a work activity except by a competent and authorised person
- where work equipment is moving around in a work area, appropriate traffic rules must be drawn up and followed for the safety of seafarers and others
- seafarers on foot should, so far as is reasonably practicable, be prevented from entering the area of operation of self-propelled work equipment
- where work cannot be done properly unless seafarers on foot are present, appropriate measures must be in place to prevent them from being injured by the work equipment.

📖 *PUWER Reg 30*

18.22.2 Seafarers should be carried on mobile work equipment only when safe facilities are provided for this purpose. Adjust the speed of the work equipment as necessary for the safety of the seafarers.

18.22.3 Do not use mobile work equipment fitted with a combustion engine in working areas unless there is sufficient ventilation, so the combustion engine does not put the health or safety of seafarers at risk.

18.23 Carrying of seafarers on mobile work equipment

18.23.1 No seafarer is to be carried on any mobile work equipment unless it is designed for that purpose.

PUWER Reg 28

18.23.2 This means it must be fitted out in such a way as to minimise risks to the safety of any seafarer, including any risks from wheels or tracks. Such equipment must also incorporate measures to prevent it rolling over or, where that is not possible, reduce the risks to health or safety of seafarers should it roll over whilst being used. Measures could include:

- stabilising the work equipment to prevent it rolling over
- providing a protection structure so that the work equipment cannot fall on its side
- providing a structure that gives sufficient clearance around the seafarers being carried if the work equipment can overturn further than that
- using any device that is equally effective in protecting the seafarers being carried.

18.23.3 Where there is a risk that any seafarer being carried by mobile work equipment could be crushed if it rolls over, provide a restraining system for the person. (This does not apply to a forklift truck with a structure as described in section 18.23.2, points 2 and 3).

18.24 Overturning of forklift trucks

18.24.1 Any forklift truck to which section 18.23.2 applies and which carries a seafarer, must be adapted or equipped to minimise the risk to health or safety from its overturning. Consider the manner and conditions in which the forklift truck is being used.

PUWER Reg 29

18.24.2 Any seafarer operating a forklift truck must have received appropriate safety training including that for the type of truck.

18.25 Self-propelled work equipment

18.25.1 When any self-propelled work equipment could present a hazard to health and safety while in motion:

- fit it with a means for preventing it from being started by an unauthorised person (eg a key-operated switch)
- where more than one item of rail-mounted work equipment is moving at the same time, fit it with appropriate facilities for minimising the consequences of a collision
- fit it with braking and stopping devices
- fit it with emergency facilities operated by a readily accessible control or automatic system for braking and stopping if the main device fails
- where the driver's direct field of vision is not wide enough to ensure safety, ensure there are adequate devices for improving their vision
- if the equipment is used in the dark:
  - fit it with lighting appropriate to the work to be carried out
  - check that it is sufficiently safe for such use
- if it, or anything carried or towed by it, involves a risk from fire and is liable to injure seafarers, it should carry appropriate firefighting appliances, unless such appliances are kept sufficiently close to it.

PUWER Reg 31

18.26 Remote-controlled self-propelled work equipment

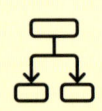

18.26.1 Where any remote-controlled self-propelled equipment could endanger the safety of seafarers while it is in motion, set it up in such a way that it stops automatically once it leaves its control range. Additionally, incorporate features to guard against the risk of crushing or other impact.

PUWER Reg 32

18.27 Drive units and power take-off shafts

18.27.1 Where the seizure of a drive unit or power take-off could present a risk to seafarers, take appropriate measures including providing guards or other protection devices referred to in section 18.4.1.

📖 *PUWER Reg 33*

18.28 Ropes and wires

18.28.1 Choose the correct rope or wire for every purpose. Use and maintain it properly to ensure the safety of life, the ship and the environment. For further information see BTA Rope Selection Guidance – Appendix 2 at https://britishtug.com/bta-produced-tow-rope-guidance/

18.28.2 When choosing a rope or wire ensure that its designated use is compatible with the manufacturer's design purpose. It should be appropriate for its intended use and expected/foreseeable operating conditions. This includes being compatible with the vessel's fittings. Relevant criteria may include minimum bend radius, minimum breaking load, material and construction type.

18.28.3 There are many different techniques to construct a rope; for example, differing numbers of strands, use of jackets and proprietary coatings. Each construction may have unique characteristics and may fail through different methods of use. Follow the manufacturer's guidance on the use and maintenance of low-twist jacketed synthetic fibre ropes and monitor their performance in dynamic loading environments.

18.28.4 The manufacturer's and specialist industry guidance will help you determine the safety factor and life expectancy of the rope or wire.

18.28.5 Man-made (synthetic) fibre ropes are relatively stronger than those of natural fibre, so for any given breaking strain they have smaller diameters. However, wear or damage will weaken them more than the same amount of wear or damage would on a natural fibre rope. High-modulus synthetic fibre (HMSF) ropes, including high-modulus polyethylene (HMPE), have a particularly high strength-to-weight ratio.

18.28.6 Table 18.1 shows recommendations for the substitution of natural fibre ropes with synthetic ones. The diameters given are for three-strand ropes and the size numbers are for eight-strand plaited ropes.

Table 18.1   Recommendations for substitution of natural fibre ropes by synthetic fibre ropes

Manila		Polyamide (eg nylon)		Polyester (terylene)		Polypropylene	
Diameter (mm)	Size	Diameter (mm)	Size	Diameter (mm)	Size	Diameter (mm)	Size
48	6	48	6	48	6	48	6
56	7	48	6	48	6	52	6.5
64	8	52	6.5	52	6.5	56	7
72	9	60	7.5	60	7.5	64	8
80	10	64	8	64	8	72	9
88	11	72	9	72	9	80	10
96	12	80	10	80	10	88	11
112	14	88	11	88	11	96	12

18.28.7   Take new rope, three-strand fibre rope and wire out of a coil very carefully to avoid disturbing the lay of the rope.

18.28.8   Rope of synthetic material stretches under load to a varying extent depending on the material. The stretch of synthetic fibre rope, which may be up to double that of natural fibre rope, usually recovers almost as soon as tension is released.

> ⚠️ **18.28.9 Warning**
>
> When a synthetic rope breaks it may snap back in a dangerous manner. If an item of running gear breaks loose it may be projected with lethal force, endangering people's lives. Avoid snatching this type of rope; this may happen inadvertently and if so personnel should stand well clear of the danger areas. HMSF ropes have very specific stretch and snap-back characteristics that will need to be considered.

18.28.10   The frictional heat generated during use can easily damage synthetic fibre ropes. Too much friction on a warping drum may fuse the rope, causing it to stick and jump when turning, which can be dangerous. Polypropylene is more liable to soften than other materials. To avoid fusing, do not surge ropes on winch barrels unless necessary and use a minimum number of turns on the winch barrel. Three are usually enough but, on **whelped** 🔍 drums, one or two extra turns may be needed to ensure a good grip; remove these as soon as practicable.

18.28.11 Choose the method of making eye splices in ropes of synthetic fibres according to the material of the rope:

- Polyamide (nylon) and polyester fibre ropes need four full tucks in the splice each with the completed strands of the rope, followed by two tapered tucks for which the strands are halved and quartered for one tuck each respectively. The length of the splicing tail from the finished splice should be equal to at least three rope diameters. The portions of the splice containing the tucks with the reduced number of filaments should be securely wrapped with adhesive tape or other suitable material.
- Polypropylene ropes should have at least three but not more than four full tucks in the splice. The protruding spliced tails should be at least equal to three rope diameters.
- Polythene ropes should have four full tucks in the splice with protruding tails of at least three rope diameters.

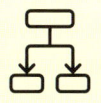

18.28.12 Certain rope constructions, such as those using high-efficiency splicing methods, may require specialist assistance in splicing to maintain the strength of the rope. Inspect the splices in ropes and wires regularly to check that they are intact. Where wire rope is joined to fibre rope, insert a thimble or other device into the eye of the fibre rope. Both wire and fibre rope should have the same direction of lay.

18.28.13 Do not use mechanical fastenings in place of splices on synthetic fibre ropes because strands may be damaged during application of the mechanical fastening. The grip of the fastenings may also be affected by slight, unavoidable fluctuations in the diameter of the strands.

18.28.14 Use synthetic fibre stoppers of similar material (but not polyamide) on synthetic fibre mooring lines. The 'West Country' method (**double and reverse stoppering** Q) is preferable, as shown in Figure 18.1.

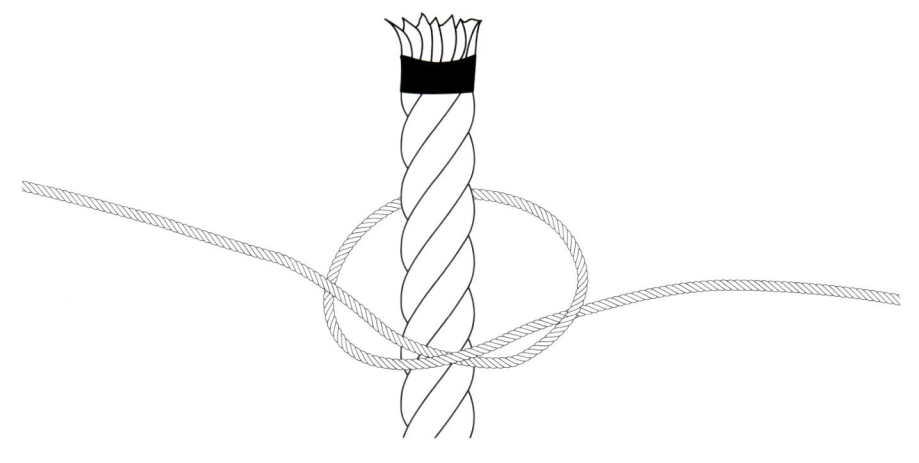

Figure 18.1   The 'West Country' method of double and reverse stoppering

18.28.15 Form eyes in wire ropes by eye-splicing or using appropriate compression fittings (using swages or ferrules). Avoid using bulldog grips if possible, and do not use them on lifting wires or mooring wires. Annex 18.2 gives further information on the use of bulldog grips.

18.28.16 Many types of rope of both synthetic and natural fibre are available, each with different properties. Table 18.2 is a guide to the chemical resistance of the main rope types, but it indicates only the possible extent of deterioration of rope. In practice much depends on the precise formulation of the material, the amount of contamination the rope receives and the length of time and the temperature at which it is exposed to contamination. In some cases, damage may not be apparent even on close visual inspection.

Table 18.2 Chemical resistance of different types of rope

Substance	Manila or sisal	Polyamide (nylon)	Polyester	Polypropylene	Polythene (HMPE)	Aramid
Sulphuric (battery) acid	None	Poor	Good	Very good	Good	Poor
Hydrochloric acid	None	Poor	Good	Very good	Very good	Good
Typical rust remover	Poor	Fair	Good	Very good		
Caustic soda	None	Good	Fair	Very good	Very good	Good
Liquid bleach	None	Good	Very good	Very good	Very good	Good
Creosote, crude oil	Fair	None	Good	Very good		
Phenols, crude tar	Good	Fair	Good	Good	Very good	Good
Diesel oil	Good	Good	Good	Good		
Synthetic detergents	Poor	Good	Good	Good		
Chlorinated solvents (eg trichloroethylene used in some paint and varnish removers)	Poor	Fair	Good	Poor	Very good	Good
Other organic solvents	Good	Good	Good	Good	Very good	Very good

18.28.17 Synthetic fibre ropes have high durability and low water absorption and are resistant to rot. Mildew does not attack synthetic fibre ropes but moulds can form on them. This will not normally affect their strength.

18.28.18 Store ropes away from heat, sunlight and extreme cold, if possible in a separate compartment that is dry and well ventilated, away from containers of chemicals, detergents, rust removers, paint strippers and other substances capable of damaging them. Cover mooring ropes with tarpaulins or, if the ship is on a long voyage, stow them away. Report any accidental contamination immediately for cleansing or other action.

18.28.19 Unlike natural fibre ropes, synthetic fibre ropes give little or no audible warning of approaching breaking point.

18.28.20 Keep mooring ropes, wires and stoppers in good condition. The user should establish maintenance procedures and records. Establish objective criteria by which to retire and replace in-service ropes and wires to minimise the risk of failure. Keep certifications for ropes and wires on board and link them positively to identifiable equipment. Include records of inspection in the vessel's planned maintenance system, maintenance plan or management system, in line with the manufacturer's guidance. Further guidance is available in IMO MSC 1/Circ 1620 – Guidelines for Inspection and Maintenance of Mooring Equipment including Lines.

MSC 1/Circ 1620

18.28.21 Inspect ropes frequently for external and, where achievable, internal wear between strands. A high degree of powdering between strands indicates excessive wear and reduced strength. Ropes with high stretch suffer greater inter-strand wear than others. Hardness and stiffness in some ropes, polyamide (nylon) in particular, may also indicate overworking.

18.28.22 Treat wires regularly with suitable lubricants. Inspect them internally for deterioration and externally for broken strands. Apply lubricants thoroughly to prevent internal as well as external corrosion, and never allow wires to dry out.

18.28.23 Ensure that the inspection and discard criteria take into account the fact that the condition of the load-bearing core of jacketed ropes cannot be adequately assessed on board ship, and therefore it is necessary to follow inspection procedures.

18.28.24 A rope or wire may lose strength if bent around deck fittings with radii below those of its specified minimum bend radius. Carefully inspect ropes that are regularly exposed to a bend radius less than the stated minimum as the service life of the rope may be less than that recommended by the manufacturer.

18.28.25 The risk of a rope or wire parting under the load is reduced by proper care, inspection and maintenance and by using it properly in service. However, ropes may progressively weaken in service. For this reason, consider their anticipated life expectancies taking specialist industry guidance into account.

- Seafarers and workers must follow training and manufacturer's instructions when using work equipment.
- See also Chapters 8 and 12.

- Work equipment must comply with the relevant regulation for use and its standard.
- PPE must be provided and worn as instructed.

Annex 18.1 Conformity requirements

Table 18.3 lists the UK instruments and all work equipment that should conform to the appropriate product standards, apart from equipment that pre-dates any relevant standards.

CE marking transferred to UK conformity assessed (UKCA) marking in Great Britain on 31 December 2020.

A UKCA or CE marking is relevant for the purpose for which the equipment is to be used. In this context, a marking signifies compliance with basic requirements of design of, manufacture of, and the specifications and test methods applicable to, a piece of work equipment, which have been adopted by the appropriate authorities. Reference to a UKCA or CE marking also includes the marking for an alternative standard that provides equivalent levels of safety, suitability and fitness for purpose.

Table 18.3 Statutory Instruments concerning the safety of work equipment

Title	Reference
The Low Voltage Electrical Equipment (Safety) Regulations 1989	SI 1989/728, amended by SI 1994/3260
The Simple Pressure Vessels (Safety) Regulations 1991	SI 1991/2749, amended by SI 1994/3098, SI 2016/1092
The Personal Protective Equipment (EC Directive) Regulations 1992	SI 1992/3139, amended by SI 1993/3074, SI 1994/2326 and SI 1996/3039
The Equipment and Protective Systems Intended for Use in Potentially Explosive Atmospheres Regulations 1996	SI 1996/192, amended by SI 1998/81
The Lifts Regulations 1997	SI 1997/831, amended by SI 2004/693 and SI 2005/831
The Pressure Equipment Regulations 1999	SI 1999/2001, amended by SI 2002/1267, SI 2004/693 and SI 2015/399
The Merchant Shipping and Fishing Vessels (Personal Protective Equipment) Regulations 1999	SI 1999/2205
The Radio Equipment and Telecommunications Terminal Equipment Regulations 2000	SI 2000/730, amended by SI 2003/1903, SI 2003/3144 and SI 2005/281
The Noise Emission in the Environment by Equipment for Use Outdoors Regulations 2001	SI 2001/1701, amended by SI 2001/3958 and SI 2005/3525, SI 2015/98
The Medical Devices Regulations 2002	SI 2002/618, SI 2003/1697, SI 2005/2909, SI 2007/400, SI 2008/2936 and SI 2013/2327, SI 2019/791, SI 2020/1478, SI 2021/873
The Electromagnetic Compatibility Regulations 2005	SI 2005/281
The Supply of Machinery (Safety) Regulations 2008	SI 2008/1597, SI 2011/2157
The Construction Products Regulations 2013	SI 2013/1387, SI 2019/465, SI 2020/1359, SI 2022/712
The Merchant Shipping (Marine Equipment) Regulations 2016	SI 1999/1957, amended by SI 2001/1638, SI 2004/302, SI 2004/1266, SI 2009/2021, SI 2016/1025

Annex 18.2 Bulldog grips

Figure 18.2 shows a typical bulldog grip.

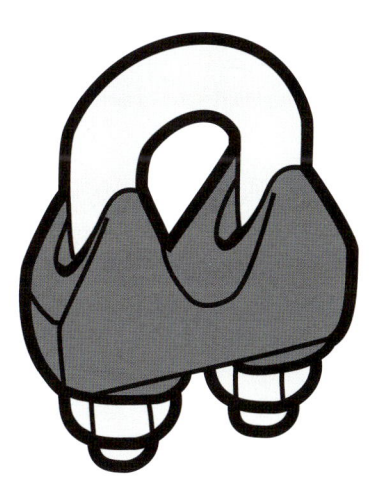

Figure 18.2   Example of a bulldog grip

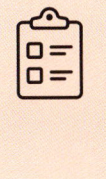

- Avoid using bulldog grips if possible, and do not use them on lifting or mooring wires.
- Do not use bulldog grips where the rope is likely to be subjected to very strong vibrations.
- Do not use bulldog grips with plastic-coated wire rope.
- Where bulldog grips are used, install them correctly following the manufacturer's instructions.
- Place the 'U' of the grip on the dead end of the rope as shown in Figure 18.3. The distance between grips should be approximately six rope diameters. The minimum number of grips depends on the rope diameter. Retighten the grips when they have been in service for several hours and recheck their tightness periodically. Correctly fitted grips should hold at least 80% of the minimum breaking load of the rope.

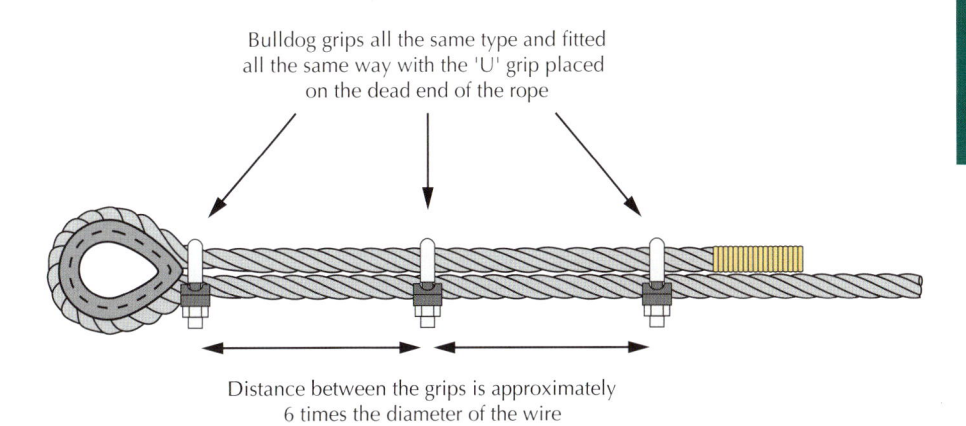

Bulldog grips all the same type and fitted all the same way with the 'U' grip placed on the dead end of the rope

Distance between the grips is approximately 6 times the diameter of the wire

Figure 18.3   Use of bulldog grips

Annex 18.3 Standards for work equipment

1. Suitability of work equipment

Work equipment should be:

- suitable for the work to be carried out
- properly adapted for that purpose
- capable of being used without any significant risks to the health and safety of any seafarer.

PUWER Reg 6(1); SI 2006/2183

2. Electrical equipment

All ship's electrical equipment and installations should be operated and maintained in such a way that there is no electrical hazard to the ship or any person.

PUWER Reg 14

3. Controls for starting or making a significant change in operating conditions

Where any work equipment could constitute a risk to the health or safety of seafarers because it contains moving parts or is mobile, it must be fitted with one or more controls for the purposes of starting it and controlling any change in its speed, pressure or other operating conditions. Additionally, it must be possible to start the machine or change its speed, etc. only by operation of the relevant control.

The requirements in the paragraph above do not apply to any automatic restarting or other changes in the operating conditions that occur as a result of the normal operating cycle of any work equipment.

PUWER Reg 17

4. Controls

All operational controls for work equipment should be clearly visible and identifiable, including the provision of appropriate marking where necessary. No control should be placed in a position where any seafarer operating it is exposed to any significant risk to their health and safety.

PUWER Reg 20

5. Stop controls

In addition to the requirements in section 4, where any work equipment could constitute a risk to health and safety, one or more readily accessible controls must be provided to either bring it to a stop or otherwise render it safe.

Any stop control must override any control required by section 3.

📖 *PUWER Reg 18*

6. Emergency stop controls

In addition to the requirements for stop controls in section 5, where any work equipment could constitute a risk to health and safety, one or more readily accessible emergency stop controls should be provided. An emergency stop control must override any controls required by sections 3 and 5.

📖 *PUWER Reg 19*

7. Control systems

Any control systems for work equipment should be safe and take account of any risks to health and safety that might result from damage to or breakdown of that control system. In this context, a control system cannot be considered safe unless:

- its operation does not create any increased risk to health or safety
- any fault in, or damage to, any part of the control system, or the loss of power supply to it, does not result in additional or increased risk to health or safety
- it does not impede the operation of any stop control required by sections 5 and 6.

📖 *PUWER Reg 21*

8. Isolation from sources of energy

Where the risk assessment indicates the need, work equipment should be provided with a suitable system for isolating it from all its sources of energy. Any isolating system should be clearly identified, capable of being locked off and indicated in the appropriate permit to work.

Suitable measures must also be in place to ensure that reconnection of any energy source to work equipment does not expose the seafarer using the equipment to any significant risk to their health and safety. Such measures must also be identified in the risk assessment and identified on the permit to work.

📖 *PUWER Reg 22*

19 Lifting equipment and operations

19.1 Introduction

19.1.1 The general principles on provision, care and use of work equipment set out in Chapter 20 also apply to lifting equipment. This chapter gives additional information on lifting. Where there is any overlap, the more stringent regulations apply.

19.1.2 'Lifting equipment' means work equipment used for lifting or lowering loads and includes the attachments used for anchoring, fixing or supporting it.

19.1.3 'Loose gear' means any gear by means of which a load can be attached to lifting equipment but does not form an integral part of either the lifting equipment or the load.

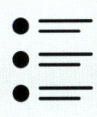

Key points
- During lifting operations ensure good and consistent communications between the personnel operating lifting equipment and the shoreside, deck and bridge.
- Store accessories for lifting in conditions that will not result in damage or degradation.

Your organisation should
- Ensure that all lifting equipment fitted on board is appropriate for its intended purpose and is safe to use.
- Ensure a valid certificate of testing and **thorough examination** 🔍 by a **competent person** 🔍 is in force for every item of lifting equipment, accessory for lifting and loose gear. All items should be tested, and then thoroughly examined and certificated for use either:
    - after manufacture or installation; or
    - after any repair or modification that is likely to alter the safe working load (SWL) or affect the strength or stability of the equipment.

A certificate for a ship's lifting equipment is valid for no more than five years.

Annex 19.1 shows the format for such certificates.

📖 *SI 2006/2184 and MGN 332 (M+F) Amendment 2*

19.1.4 In addition to the strength and stability of the lifting equipment, also consider the stability, angle of heel and potential **down-flooding** 🔍 of any vessel as a result of using a crane, davit, derrick or other lifting device fitted on it. This is especially important when fitting a crane on a workboat or other small vessel. Get advice from the crane manufacturer in such cases before fitting the crane.

19.1.5 A suitably qualified person should check the vessel's stability before installing a crane, and following any modification to it, to ensure that the vessel can operate safely with the crane fitted and in use. Failure to do this could endanger the vessel and the workers on it. See BS 7121: Part 2: 2013 Code of practice for safe use of cranes, Part 2 Inspection, testing and examination.

19.1.6 Any welding of material should be to an approved, acceptable standard because any fitting is only as strong as the weld that connects it to the vessel's structure.

19.1.7 If counterbalance weights are moveable, take effective precautions to ensure that the lifting equipment is not used for lifting in an unstable condition. In particular, all weights should be correctly installed and positioned.

19.1.8 Do not use lifting equipment with pneumatic tyres unless the tyres are in a safe condition and inflated to the correct pressures. Provide the means to check this.

19.1.9 The operator should check safety devices fitted to lifting equipment before work starts and at regular intervals thereafter to ensure that they are working properly.

19.1.10 ## Accessories for lifting

When selecting accessories for lifting take into account:

- the loads to be handled
- the gripping points
- the loose gear for attaching the load, and for attaching the accessories to the lifting equipment
- the atmospheric and environmental conditions
- the mode and configuration of slinging
- vessel motions
- stability issues.

Controls

19.1.11 Controls of lifting equipment should be permanently and legibly marked showing their function and their operating directions by arrows or other simple means. Marks should indicate the position or direction of movement (eg hoisting, lowering, slewing or luffing).

19.1.12 Do not fit makeshift extensions to controls or make any unauthorised alterations to them. Foot-operated controls should have slip-resistant surfaces.

19.1.13 Do not use any lifting device with an inoperative locking pawl, safety attachment or device. If, in exceptional circumstances, limit switches need to be isolated to lower a crane to its stowage position, take care to complete the operation safely.

19.2 Regular maintenance

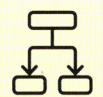

19.2.1 To ensure that all parts of lifting equipment and related equipment are kept in good repair and working order, do preventive maintenance work regularly. Maintenance should include regular examinations by a competent person. Do the examinations as required by the regulations but in any event at least once annually. Look for general material defects such as cracks, distortion, corrosion and wear and tear that could affect SWL and overall strength.

19.2.2 When it is suspected that any lifting equipment, or any part of that equipment, may have been subjected to loads exceeding the SWL, or treated in a way likely to cause damage, take it out of service until a competent person can examine it thoroughly.

19.2.3 Some suggested maintenance tasks are as follows:
- Grease equipment thoroughly and frequently because dry bearings impose additional loads that can lead to failure.
- Check all ropes, wires and chains regularly for wear, damage and corrosion and replace them as necessary. Examine ropes thoroughly, including lengths that remain static in use; these may also be in areas difficult to access.
- Renew shackles, links and rings when wear or damage is evident.
- Examine structures frequently for corrosion, cracks, distortion and wear of bearings, and securing points.
- Check hollow structures such as gantries or masts for water trapped inside. If water is found, drain the structure, treat it appropriately, then seal it.
- Carry out regular function tests of controls, stops, brakes, and safety devices for hoisting gear, preferably before the start of operations.

This list is illustrative only and additional items may be appropriate dependent upon the equipment fitted to an individual vessel.

19.2.4 Any replacement parts must comply with the manufacturer's instructions and be of an equivalent construction to the original part. Replacing parts with incorrect or counterfeit parts of inferior quality can seriously affect the safety of lifting equipment.

19.2.5 After repairing or altering any lifting equipment a competent person should examine it and retest it if appropriate. This also applies if any significant changes are made or noticed to the general condition of the equipment.

19.3 Thorough examination and inspection

19.3.1 Where the safety of lifting equipment depends on the installation conditions, a competent person should inspect it before using it for the first time. Inspect it on initial installation or after re-assembly at another location, to ensure that it has been installed correctly, in line with any manufacturer's instructions, and is safe for workers to operate as well as functioning safely.

19.3.2 Examine thoroughly any lifting equipment or accessory that is regularly in service, and has been exposed to conditions that could cause deterioration that is likely to result in dangerous situations:

- at regular intervals (at least either every 6 months or every 12 months depending on whether the lifting equipment is for lifting people) or
- in line with an examination scheme drawn up by a competent person and
- whenever exceptional circumstances have occurred (eg modification work, accidents, natural phenomena and prolonged periods of inactivity) which are liable to jeopardise the safety of the lifting equipment, to ensure that health and safety conditions are maintained and that any deterioration can be detected and remedied in good time.

Table 19.1 shows the recommended in-service examination periods.

SI 2006/2184; Reg 12(2); INDG422

Table 19.1 Recommended in-service examination periods

Type of equipment	6 months	12 months	Examination scheme
Accessory for lifting	✔		✔
Equipment used to lift people	✔		✔
All other lifting equipment		✔	✔

19.3.3 Do not use accessories for lifting, other than those subject to section 19.4.2, first bullet point, unless they have been thoroughly examined within the 12 months immediately before such use.

19.4 Defect reporting and testing: advice to competent persons

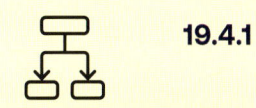

19.4.1 By law lifting equipment must be tested every five years. This section advises the competent person doing the test.

19.4.2 The requirements for testing a piece of lifting equipment will be met if one of the following appropriate tests is done before use:

- proof loading the equipment concerned
- where appropriate, testing a sample to destruction
- when retesting after repairs or modifications, a test that satisfies the competent person who subsequently examines the equipment (a ship's lifting equipment may be retested by means of a static test, eg by dynamometer where appropriate).

19.4.3 Where proof loading is part of a test, apply a test load in excess of the SWL as specified in the relevant standard or, in other cases, by at least the amount set out in Table 19.2.

Table 19.2   Proof load (tonnes)

SWL (tonnes)	Lifting equipment	Single-sheave cargo and pulley blocks	Multi-sheave cargo and pulley blocks	Lifting beams and frames, etc.	Other lifting gear
0–10	SWL × 1.25	SWL × 4	SWL × 2	SWL × 2	SWL × 2
11–20	SWL × 1.25	SWL × 4	SWL × 2	SWL × 1.04 + 9.6	SWL × 2
21–25	SWL + 5	SWL × 4	SWL × 2	SWL × 1.04 + 9.6	SWL × 2
26–50	SWL + 5	SWL × 4	SWL × 0.933 + 27	SWL × 1.04 + 9.6	SWL × 1.22 + 20
51–160	SWL × 1.1	SWL × 4	SWL × 0.933 + 27	SWL × 1.04 + 9.6	SWL × 1.22 + 20
161+	SWL × 1.1	SWL × 4	SWL × 1.1	SWL × 1.1	SWL × 1.22 + 20

Note: Where lifting equipment is normally used with a specific removable attachment and the weight of that attachment is not included in the marked SWL, then for the purposes of using Table 19.2 the SWL of that equipment is the marked SWL plus the weight of the attachment.

19.4.4 Report any defect found in any lifting equipment, including that provided by a shore authority, immediately to the master or to another responsible person who should take appropriate action. Keep certificates of a test or thorough examination in a secure place on board for at least two years.

19.4.5 Similar principles apply to cargo-securing devices as to lifting equipment. Instruct the crew and persons employed for the securing of cargoes in the correct way to apply and use the cargo-securing gear on board the ship. The ship's approved cargo-securing manual gives guidance on the securing of cargoes and handling of security devices.

19.5 Certificates

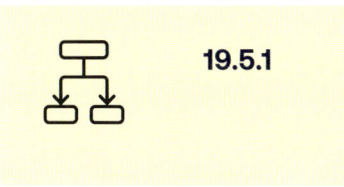

19.5.1 The company must obtain a certificate no later than 28 days after any test and thorough examination of any lifting equipment. Work should not proceed without a valid certificate.

19.6 Records of lifting equipment

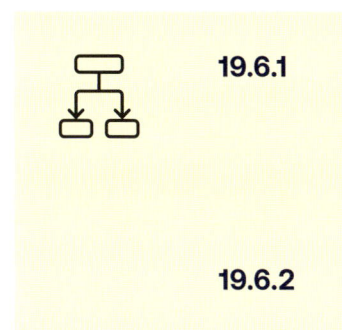

19.6.1 All vessels must keep records of manufacture, examination, inspection and testing of lifting equipment. Keep records and service history of the equipment, dates of when and where it is brought into use, its safe working load, plus any repairs, modifications, tests and examinations carried out.

19.6.2 Annex 19.2 provides a form to register lifting appliances and loose gear used for cargo handling, based on the model recommended by the International Labour Organization.

19.7 Positioning and installation

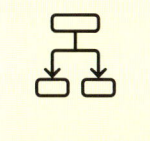

19.7.1 Do not use permanently installed lifting equipment unless it has been positioned or installed to minimise the risk of:

- the equipment or a load striking a worker
- a load drifting dangerously or falling freely
- a load being released unintentionally.

Reg 8

19.8 Lifting operations

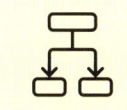

19.8.1 Every lifting operation must be:

- subject to risk assessment
- properly planned
- appropriately supervised
- carried out in a safe manner.

19.8.2 No lifting operation should begin using lifting equipment that is mobile or can be dismantled unless the company is satisfied that it will remain stable during use under all foreseeable conditions, taking into account the nature of the surface on which it stands.

19.8.3   All lifting operations must be properly planned, appropriately supervised and carried out to protect the safety of workers. While this applies to all vessels, it is particularly important when using cranes on workboats and other small vessels due to their impact on the stability of the vessel. Overloading a crane or attempting to lift at the wrong angle could, in some circumstances, result in down-flooding and the vessel possibly sinking. Always observe any restrictions relating to the use of lifting appliances as stated in the vessel's stability book.

19.8.4   Weather conditions can play a significant part in lifting operations. High winds or wave action may cause suspended loads to swing dangerously or mobile equipment to topple. Consider the effects of weather conditions on all lifting operations, whether inside the ship or outside on deck. Suspend such operations before conditions deteriorate to the extent that lifting becomes dangerous.

19.8.5   No person should be lifted except where the equipment is designed and certified for that use or specially adapted and equipped for that purpose, or for rescue or in emergencies.

19.8.6   Minimise contact between bare ropes or warps and moving parts of the equipment by installing appropriate protective devices.

19.8.7   Personnel should never stand on, stand below or pass beneath a load that is being lifted. Establish a safe system of work and supervise to ensure that loads are not lifted over any access way.

19.8.8   All loads should be properly slung and properly attached to lifting gear, and all gear properly attached to equipment.

19.8.9   Any lifts by two or more appliances simultaneously can create hazardous situations; do this only when unavoidable. Conduct lifts properly under the close supervision of a responsible person, after thorough planning of the operation.

19.8.10 Do not use lifting equipment in a manner likely to subject it to excessive overturning movements.

19.8.11 Do not knot ropes, chains and slings.

19.8.12 A thimble or loop splice in any wire rope should have at least three tucks, with a whole strand of rope and two tucks, with one half of the wires cut out of each strand. Tuck the strands against the lay of the rope. Any other form of splice that can be shown as equivalent can also be used.

19.8.13 Do not pass lifting gear around edges liable to cause damage without using a **chafe guard** Q.

19.8.14		Where a particular type of load is normally lifted by special gear, such as plate clamps, substitute other arrangements only if they are equally safe.
19.8.15		The manner of use of natural and synthetic fibre ropes, magnetic and vacuum lifting devices and other gear should take proper account of the particular limitations of the gear and the nature of the load to be lifted.
19.8.16		Inspect wire ropes regularly. When necessary treat them with the correct lubricants, as recommended by the manufacturer, to prevent water being trapped inside the wire. Apply lubricants thoroughly to prevent internal as well as external corrosion. Never allow the ropes to dry out.
19.8.17		Cargo-handling equipment that is lifted onto or off ships by crane or derrick should be provided with suitable points for the attachment of lifting gear, so designed as to be safe in use. The equipment should also be marked with its own gross weight and SWL.
19.8.18		Before trying to free equipment that has become jammed under load, first try to take the load off safely. Take precautions against sudden or unexpected freeing. People who are not directly engaged in the operation should stay in safe or protected positions.
19.8.19		When lifting machinery and, in particular, pistons by means of screw-in eye-bolts, check the eye-bolts to ensure that they have collars, that the threads are in good condition and that the bolts are screwed hard down on to their collars. Clean the screw holes for lifting bolts in piston heads and check that the threads are not wasted before inserting the bolts.

19.9 Safe working load

19.9.1 Do not lift a load greater than the SWL unless:

- a test is required by regulation
- the weight of the load is known and is the appropriate proof load
- the lift is a straight lift by a single appliance
- the lift is supervised by a competent person who would normally supervise a test and carry out a thorough inspection
- the competent person specifies in writing that the lift is appropriate (in weight and other respects) to act as a test of the equipment, and agrees to the detailed plan of the lift
- no person is exposed to danger.

19.9.2 Any grab fitted to lifting equipment should be of an appropriate size, taking into account the SWL of the equipment, the additional stresses on the equipment likely to result from the operation, and the material being lifted.

19.9.3 For a single sheave block used in double purchase, assume that the working load applied to the wire equals half the load suspended from the block.

19.9.4 The SWL of a lift truck means its actual lifting capacity. In the case of a forklift truck this relates to the load that can be lifted, based on the distance from the centre of gravity of the load from the heels of the forks. It may also specify lower capacities in certain situations, such as for lifts beyond a certain height.

19.10 Operational safety measures

19.10.1 Warning
Powered lifting equipment should always have a person at the controls while it is in operation; never leave it to run with a control secured in the 'ON' position.

Regulation 10

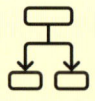

19.10.2 When leaving any powered lifting equipment unattended with the power on, take loads off and put controls in a 'NEUTRAL' or 'OFF' position. Where practical, lock controls or inactivate them to prevent accidental restarting. When work is completed, shut the power off.

19.10.3 The person operating any lifting equipment should have no other duties that might interfere with their primary task. They should be in a proper and protected position, facing the controls and, so far as is practicable, have a clear view of the whole operation.

19.10.4 Where the operator of the lifting equipment does not have a clear view of the whole path of travel of any load carried by that equipment, take precautions to prevent danger. Generally this requirement should be met by employing a competent and properly trained signaller to give instructions to the operator. A 'signaller' includes any person who gives directional instructions to an operator while they are moving a load, whether by manual signals, radio or otherwise.

19.10.5 The signaller should have a clear view of the path of travel of the load where the operator of the lifting equipment cannot see it.

19.10.6	Where necessary, employ additional signallers to give instructions to the first signaller.
19.10.7	Every signaller should be in a position that is:

- safe
- in plain view of the person to whom they are signalling, unless an effective system of radio or other contact is in use.

19.10.8	All signallers should be instructed in, and should follow, a clear code of signals, agreed in advance and understood by all concerned in the operation. Annex 19.3 shows examples of hand signals recommended for use with lifting equipment on ships.
19.10.9	Signallers are not necessary if a load can be guided by fixed guides, or by electronic means, or in some other way, so that it is as safely moved as if it was being controlled by a competent team of driver and signallers.

Additional measures for small vessels

19.10.10	Provide an inclinometer or other efficient device to display the heel angle on board to guide the operator when controlling the lifting of items of unknown weight.
19.10.11	Consider which openings below deck should be secured weathertight during lifting operations. All personnel should be above deck before a lifting operation commences. Post notices with this information on or near the lifting equipment.

19.11 Use of winches and cranes

19.11.1	Secure the drum end of wire runners or falls to winch drums or crane drums by proper clamps or U-bolts. The runner or fall should be long enough to leave at least three turns on the drum at maximum normal extension. Avoid making slack turns of wire or rope on a drum because they are likely to pull out suddenly under load.
19.11.2	When changing a winch from single to double gear or vice versa, first release any load and secure the clutch so that it cannot disengage when the winch is working.
19.11.3	Maintain steam winches so that the operator is not exposed to the risk of scalding by leaks of hot water and steam.
19.11.4	Before operating a steam winch clear the cylinders and steam pipes of water by opening the appropriate drain cocks. Keep the stop valve between winch and deck steam line unobstructed. Take adequate measures to prevent steam obscuring the driver's vision in any part of a working area.

19.11.5 Maintain and operate ships' cranes properly in line with the manufacturers' instructions. The company and the master, as appropriate, should ensure that sufficient technical information is available, including the:

- length, size and SWL of falls and topping lifts
- SWL of all fittings
- boom lifting angles
- manufacturers' instructions for replacing wires, topping up hydraulics and other maintenance as appropriate.

19.11.6 Power-operated rail-mounted cranes should incorporate in their control systems:

- facilities to prevent unauthorised start-up
- an efficient braking mechanism that will arrest the motion along the rails and, where safety constraints require, emergency facilities operated by readily accessible controls or automatic systems for braking or stopping equipment in the event of failure of the main facility
- guards that reduce as far as possible the risk of the wheels running over people's feet, and that remove loose materials from the rails.

19.11.7 When moving a travelling crane replace any necessary holding bolts or clamps before resuming operations.

19.11.8 Access to a crane should always be by the proper means provided. Cranes should be stationary while they are being accessed.

19.12 Use of derricks

19.12.1 Ships' derricks should be properly rigged. The company and the master should ensure that rigging plans are available including information on the:

- position and size of deck eye-plates
- position of inboard and outboard booms
- maximum headroom: the permissible height of cargo hook above hatch coaming
- maximum angle between runners
- position, size and SWL of blocks
- length, size and SWL of runners, topping lifts, guys and preventers
- SWL of shackles
- position of derricks producing maximum forces
- optimum position for guys and preventers to resist maximum forces
- combined load diagrams showing forces for a load of 1 tonne or the SWL
- guidance on the maintenance of the derrick rig.

19.12.2 The operational guidance in the remainder of this section applies generally to the conventional type of ship's derrick. Follow the manufacturers' instructions for other types, such as 'Hallen' and 'Stulken' derricks.

19.12.3 Fit runner guides to all derricks so that when the runner is slack, the bight is not a hazard to people walking along the decks. Where rollers are fitted to runner guides, they should rotate freely.

19.12.4 Before a derrick is raised or lowered, warn everyone on deck nearby not to stand in, or be in danger from, bights of wire and other ropes. Flake out all necessary wires.

19.12.5 When raising, lowering or adjusting a single span derrick secure the hauling part of the topping lift or bull wire (winch-end whip) adequately to the drum end.

19.12.6 The winch driver should raise or lower the derrick at a speed consistent with the safe handling of the guys.

19.12.7 Before raising, lowering or adjusting a derrick with a topping lift purchase, first flake out the hauling part of the span for its entire length in a safe manner. Someone should be available to assist the person controlling the wire on the drum, keeping the wire clear of turns and making fast to the bitts or cleats. Where the hauling part of a topping lift purchase is led to a derrick span winch, handle the bull wire in the same way.

19.12.8 To fasten the derrick in its final position, secure the topping lift purchase to bitts or cleats. First put on three complete turns, followed by four crossing turns. Finally, secure the whole with a lashing to prevent the turns jumping off due to the wire's natural springiness.

19.12.9 When lowering a derrick on a topping lift purchase, employ someone to lift and hold the pawl bar, and be ready to release it should the need arise. They should engage the pawl fully before releasing the topping lift purchase or bull wire. They should not attempt, or be given, any other task until this operation is complete; in no circumstances should they wedge or lash up the pawl bar.

19.12.10 A derrick with a topping winch, and particularly one that is self-powered, should not be topped hard against the mast, table or clamp causing undue strain on the topping lift purchase and its attachments in such a way that the initial heave required to free the pawl bar prior to lowering the derrick cannot be achieved.

19.12.11 Secure a heel block additionally by means of a chain or wire so that the block will be pulled into position under load but will not drop when the load is released.

19.12.12 Lower the derrick to the deck or crutch and secure it properly whenever repairs or changes to the rig are to be carried out.

19.12.13 If heavy cargo is to be dragged under deck with a ship's winches, the runner should be led directly from the heel block to avoid overloading the derrick boom and rigging. Where a heavy load is to be removed, use a snatch block or bull wire to provide a fairlead for the runner and to keep the load clear of obstructions.

19.13 Use of derricks in union purchase

19.13.1 To avoid excessive tensions when using union purchase:

- the angle between the married runners should not normally exceed 90°, and should never exceed 120°
- keep the cargo sling as short as possible so as to clear the bulwarks without the angle between the runners exceeding 90° (or 120° in special circumstances)
- top the derricks as high as practicable, consistent with safe working practices
- do not rig the derricks further apart than necessary.

19.13.2 The following examples show how rapidly loads increase on derricks, runners and attachments as the angle between runners increases:

- At a 60° included angle, the tension in each runner is just over half the load.
- At 90°, the tension is nearly three-quarters of the load.
- At 195°, the tension is nearly 12 times the load.

19.13.3 When using union purchase, winch operators should wind in and pay out in step; otherwise, dangerous tensions may develop in the rig.

19.13.4 Always rig an adequate preventer guy on the outboard side of each derrick when used in union purchase. Loop the preventer guy over the head of the derrick, and as close to and parallel with the outboard guy as available fittings permit. Secure each guy to individual and adequate deck or other fastenings.

19.13.5 Narrow angles between derricks and outboard guys and between outboard guys and the vertical should be avoided in union purchase because these materially increase the loading on the guys. The angle between the outboard derrick and its outboard guy and preventer should not be too large and it may cause the outboard derrick to jack-knife. In general, the inboard derrick guys and preventers should be secured as close as possible to an angle of 90° to the derrick.

19.14 Use of stoppers

19.14.1 Use mechanical topping lift stoppers whenever they are fitted. Where chain stoppers are used, always apply them by two half-hitches in the form of a cow hitch, suitably spaced with the remaining chain and rope tail backed round the wire and held taut to the wire.

19.14.2 Shackle a chain stopper as near as possible in line with the span downhaul and always to an eye-plate. Do not pass it round on a bight because this would induce bending stresses similar to those in a knotted chain.

19.14.3	Do not shackle a stopper to the same eye-plate as the lead block for the span downhaul. This is particularly hazardous when the lead block has to be turned to take the downhaul to the winch or secure it to bitts or cleats.
19.14.4	Always ease the span downhaul to a stopper. The stopper should take the weight before removing turns from the winch, bitts or cleats.

19.15 Overhaul of cargo gear

19.15.1	When replacing a cargo block or shackle ensure that the replacement is of the correct type, size and SWL necessary for its intended use.
19.15.2	All shackles should have their pins effectively secured or seized with wire.
19.15.3	On completion of the work check that all the split pins in blocks and so on have been replaced and secured.
19.15.4	On completion of the gear overhaul clean all working places of oil and grease.

19.16 Trucks and other vehicles/appliances

19.16.1	Do not carry personnel other than the driver on a truck unless it has been constructed or adapted for the purpose. Riding on the forks of a forklift truck is not permitted. The driver should be careful to keep all parts of their body within the limits of the width of the truck or load.
19.16.2	Only competent people should use trucks for lifting and transporting, and only when the ship is in still water; never in a seaway.
19.16.3	Do not use appliances powered by internal combustion engines in enclosed spaces unless they are adequately ventilated. Do not leave the engine running when the truck is idle.
19.16.4	When trucks for lifting or transporting are not in use or are left unattended while the vessel is in port, align them alongside the ship with brakes on, operating controls locked and, where applicable, the forks tilted forwards flush with the deck and clear of the passageway. If the trucks are on an incline **chock** Q their wheels. If they are not to be used for some time, and always while at sea, secure them properly to prevent movement.
19.16.5	Do not try to handle a heavy load by using two trucks at the same time. Do not use a truck to handle a load greater than its marked capacity or to move insecure or unsafe loads.
19.16.6	Do not lift tank containers directly with the forks of forklift trucks because of the risks of instability and of damaging the container with the ends of the forks. Tank containers may be lifted using forklift trucks fitted with suitably designed side or top lifting attachments, but take care because of the free surface effect in partly filled tanks.

19.17 Personnel-lifting equipment, lifts and lift machinery

19.17.1 Except under the conditions required by section 19.17.2, do not use lifting equipment to lift people unless it is designed for the purpose.

19.17.2 If in exceptional circumstances it is necessary to lift people using lifting equipment that has not been specifically designed for the purpose:

- the control position of the lifting equipment must be manned at all times
- the people being lifted must have a reliable means of communication, whether direct or indirect, with the operator of the lifting equipment.

19.17.3 Lifting equipment that is designed for lifting people must not be used for that purpose unless it has been constructed, maintained and operated such that a worker may use it or do work activities from the carrier without risk to their health and safety, and in particular so that:

- the worker will not be crushed, trapped or struck, especially through inadvertent contact with objects
- the lifting equipment is so designed or has suitable devices:
  - to prevent any carrier falling or, if that cannot be prevented for reasons inherent in the site and height differences, the carrier has an enhanced safety coefficient suspension rope or chain
  - to prevent the risk of any person falling from the carrier
- any person trapped in the carrier in the event of an incident is not thereby exposed to danger and can be freed.

19.17.4 A competent person should inspect any rope or chain provided under section 19.17.3, first part of second bullet point, every working day.

19.17.5 Chapter 31 and MGN 332 (M+F) Amendment 2 provide guidelines on the transfer of personnel.

📖 *MGN 332 (M+F) Amendment 2*

19.18 Maintenance and testing of lifts

19.18.1 Before a lift is put into normal service, a competent person must test and examine it and issue a certificate or report. Details of the tests and examinations required for the issue of a certificate are given in British Standards and other equivalent standards.

> *Thorough examination and testing of lifts: Simple guidance for lift owners (INDG339) is also available from the Health and Safety Executive (HSE).*

19.18.2 A competent person should carry out a regular, thorough examination at least every six months, or in line with an examination scheme, and issue a certificate or report. A third party must do a more detailed examination and test of parts of the lift installation at periodic intervals determined by the manufacturer or their representative, or at least every 12 months.

19.18.3 A person chosen to act as a competent person must be over 18 years old. They should have enough practical and theoretical knowledge and actual experience of the type of lift that they have to examine to be able to detect defects or weaknesses and assess their importance in relation to the safety of the lift. Specialist lift maintenance courses are available and recommended.

19.18.4 Only authorised people who are familiar with the work and the appropriate safe working procedures can work on lifts. Procedures must include provision for the safety of people working on a lift and others who may also be at risk.

19.18.5 Do an initial risk assessment to identify the hazards associated with work on each lift installation, including work requiring access to the lift shaft. Draw up safe working procedures for each lift installation. People who are to be authorised to work on or inspect a lift installation must comply with these procedures.

19.18.6 The risk assessment should include, as appropriate:

- whether there are safe clearances above and below the car at the extent of its travel
- whether a car-top control station is fitted and its means of operation
- the working conditions in the machine and pulley rooms.

19.18.7 Based on the findings of the risk assessment use a permit-to-work system, as described in Chapter 14, when personnel need to enter the lift shaft or override the control safety systems. No person should work alone on lifts.

19.18.8 Display appropriate safety signs in the area and on control equipment such as call lift buttons. Use barriers when lift landing doors need to remain open to the lift shaft.

19.18.9 The most important single factor in minimising risk of accidents is to avoid misunderstandings between personnel. Establish a means of communication to the **authorising officer** 🔍 and between people involved in working on a lift and always maintain it. This might be by telephone, portable hand-held radio or a person-to-person chain. Whatever the arrangement, act only as a result of the positive receipt of confirmation that the message is understood.

19.18.10 **Isolate** 🔍 the lift before attempting to access the lift shaft. Lock the mains switch in the 'OFF' position (or withdraw the fuses and keep them in a safe place) and display an appropriate safety sign at the point of isolation. This should include both main and emergency supplies. Also do not allow the landing doors to remain open longer than necessary. Protect the machine room against unauthorised entry and, after completion of work, check that all equipment used in the operation has been cleared from the well.

19.18.11 When personnel need to travel on top of a car, it is much safer to use the car-top control station (comprising a stopping device and an inspection switch/control device) in line with British Standards or an equivalent standard. Consider the arrangement and location of the control station: whether the stopping device can be operated before stepping onto the car on top of the lift car if no stopping device is fitted.

☑ When lifting equipment power is on, there must always be personnel at the controls.

💡
- A certificate for a ship's lifting equipment is valid for no more than five years.
- If there are concerns that lifting equipment may be damaged take it out of service until a competent person can examine it thoroughly.
- Risk-assess the stability of the lifting equipment of the vessel before starting any lifting operations.

Annex 19.1 Certificates of testing and thorough examination of equipment

Annex 19.1.1 Certificate of test and thorough examination of lifting appliances

Name of ship: Certificate no: ....

Official number:

Call sign:

Port of registry:

Name of owner:

(1) Situation and description of derricks used in union purchase (with distinguishing numbers or marks, if any), which have been tested and thoroughly examined	(2) Maximum height of triangle plate above hatch coaming (m) or maximum angle between runners	(3) Test load (tonnes)	(4) Safe working load, SWL (U), when operating in union purchase (tonnes)

Position of outboard preventer guy attachments: (a) forward/aft* of mast and (m)

(b) from ship's centre line (m)

Position of inboard preventer guy attachments: (a) forward/aft* of mast

and

(b) from ship's centre line (m)

* Delete as appropriate

Name and address of the firm or competent person who witnessed testing and carried out thorough examination:

....
....
....

I certify that on the date to which I have appended my signature, the gear shown in column (1) was tested and thoroughly examined and no defects or permanent deformation were found; and that the safe working load is as shown.

Date: ....

Signature: ....

Place: ....

Note: This certificate is the standard international form as recommended by the International Labour Office in accordance with ILO Convention No. 152.

Annex 19.1.2 Certificate of test and thorough examination of derricks used in union purchase

Name of ship: Certificate no: ....

Official number:

Call sign:

Port of registry:

Name of owner:

(1) Situation and description of derricks used in union purchase (with distinguishing numbers or marks, if any), which have been tested and thoroughly examined	(2) Maximum height of triangle plate above hatch coaming (m) or maximum angle between runners	(3) Test load (tonnes)	(4) Safe working load, SWL (U), when operating in union purchase (tonnes)

Position of outboard preventer guy attachments: (a) forward/aft* of mast and   (m)

                                                        (b) from ship's centre line   (m)

Position of inboard preventer guy attachments: (a) forward/aft* of mast   (m)

                                                        and

                                                        (b) from ship's centre line   (m)

*Delete as appropriate

Name and address of the firm or competent person who witnessed testing and carried out thorough examination:

....
....
....

I certify that on the date to which I have appended my signature, the gear shown in column (1) was tested and thoroughly examined and no defects or permanent deformation were found; and that the safe working load is as shown.

Date: ....

Signature: ....

Place: ....

Note: This certificate is the standard international form as recommended by the International Labour Office in accordance with ILO Convention No. 152.

Annex 19.1.3 Certificate of test and thorough examination of loose gear

Name of ship:
Official number:
Call sign:
Port of registry:
Name of owner:

Certificate no: ....

Distinguishing number or mark	Description of loose gear	Number tested	Date of test	Test loaded (tonnes)	Safe working load (SWL) (tonnes)

Name and address of makers or suppliers:

....
....
....

Name and address of the firm or competent person who witnessed testing and carried out thorough examination:

....
....
....

I certify that the above items of loose gear were tested and thoroughly examined and no defects affecting their SWL were found.

Date: ....

Signature: ....

Place: ....

Note: This certificate is the standard international form as recommended by the International Labour Office in accordance with ILO Convention No. 152.

Annex 19.1.4 Certificate of test and thorough examination of wire rope

Name of ship: Certificate no. ....

Official number:

Call sign:

Port of registry:

Name of owner:

Name and address of makers or suppliers:

....
....
....

Nominal diameter of rope (mm):

Number of strands:

Number of wires per strand:

Core:

Lay:

Quality of wire (N/mm^2):

Date of test of sample:

Load at which sample broke (tonnes):

Safe working load of rope (tonnes):

Intended use:

....

Name and address of the firm or competent person who witnessed testing and carried out thorough examination:

....
....
....

I certify that the above particulars are correct, and that the rope was tested and thoroughly examined and no defects affecting its SWL were found.

Date: ....

Signature: ....

Place: ....

Note: This certificate is the standard international form as recommended by the International Labour Office in accordance with ILO Convention No. 152.

Annex 19.2 Register of ships' lifting appliances and cargo-handling gear

Name of ship:
Official number:
Call sign:
Port of registry:
Name of owner:
Register number:
Date of issue:
Issued by:
Signature and stamp:

Note: This Register is the standard international form as recommended by the International Labour Office (ILO) in accordance with ILO Convention No. 152.

Part 1 Thorough examination of lifting appliances and loose gear

(1) Situation and description of lifting appliances and loose gear (with distinguishing numbers or marks, if any) which have been thoroughly examined (see note 1)	(2) Certificate numbers	(3) Examination performed (see note 2)	(4) I certify that on the date to which I have appended my signature, the gear shown in column (1) was thoroughly examined and no defects affecting its safe working condition were found other than those shown in column (5) (date and signature)	(5) Remarks (to be dated and signed)

Note 1: If all the lifting appliances are thoroughly examined on the same date it will be sufficient to enter in column (1) 'All the lifting appliances and loose gear'. If not, the parts which have been thoroughly examined on the dates stated must be clearly indicated.

Note 2: The thorough examinations to be indicated in column (3) include:

- initial
- 12-monthly
- five-yearly
- repair/damage
- other thorough examinations, including those associated with heat treatment.

Part 2   Regular inspections of loose gear

(1) Situation and description of loose gear (with distinguishing numbers or marks, if any) which has been inspected (see note 1)	(2) Signature and date of the responsible person carrying out the inspection	(3) Remarks (to be dated and signed)

Note 1: All loose gear should be inspected before use. However, entries need only be made when the inspection discloses a defect.

Annex 19.3 Code of hand signals

Preliminary remark: The following sets of coded signals are examples of those implemented by the EU Directive 92/58/EEC, but where there are accepted national signals in common use (as indicated *) these too are acceptable.

Visit the Health and Safety Executive (HSE) website to see these signs demonstrated in a video clip: https://www.hse.gov.uk/workplacetransport/safetysigns/banksman/index.htm

Meaning	Description	Illustration
	A. General hand signals	
START Attention Start of command	Both arms are extended horizontally with the palms facing forward.	
TAKING THE STRAIN or INCHING THE LOAD	The right arm points upwards with the palm facing forwards. The fingers are clenched and then unclenched.	*
STOP Interruption End of movement	The right arm points upwards with the palm facing forwards.	
END of the operation (operations cease)	Both hands are clasped at chest height.	
	OR Both arms extended at 45º downwards and lower arms crossed back and forth sharply across torso.	*
	B. Vertical movements	
RAISE	The right arm points upwards, with the palm facing forward, and slowly makes a circle.	
LOWER	The right arm points downwards, with the palm facing inwards, and slowly makes a circle.	

DERRICKING THE JIB	Signal with one hand. Other hand on head.	Jib up * Jib down *
TELESCOPING THE JIB	Signal with one hand. Other hand on head.	Extend jib * Retract jib *
VERTICAL DISTANCE	The hands indicate the relevant distance.	
	C. Horizontal movements	
MOVE FORWARDS (Travel to me)	Both arms are bent with the palms facing upwards and the forearms make slow movements towards the body.	
MOVE BACKWARDS (Travel from me)	Both arms are bent with the palms facing downwards and the forearms make slow movements away from the body.	
RIGHT to the signaller's (in the direction indicated)	The right arm is extended more or less horizontally with the palm facing downwards and slowly makes small movements to the right.	
LEFT to the signaller's (in the direction indicated)	The left arm is extended more or less horizontally with the palm facing downwards and slowly makes small movements to the left.	

HORIZONTAL DISTANCE	The hands indicate the relevant distance.	
SLEWING (in the direction indicated)	Both arms close to side, extending one arm 90º from elbow.	* *
	D. Danger	
DANGER EMERGENCY STOP	Both arms point upwards with the palms facing forwards.	
	E. Other	
SECURE	Secure the load: both arms are crossed closely to the chest with hands clenched.	
TWISTLOCKS Twistlocks on/off	The left arm points upwards. Rotate wrist of left hand clockwise for signalling twist on, and anticlockwise for signalling twist off.	or
	F. Operating instructions	
QUICK	All movements faster.	
SLOW	All movements slower.	

Annex 19.4 Standards

The Merchant Shipping and Fishing Vessels (Lifting Operations and Lifting Equipment) Regulations 2006 introduce measures intended to protect workers from risks arising from the provision and use of lifting equipment. Full guidance is given in marine guidance note MGN 332 (M+F) Amendment 2.

SI 2006/2184; MGN 332 (M+F) Amendment 2

Lifting equipment

Regulation 6 requires lifting equipment to be:

- of adequate strength and stability for each load, having regard in particular to the stress induced at its mounting or fixing points; and
- securely anchored; or
- adequately ballasted or counterbalanced; or
- supported by outriggers, as necessary to ensure its stability when lifting.

Lifting equipment should be of steel or other acceptable material and securely fastened to the vessel's structure. The maximum safe working load (SWL) and maximum radius of operation of all derricks and lifting equipment are required to be part of the specification on all new constructions with associated ropes, wires and guys, eye-plates, shackles and blocks designed to meet these loads.

The vessel's structure, crane, davit, derrick or other lifting device and the supporting structure should be of sufficient strength to withstand the loads that will be imposed when operating at its maximum load moment.

Every part of a load that is used in lifting it, as well as anything attached to the load and used for that purpose, should be of good construction, of adequate strength for the purpose for which it is to be used and free from defects.

Marking of equipment

Lifting equipment must be clearly marked to indicate its safe working loads.

Where the safe working load depends on the configuration of the equipment:

- the work equipment is clearly marked to indicate the SWL for each configuration of the equipment
- information that clearly indicates the SWL for each configuration of the work equipment is kept with the equipment.

Any lifting equipment where the SWL varies with its operating radius is fitted with an accurate indicator, clearly visible to the operator, showing the radius of the load lifting attachment at any time and the safe working load corresponding to that radius.

Lifting equipment that is designed for lifting persons is appropriately and clearly marked.

Lifting equipment that is not designed for lifting persons but which may be so used in error is appropriately and clearly marked to the effect that it is not designed for lifting persons.

Loose gear must be clearly and legibly marked with its safe working load or otherwise marked in such a way that it is possible for any user to identify the characteristics necessary for its safe use including, where appropriate, its SWL.

Loose gear that weighs a significant proportion of the SWL of any lifting equipment with which it is intended to be used must be clearly marked with its own weight.

Trucks and other vehicles/appliances

When vehicles/work trucks or other mechanical appliances are used aboard a vessel to carry personnel, they should where possible be constructed so as to prevent them overturning, or they should be equipped or adapted to limit the risk to those carried, by one or more of the following protection measures:

- An enclosure for the driver.
- A structure ensuring that, should the vehicle overturn, safe clearance remains between the ground and the parts of the vehicle where people are located when it is in use.
- A structure restraining the workers on the driving seat so as to prevent them from being crushed. These protection structures may be an integral part of the vehicle/work equipment. They are not required when the work equipment is stabilised or where the equipment design makes rollover impossible.

20 Work on machinery and power systems

20.1 Introduction

20.1.1 Work on machinery and power systems can be dangerous. It will often require careful assessment of the risks involved, permits to work and clear understanding of responsibilities. Close coordination is also necessary between the **competent person** who does the work and the authorising person who is responsible for the workers, and for the work being done safely and efficiently.

20.1.2 Based on the hazards and findings identified from risk assessment, before any maintenance work begins, put appropriate control measures in place to protect the seafarers and others who may be affected.

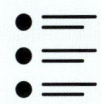

Key points
- Do risk assessments to identify and mitigate risk before work begins.
- Put in place appropriate control measures to protect the seafarers involved and others who may be affected.
- Use permits to work whenever isolations are required for work on machinery and power systems.
- Keep safe means of access and escape clear at all times when working on machinery and power systems.
- Consider using **lock out tag out (LOTO)** systems to effectively control isolations required on machinery during maintenance, repair or inspection.

Your organisation should
- assess the identified risk to ship, personnel and environment and put in place appropriate safeguards
- provide suitable tools and equipment to do the work safely
- provide manufacturers' guidance and machinery manuals for maintenance and repair
- provide chemical data sheets for all substances that seafarers might be exposed to while working
- provide appropriately sized personal protective equipment (PPE) that is fit for purpose and minimises the risk of injuries while working.

20.2 General

20.2.1 Seek the authority of the master and **chief engineer** 🔍 before doing repair or maintenance work that may affect the supply of water to the fire main or sprinkler systems and before isolating any alarm system.

20.2.2 Always keep access to firefighting equipment, emergency escape routes and watertight doors free from obstruction.

20.2.3 Remove the safety guards on machinery or equipment only when the machinery is **isolated** 🔍 (prevented from operating). If removal is essential for maintenance or examination, take the following precautions:

- A responsible person should authorise the removal, and a competent person should do the work or examination.
- There should be adequate clear space and lighting for the work.
- Anyone working close to the machinery should be told about the risks, safe systems of work and precautions to take.
- Post a conspicuous warning notice in the immediate area.

20.2.4 When removing floor plates or handrails post warning notices, guard or fence the openings, and ensure the area is well lit. Resecure the floor plates and handrails in place once the work is done.

20.2.5 Use lifting handles when removing or replacing a floor plate. When there are no lifting handles lever up the plate with a suitable tool and insert a **chock** 🔍 before lifting. Never use fingers to prise up the edges.

20.2.6 Solvents used for cleaning can be toxic; always follow the manufacturer's instructions. The area should be well ventilated, and do not allow smoking.

20.2.7 When working on any powered machinery or equipment, isolate any parts that may present a risk to personnel from potential uncontrolled energy sources during repair, service, or maintenance work.

20.2.8 The basic rules are:

- isolate all power sources (usually, but not always, electrical energy)
- lock the isolator in the 'off' position (eg with a padlock)
- post a warning sign that maintenance work is in progress.

20.2.9 Dissipate any stored energy (eg electrical, hydraulic or pneumatic power) before the work starts. Before anyone enters or works on the equipment, a competent person must verify and confirm that the isolation is effective.

20.3 Work in machinery spaces

20.3.1 Dangerous parts of a ship's machinery or other equipment should have guards or protection devices to prevent access to danger zones or to halt movements of dangerous parts before people reach the danger zones. Guidance is given in Marine Guidance Note MGN 331 (M+F) Amendment 2.

SI 2006/2183; MGN 331 (M+F) Amendment 2

20.3.2 Lag or otherwise shield all steam pipes, exhaust pipes and fittings, which because of their location and temperature present a hazard. The insulation of hot surfaces should be properly maintained, particularly near oil systems.

SOLAS II-2 Reg 4.2.2.6

20.3.3 Seafarers working in noisy machinery spaces should wear suitable hearing protection (see section 8.5).

20.3.4 Where a high noise level in a machinery space, or the wearing of ear protectors, may mask an audible alarm, where practicable provide a visual alarm of suitable intensity, to attract attention and indicate that an audible alarm is sounding. This should preferably be a light or lights with rotating reflectors.

Guidance is available in the International Maritime Organization (IMO) Code on Alerts and Indicators.

20.3.5 Find the source of any oil leakage and repair it as soon as possible.

20.3.6 Do not allow waste oil to accumulate in the bilges or on tank tops. Dispose of any leakage of fuel, lubricating or hydraulic oil in line with The Merchant Shipping (Prevention of Oil Pollution) Regulations 2019 at the earliest opportunity.

SI 2019/42

20.3.7 Take extra care when filling any settling or other oil tank to prevent it overflowing, especially in an engine room where exhaust pipes or other hot surfaces are directly below. Secure manholes or other openings in the tanks so that if a tank is overfilled the overflow arrangements will direct the oil to a safe place.

20.3.8	When filling tanks that have their sounding pipes in the machinery spaces, ensure that weighted cocks are closed. Never secure a weighted cock on a fuel or lubricated oil tank sounding pipe or on a fuel, lubricating or hydraulic oil tank gauge in the 'open' position.
20.3.9	Keep engine room bilges clear of rubbish and other substances so that mud-boxes are not blocked, and so the bilges may be readily and easily pumped.
20.3.10	Regularly test remote controls fitted for stopping machinery or pumps, or for operating oil-tank quick-closing valves in the event of fire to ensure that they are fully functional. This also applies to the controls on fuel storage daily service tanks (other than double bottoms) and lubricating oil tanks.
20.3.11	Use cleaning solvents in line with the manufacturers' instructions and in a well-ventilated area.
20.3.12	Secure spare gear, equipment, and machinery under repair/maintenance against movement and stow it so as to prevent injury or damage in any and all weather and operational conditions.
20.3.13	Procedures should be in place to identify defects caused by vibration, fatigue, poor components and poor fitting of the fuel system and to keep hot surfaces protected.
	MSC 1/Circ 1321
20.3.14	Keep a supply of the necessary tools for personnel working in the engine room in a convenient place. This should minimise the distance a loaded toolbox needs transporting and, as much as possible, avoid the need to carry tools up and down ladders.
20.3.15	Keep a supply of PPE and consumables (eg light bulbs, flashlights, batteries, rags, log books and stationery) close to the engine room for the personnel working there.
	MSC/Circ 834

20.4 Unmanned machinery spaces

20.4.1 Seafarers should enter or remain in an unmanned machinery space alone only if they have the permission of, or have been instructed by, the engineer officer in charge at the time. They may go there only to carry out a specific task that should be done in a comparatively short time. Before entering the space, at regular intervals while in the space, and on leaving it, they must report by telephone (or other means provided) to the duty deck officer (see also section 20.4.4). Before they enter the space, explain clearly how they should report. Consider using a **permit to work** where appropriate (see section 14.2) and note the following:

- The permit to work should be signed by an **authorising officer** and given to the competent person in charge of the work to be carried out on or close to the high-voltage apparatus.
- Inform the competent person of all permit to work details, including as a minimum:
  - the extent of the work
  - exactly what apparatus is **dead**
  - that it is isolated from all live conductors
  - that it is discharged and earthed.
- Confirm that electrical hazards have been assessed and that it is safe to work.

20.4.2 If the engineer officer in charge enters the machinery space alone, they too should report to the deck officer before entry, at regular intervals while in the space, and on leaving it.

20.4.3 A notice of the safety precautions that seafarers working in unmanned machinery spaces must take should be clearly posted at all entrances to the space. Warn them that in unmanned machinery spaces there is a likelihood of machinery suddenly starting up.

20.4.4 If there is a personnel alarm system in place it might not be necessary to report at regular intervals. A personnel alarm is a system that will indicate a person's presence and their well-being in unmanned machinery spaces. Vessels without a personnel alarm system should have additional guidance recorded in the safety management system.

20.4.5 Unmanned machinery spaces should be adequately lit at all times.

20.4.6 Engine room staff should tell the bridge about any changes they are considering to machinery under bridge control. They should tell the bridge when machinery under bridge control is being changed to engine room control.

20.5 Maintenance of machinery

20.5.1 Before servicing or repairing machinery, prevent it being turned on or started automatically or from a remote-control system, as follows:

- Isolate electrically operated machinery from the power supply.
- Close both the steam and exhaust valves on steam-operated machinery. Lock the valves or tie them shut, or use some other means to indicate that the valves should not be opened. Take the same care when dealing with heated water under pressure as when working on steam-operated machinery or pipework.
- Hydraulic-operated machinery should have its own oil supply valve isolated as well as the oil return valve isolated, if fitted.
- In all cases, post notices at or near the controls warning that this machinery must not be used.

20.5.2 The cleaning or replacement of fuel or lubricating filter elements on engines or turbines should, so far as practicable, only be done with the engine or turbine in the stopped condition. Where valves or filter covers have to be removed or similar operations have to be performed on pressurised systems, isolate that part of the system by closing the appropriate valves. The position of a duplex filter changeover cock does not guarantee that the 'out of service' filter chamber has been isolated. Open the drain and/or vent cocks gradually to depressurise the system before slackening off any other fastenings or bolts.

20.5.3 When breaking joints of pipes, fittings, etc. do not completely remove the fastenings until the joint has been broken and it has been established that no pressure remains within.

20.5.4 Before opening a section of a steam pipe system to the steam supply, open all drains. Let out the steam very slowly and keep the drains open until all the water has been expelled.

20.5.5 Maintenance or repairs to, or close to, moving machinery should be permitted only where no danger exists or where it is impracticable to stop the machinery. Workers should wear close-fitting clothing and cover long hair (see section 8.4.5). The officer in charge should consider whether it is necessary in the interests of safety for a second person to be in close attendance whilst the work is being carried out.

20.5.6 Firmly secure any heavy parts of dismantled machinery that have been temporarily put aside against movement in a seaway and, as far as practicable, keep them clear of walkways. Cover sharp edges/projections when reasonably practicable.

20.5.7		Secure spare gear, tools and other equipment or material appropriately after use, especially near stabiliser or steering gear rams, switchboards and batteries.

20.5.8 Use a marlin spike, steel rod or other suitable device to align holes when reassembling or mounting machinery; never use your fingers.

20.5.9 If guards or other safety devices have been removed from machinery, replace them immediately once the work is completed and before testing the machinery or equipment.

20.5.10 Use an approved safety lamp for lighting spaces where oil or oil vapour is present. Disperse vapour by ventilation before the work is done.

20.6 Boilers and thermal oil heaters

20.6.1 Open boilers only under the direction of an engineering officer:

- After emptying check that the vacuum is broken before removing manhole doors.
- Even if an air cock has been opened to break the vacuum, always loosen the manhole door nuts and break the joint before removing the dogs and knocking in the doors.

20.6.2 First remove the top manhole doors. Seafarers should stand clear of hot vapour when doors are opened. Seafarers should not enter any boiler, boiler furnace or boiler flue until it has cooled enough to make work safe.

20.6.3 Before allowing entry to a boiler that is part of a range of two or more boilers, the engineer officer in charge should ensure that either:

- all inlets through which steam or water might enter the boiler from any other part of the range have been disconnected, drained and left open to the atmosphere, or
- where that is not practicable, all valves or cocks, including blowdown valves controlling entry of steam or water, have been closed and securely locked, and notices posted to prevent them being opened again until authorisation is given.

Keep the above precautions as long as people remain in the boiler.

20.6.4 Seafarers cleaning tubes, descaling boilers and cleaning backends should wear appropriate PPE including respirators. The company should ensure that seafarers read the accompanying data sheet to any chemical agents they may use in their work.

20.6.5 Warning

Seafarers should be aware of any potentially hazardous gaseous by-products that may be produced from the reaction of the cleaner/descaling product and the object itself, or from products used together. This may result in an asphyxiating, explosive or otherwise hazardous atmosphere.

20.6.6 A boiler is an enclosed and therefore potentially a dangerous space. Take care before entering a boiler that has not been used for some time or where chemicals have been used to prevent rust forming. The atmosphere may be deficient in oxygen so test it before allowing any person to enter (see Chapter 15).

Post a notice at each boiler setting out operating instructions. In the boiler room post information provided by the manufacturers of the oil-burning equipment..

20.6.7 To avoid the danger of a blowback when lighting boilers, follow the correct flashing-up procedure:

- Check that the furnace floor is free from loose oil.
- Check that the oil is at the correct temperature for its grade; if not, regulate the oil temperature before trying to light it.
- Blow the furnace through with air to clear any oil vapour.
- Use the torch provided for the purpose, or a manufacturer-approved ignitor, for lighting a burner unless an adjacent burner in the same furnace is already lit. Do not use other means of ignition, such as putting loose burning material into the furnace.
- If all is in order, the operator should stand to one side, insert the lighted torch and turn the fuel on. Check that there is not too much oil on the torch that could drip and possibly cause a fire.
- If the oil does not light immediately, turn off the fuel supply. Ventilate the furnace by allowing air to blow through for two or three minutes to clear any oil vapour. During this time remove the burner and the atomiser and check the tip to ensure that they are in good order. Then try again.
- If there is a total flame failure while the burner is alight, turn off the fuel supply.

20.6.8 Keep the means of escape from the boiler fronts and firing spaces clear at all times.

20.6.9 If a gauge glass cover is used it should always be in place when the glass is under pressure. If a gauge glass or cover needs replacing or repairing, shut off the gauge and drain it before removing the cover.

20.6.10		Apply the same isolating and maintenance principles to thermal oil heaters and systems as those for boilers. However, because venting systems are closed to the thermal oil header tank, and drainage systems are closed to the thermal oil drain tank, take additional care when isolating heaters to ensure that the system is fully drained. Check that no residual pressure remains before removing fittings or disconnecting pipes.
20.6.11		Once work on thermal systems is completed prevent water and moisture getting into the system. The steam produced, its sudden expansion and the significantly greater volume occupied may damage equipment and cause significant disruption to the whole system. Before refilling the system from the thermal oil header tank, test the header tank drain for the presence of water. Do this also whenever the header tank is refilled from the thermal oil drain tank or the thermal oil storage tank.

> **20.6.12** **Warning**
> There may be an explosion if you try to relight a burner from the hot brickwork of the furnace.

20.7 Auxiliary machinery and equipment

20.7.1	Before starting work on an electric generator or auxiliary machine, stop the machine and secure the starting air valve or similar device so it cannot be operated. Post a notice warning that the machine must not be started or the turning gear used. To avoid the danger of motoring and electric shock to any person working on the machine, isolate it electrically from the switchboard or starter before starting work. Open the circuit breaker and post a notice at the switchboard warning seafarers that the breaker must not be closed. Where possible lock the circuit breaker open and/or prevent access.
20.7.2	Before starting a diesel engine, turn the engine with the indicator cocks open. Disconnect the turning gear and secure it before trying to start the engine.
20.7.3	Keep diesel engine relief valves, crankcase explosion doors and scavenge belt safety discs free from oily deposits and flammable materials.
20.7.4	Never apply flammable coatings to the internal surfaces of air starting reservoirs.
20.7.5	When testing a diesel engine fuel injector or other high-pressure parts of injection equipment, contain the jets so they cannot spray onto any part of the body.

> **20.7.6 Warning**
> Never use oxygen for starting engines. Doing this would probably cause a violent explosion.

20.7.7 Use assistive-start substances (typically volatile, low flash point aerosols) only in line with both engine and substance manufacturer guidance and where necessary:

- Use of these substances is potentially dangerous, especially in engines that utilise compressed air for starting arrangements; uncontrolled explosions and engine damage have been documented.
- Never use these substances on hot engines or when manifold heater plugs (glow plugs) are being used.

20.8 Main engines

20.8.1 Where necessary use suitable staging, adequately secured, to provide a working platform.

20.8.2 Before allowing anyone to enter or work in the main engine crankcase or gear case:

- the engine-starting system must be in local control and fully isolated with starting air drains opened to the atmosphere
- turning gear should be engaged and warning notices posted at the start position and turning gear local control
- turning gear should be under the control of the person doing the work
- the spaces should be well ventilated and the atmosphere tested before a seafarer enters.

20.8.3 Before using the main engine turning gear check that all seafarers are clear of the crankcase and any moving part of the main engine. Also check that the duty deck officer has confirmed that the propeller is clear.

20.8.4 If a hot bearing has been detected in a closed crankcase, do not open the crankcase until sufficient time has been allowed for the bearing to cool; otherwise the entry of air could create an explosive air/oil vapour mixture.

20.8.5		The opened crankcase or gear case should be well ventilated to expel any flammable gases before bringing any source of ignition near to it, such as a portable lamp (unless of an approved safety type).
20.8.6		Before restarting the main engine a responsible engineer officer should check that the shaft is clear and inform the duty deck officer, who should confirm that the propeller is clear.

20.9 Refrigeration machinery and refrigerated compartments

20.9.1 No one should enter a refrigerated chamber for maintenance activities without first informing a responsible officer or having completed a permit to work, as part of the risk assessment process.

20.9.2 Seafarers charging or repairing refrigeration plants should know the precautions to take when handling the refrigerant. They should wear appropriate PPE when handling these chemicals. Adequate information should be available on each vessel, laying down the operation and maintenance safeguards of the refrigeration plant, the particular properties of the refrigerant and the precautions for its safe handling.

20.9.3 The compartment or flat in which refrigeration machinery is fitted should be adequately ventilated and lit. Where fitted, both the supply and exhaust fans to and from compartments in which refrigeration machinery is situated should be kept running at all times. Keep inlets and outlets unobstructed. When there is any doubt as to the adequacy of the ventilation, use a portable fan or other suitable means to help remove toxic gases from around the machine.

20.9.4 If it is known or suspected that the refrigerant has leaked into any compartments, do not try to enter those compartments until a responsible officer has been advised of the situation. If it is necessary to enter the space follow the procedures for entry into enclosed spaces (see Chapter 15).

20.9.5 When charging refrigerant plants through a charging connection in the compressor suction line, it is sometimes the practice to heat the cylinder to evaporate the last of the liquid refrigerant. Do this only by placing the cylinder in hot water or by a similar indirect method; never heat the cylinder directly with a blow lamp or other flame. Advice on the handling and storage of gas cylinders is given in section 24.8.

20.9.6 If it is necessary for repair or maintenance to apply heat to vessels containing refrigerant, open the appropriate valves to prevent a build-up of pressure within the vessels.

20.10 Critical equipment

20.10.1 A risk assessment will be required before shutting down the equipment. The risk assessment should include, but not be limited to, the following:

- alternative back-up equipment/systems
- any necessary changes in operational procedures because the equipment is out of service
- any additional safety procedures, such as emergency equipment.

20.10.2 If the agreed out-of-service period for critical equipment or systems maintenance cannot be achieved, any extension or alternative actions will require a review by both the on-board and shore management. A further risk assessment may be required if circumstances change (such as environmental conditions, crew fatigue or operational parameters).

20.11 Steering gear

20.11.1 Do not work on steering gear when the vessel is making way, only when it has stopped. Immobilise the rudder by closing the valves on the hydraulic cylinders or by other appropriate means.

20.12 Hydraulic and pneumatic equipment

20.12.1 Before repairing or maintaining hydraulic and pneumatic equipment remove any load or, if this is not practical, adequately support it by other means. Release all pressure in the system. Isolate the part being worked on from the power source and post a warning notice near the isolating valve, which should be locked.

20.12.2 Take precautions against the possibility of residual pressure being released when unions or joints are broken.

20.12.3 Absolute cleanliness is essential for the proper and safe operation of the hydraulic and pneumatic system. Keep the working area, tools, system and its components clean during servicing work. Ensure that replacement units, especially fluid passages, are clean and free from any contamination.

20.12.4 Use only replacement components that comply with manufacturers' recommendations. Inspect any renewed or replacement items or test them before putting them into operation within the system.

20.12.5 Since vapours from hydraulic fluid may be flammable, keep naked lights away from hydraulic equipment that is being tested or serviced.

20.12.6 Never allow a jet of hydraulic fluid under pressure to spray onto any part of the body. If hydraulic fluid under high pressure spills onto unprotected skin, get medical help immediately and wash off the fluid thoroughly.

20.13 Electrical equipment

20.13.1 The risk of electric shock is greater on board ship due to moisture, high humidity and high temperature (including sweating), which reduce the contact resistance of the body. In those conditions, severe and even fatal shocks may be caused at 60 volts or lower. Also remember that cuts and abrasions significantly reduce skin resistance.

20.13.2 Post notices giving instructions on the treatment of electric shock in every place containing electrical equipment and switchgear. Immediate on-the-spot treatment of an unconscious patient is essential.

20.13.3 Before working on electrical equipment, ensure that:

- the energy source is isolated and any residual energy dissipated first
- fuses are removed or circuit breakers opened to ensure that all related circuits are **dead**
- switches and circuit breakers are locked open and a 'do not close' notice has been posted
- where a fuse has been removed, it has been retained by the person working on the equipment until the job is finished
- any interlocks or other safety devices are operative
- the work will be carried out by, or under the direct supervision of, a competent person with sufficient technical knowledge and a permit to work
- additional precautions are taken to ensure safety when work is to be undertaken on high-voltage equipment (designed to operate at a nominal system voltage in excess of 1000 volts).

Lock out tag out

20.13.4 Lock out tag out (LOTO) is a safety procedure to ensure that energy and power sources are properly isolated, shut off and cannot be started up again before maintenance or repair work is completed. This ensures that:

- energy sources are isolated and rendered inoperative before work is started on the equipment
- the isolated power sources are then locked and a tag is placed on the lock identifying the worker who placed it (see section 20.2.9).

The key should then be held according to company procedure. This is to prevent accidental start-up and to ensure that the lock is only removed once the work is completed and has been tested for recommissioning purposes.

20.13.5 When dual LOTO is taking place on the same energy source, the same principles apply. This prevents accidental start-up of a machine or power source while it is in a hazardous state or while a worker is in direct contact with it.

Other precautions

20.13.6 Use voltage indicators and proving units to prove successful isolation. Parts of equipment, even when switched off, may remain **live** 🔍; identify any such circuits clearly by looking at the wiring diagram.

20.13.7 Never store or leave flammable materials near switchboards.

20.13.8 Avoid work on or near live equipment if possible. When it is essential for the safety of the ship or for testing, take the following precautions:

- A second person, who should be electrically competent and trained in first aid, should be continually in attendance.
- The working position should be safe and secure to avoid accidental contact with the live parts. Wear insulated gloves where practicable.
- Avoid contact with the deck, particularly if it is wet. Footwear may not give adequate insulation if it is damp or has metal studs or rivets. Use a dry insulating mat at all times.
- Avoid contact with bare metal. A hand-to-hand shock is especially dangerous. To minimise the risk of a second contact if the working hand accidentally touches a live part, keep one hand in a trouser pocket whenever practicable.
- Remove any jewellery such as wristwatches, metal identity bracelets and rings. They provide low-resistance contacts with the skin. Metal fittings on clothing or footwear (eg buttons, zips) are also dangerous.

20.13.9 Any test meters (and their associated leads/probes) used should be rated for the voltage and/or current being tested. Meter probes should have only minimum amounts of metal exposed and the insulation of both probes should be in good condition. Take care that the probes do not short-circuit adjacent connections. When measuring voltages that are greater than 250 volts, attach and remove the probe with the circuit dead.

20.13.10 The conducting tips of probes should have a maximum dimension of 4 mm (and where possible 2 mm or less and/or fitted with a retractable shield). Leads should be flexible and long enough but not so long as to be unwieldy. Meter sockets and lead plugs should prevent finger contact being made with the conductor if the lead becomes detached from the socket.

20.13.11 All seafarers should be aware of the potential dangers in the space in which they are working. The test equipment should be suitable for the system being examined, checked for damage before use, and proved to be operational before and after use.

20.14 Main switchboards

20.14.1 The internal cleaning and maintenance of the main switchboard must be done only while the switchboard is 'dead'. A full risk assessment must have been carried out (see Chapter 1) and a formal permit to work issued (see Chapter 14).

20.14.2 The risk assessment will identify the actions and checks required to make the switchboard safe, and these will be identified in the permit to work. The major checks to be listed on the permit to work will identify the necessary interconnections to and from, and/or within, the main switchboard and verify that they are disconnected. These will include but are not limited to:

- the shore power supply
- the emergency generator
- the emergency power supply.

20.14.3 The internal cleaning and maintenance of the main switchboard would, in general, be an integral part of a ship's dry-dock programme or that of an extended maintenance programme.

20.15 High-voltage systems

20.15.1 Additional precautions are necessary to ensure safety when work is to be done on high-voltage equipment (designed to operate at a nominal system voltage in excess of 1000 volts).

20.15.2 Use a **limitation of access** instruction to give a written definition of the limits of work to be done in the vicinity of, but not on, high-voltage equipment/installations.

20.16 Work on high-voltage equipment/installations

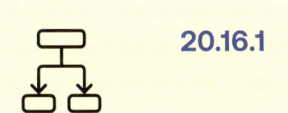

20.16.1 No work must be carried out on high-voltage equipment/installations unless an agreed **switching plan** has been developed and implemented so that the equipment/installations are:

- dead
- isolated and all practicable steps have been taken to lock off live conductors, voltage transformers (except where the connections are bolted) and dead conductors that may become live
- **earthed** at all points of disconnection of high-voltage supply and **caution notices** have been attached in English and any other working language of the vessel
- released for work by the issue of a permit to work or a **sanction for test**.

A switching plan for the safe deisolation and/or reenergising of the system should be in place. The deisolation switching plan should take into account the safe reconnection of live conductors and removal of earthed connection. A safe deisolation plan might not just be the reverse of the isolation switching plan.

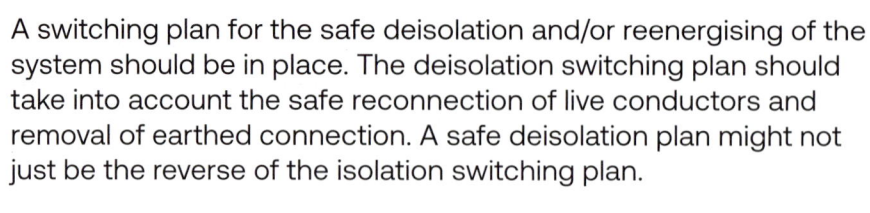

Also, the competent person designated to carry out the work should fully understand the nature and scope of the work to be carried out and have witnessed a demonstration that the equipment/installation is dead at the point of work.

Operation of switchgear

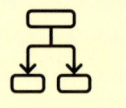

20.16.2 Routine high-voltage switching should be carried out by a competent person in the normal course of their duties, using the equipment provided for the purpose.

High-voltage switching to isolate equipment for maintenance, inspection and/or testing should be done by an **authorised person** or a competent person acting in the presence of and to the instructions of an authorised person. The sequence of switching, isolation and earthing is to be carried out in line with an agreed switching plan.

In an emergency, any competent person may carry out high-voltage switching to cut off supply.

The recipient should repeat in full any message relating to the operation of the high-voltage system that has been transmitted by telephone/radio. The sender should then confirm that the message has been accurately received.

Making live or dead by signals or a pre-arranged understanding after an agreed time interval is not permitted.

Withdrawn apparatus

20.16.3 **High-voltage apparatus** that has been isolated and removed from its normal operating position may be worked on without a permit to work or sanction for test, provided that:

- it has been discharged
- it is prevented by barriers and locking from being restored to a live position
- access to high-voltage conductors on the switchboard is prevented.

Locking off

20.16.4 Lock shut all spout (orifice) shutters not required for immediate work or operations. (Exception: on certain types of switchgear, access to the shutters is restricted while the circuit breaker is still in the cubicle. Under these circumstances, it is acceptable to lock either the cubicle door or the racking mechanism, whichever is appropriate, which must prevent further withdrawal of the circuit breaker, so long as the circuit breaker has been withdrawn from its normal operating position.)

Protective equipment

20.16.5 Do not adjust protective equipment associated with the high-voltage equipment/installations and that forms part of the system. Do not put it into or take it out of commission without the sanction of the chief engineer or **superintendent/senior electrical engineer**.

Do not commission or recommission (after major work) high-voltage equipment/installations until the protective devices have been proved to be functioning correctly.

Insulation testing

20.16.6 All high-voltage equipment/installations that are either new or have undergone substantial maintenance or alteration must undergo a high-voltage test in line with figures approved in writing by the chief engineer or superintendent/electrical engineer.

Failure of supply

20.16.7 During failures of supply, consider all apparatus, equipment and conductors as being live until isolated and proved dead.

Entry to enclosures containing high-voltage equipment/installations

20.16.8 Keep compartments and other enclosures containing high-voltage apparatus locked except when entry or exit is necessary.

The keys or **key safe** giving normal access to such enclosures shall be accessible to authorising officers only.

No person except an authorising officer, or a competent person who is under the immediate supervision of an authorising officer, who must be continuously present, must enter any enclosure in which it is possible to touch exposed high-voltage conductors.

Entry to compartments or other enclosures containing high-voltage equipment/installations is limited to authorising officers or other people only when accompanied by an authorising officer.

Compartments containing high-voltage equipment/installations that are not protected by insulated covers should be entered only when the equipment/installations are isolated and earthed.

Earthing

20.16.9 Circuit mains earths must be applied and removed only by an authorising officer or a person competent to do so in the authorising officer's presence and following their instructions.

When high-voltage equipment/installations have been made dead and isolated, the conductors to be earthed must be proved dead, if practicable, using an approved potential indicator. The potential indicator should be in date for calibration and be tested immediately before and after use, to prove it is in good working order.

Where practicable, apply **circuit main earths** through a circuit breaker or earthing switches.

Before closing to earth, make the trip features inoperative unless this is impracticable. After closing lock the circuit breaker in the earth position and make the trip features inoperative, posting a caution notice nearby.

Additional earths may be applied at the point of work after a permit to work has been issued by the competent person in charge of the work.

Circuit main earths/additional earths may also be removed/replaced at the point of work after the issue of a sanction for test by the authorised person conducting the test.

A circuit main earth applied at the point of work may be removed and replaced one phase at a time to facilitate the work, provided this instruction is recorded on the permit to work. If this is the only circuit main earth connected to the apparatus, then a person authorised to issue permits to work must remain at the point of work and be responsible for the safety of all those engaged in the work while the circuit main earth is removed. No other simultaneous work must be permitted on any part of the circuit during the validity of this permit to work.

Notices (tags)

20.16.10 Post caution notices and **danger notices** on all high-voltage equipment/installations covered by a permit to work or sanction for test drawing people's attention to non-interference or danger as appropriate.

Work on high-voltage cables

20.16.11 No person must touch the insulation that covers or supports any conductor subject to **high voltage** unless the conductor is earthed.

Before issuing a permit to work a person authorised to issue permits must identify the cable to be worked on and proven dead at the point of work. Assume that all cables are live high-voltage cables until proven otherwise.

20.16.12 Before issuing a permit to work to cut into or disturb the insulation of a high-voltage cable (except as required below) the person who is to issue the permit to work must ensure compliance with the following and, where practicable, shall involve the recipient of the permit to work:

- Check cable records.
- Visually trace the cable from the point of work to a point where the apparatus is clearly identified by permanent labelling and in such a way that there is no doubt about the cable's identity.
- Where this is not practicable, the cable shall be identified by signal injection methods; the cable shall be spiked with an approved spiking gun as near to the point of work as practicable. When practicable, the cable shall be cut with the spiking gun in position; tests shall be made to confirm the cable cut is the correct one. All this shall be carried out under a sanction for test.
- Where work is to be carried out on cables where the conductors and/or sheath may be subject to induced voltages from live equipment in close proximity, where practicable the conductors and/or sheath shall be earthed and appropriate PPE used.

Where the above procedures are not practicable a special procedure shall be written and **approved** by the chief engineer or **electro-technical officer**.

Work on transformers

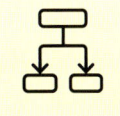

20.16.13 When work is to be carried out on any connections up to a point of isolation or the windings of a transformer, isolate all windings irrespective of voltage. Apply circuit main earths at the points of isolation from high-voltage supply. Lock open low-voltage points of isolation.

Work on ring main units

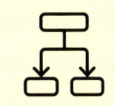

20.16.14 The design of ring main units usually prevents the use of a potential indicator, before earthing. It is therefore extremely important to isolate the appropriate remote end before applying any earth.

Before starting operations check the system diagram and note the onsite labelling on an approved switching procedure.

Do all work and switching on ring main units in strict accordance with the manufacturer's instructions.

Work within the switching chamber of the ring main unit may require the isolation and earthing of all remote ends of the ring main unit.

Work on busbars and directly connected busbar equipment

20.16.15 Before any work begins on a busbar or section of busbar, including any directly connected equipment, isolate the busbar from any point of supply, including voltage transformers. Isolate any directly connected cable and earth it at the remote end.

Withdraw all switches on the busbar or section of busbar to their isolated position.

Lock all isolating arrangements with shutters covering high-voltage contacts. Lock shut any contacts that may become alive and those where no work is to be done and post warning notices.

Prove dead the busbar or section of the busbar to be worked on using an approved potential indicator following the rules for earthing (see section 20.16.11).

Apply a circuit main earth to the busbar on at least one switch panel on the section of busbar on which work is to be done. Apply an additional circuit main earth at any remote ends of directly connected equipment.

Apply an additional circuit main earth at any other position as is necessary to ensure that the busbar remains earthed at all times while work is being carried out.

Issue a separate permit to work or sanction for test for each section of busbar. Do not issue more than one permit to work or sanction for test at the same time for any section of busbar or any electrical equipment directly connected to it.

Prove dead any orifices where work is to be done immediately beforehand by using an approved potential indicator.

20.17 Arc-flash associated with high- and low-voltage equipment

20.17.1 An arc-flash occurs when an electric current flows through an air gap. The air is the conductor and an arc can form between phase-to-ground (neutral) or phase-to-phase and is accompanied by ionisation of the surrounding air.

20.17.2 The incident energy associated with an arc-flash is measured in calories per square centimetre (cal/cm^2). It is the amount of thermal energy from an arc-flash that reaches a surface, such as a person's skin.

20.17.3 The greater the incident energy value is, the more severe the burn injury. The energy required to produce the onset of a partial-thickness burn is 1.2 cal/cm^2 and this is the benchmark for personal protection. (A partial-thickness burn affects both the outer and underlying layer of skin and causes pain and redness, swelling and blistering (NHS, 2020).)

20.17.4 Clearly mark all high- and low-voltage equipment that presents an arc-flash hazard to personnel.

20.17.5 There are various methods to reduce the risk to personnel if an arcing fault occurs within electrical equipment.

These may include:

- reducing fault levels
- tripping times by design
- maintenance settings
- arc-flash detection systems
- 'arc-proof' equipment tested in accordance with the relevant national or international standards
- removing personnel from the location of hazards (remote operation/circuit breaker racking)
- consider appropriate arc-flash protection PPE.

20.17.6 Arc-flash protection is to minimise the likelihood of burn injury by providing an adequate thermal barrier that will limit the energy exposure of a person's skin to no more than 1.2 cal/cm^2. Remember that 1.2 cal/cm^2 is where the onset of a partial-thickness burn can occur, so there is still a possibility of being injured while protected.

MGN 452 (M)

20.18 Storage batteries: general

20.18.1 As batteries may give off gases when charging, keep battery containers and compartments well-ventilated to prevent an accumulation of dangerous gas. Store damaged lithium batteries securely until they can be disposed of safely, following the battery manufacturer's instructions and according to company policy.

20.18.2 Do not allow smoking or any type of open flame in a battery compartment. Post a conspicuous notice to this effect at the entrance to the compartment.

20.18.3 Lighting fittings in a battery compartment should be of an intrinsically safe design with any protective coverings tightly fitted and maintained in accordance with the manufacturer's recommendations.

20.18.4 If the lighting fixtures are found to have defects that cannot be easily rectified, isolate the lighting fixture circuit.

20.18.5 Do not make any unauthorised modifications or additions to electrical equipment (including lighting fittings) in battery compartments.

20.18.6 All batteries require appropriate maintenance following the manufacturer's instructions.

20.18.7	The term 'maintenance free' is used to describe batteries constructed to prevent the ability to top up electrolyte levels. 'Maintenance free' batteries are manufactured with sufficient electrolyte levels for recommended service life and application. However, excess use, lack of maintenance and incorrect application can increase water loss rates and gassing, leading to lower electrolyte levels. Carry out maintenance following the manufacturer's instructions.
20.18.8	Do not use portable electric lamps and tools, or other portable power tools that might give rise to sparks, in battery compartments.
20.18.9	Do not use the battery compartment as a store for any materials or gear not associated with it.
20.18.10	A short-circuit of even one cell may produce an arc or sparks that may cause an explosion of any hydrogen present. Additionally, the very heavy current that can flow in the short-circuiting wire or tool may cause burns due to rapid overheating of the metal.
20.18.11	Maintain insulation and/or guarding of cables in battery compartments in good condition.
20.18.12	Keep all battery connections clean and tight to avoid sparking and overheating. Never use temporary clip-on connections because they may work loose due to vibration and cause a spark or short circuit.
20.18.13	Never place metal tools, such as wrenches or spanners, on top of batteries because they may cause sparks or short-circuits. The use of insulated tools is recommended.
20.18.14	Remove jewellery such as watches and rings when working on batteries. A short-circuit through any of these items will heat it rapidly and may cause a severe skin burn. If rings cannot be removed, tape them heavily in insulating material.
20.18.15	Switch off the battery chargers and all circuits fed by the battery when connecting or disconnecting leads. If a battery is in sections, it may be possible to reduce the voltage between cells in the work area, and hence the severity of an accidental short-circuit or electric shock, by removing the jumper leads between sections before beginning work. Although individual cell voltages may not prevent a shock risk, dangerous voltages can exist when numbers of cells are connected in series. A lethal shock needs a current of only tens of milliamps so take care when the voltage exceeds 50 volts.
20.18.16	Check the battery-charging systems to ensure that it is only possible to charge within the specified rate. Check battery boxes for fixing and integrity as part of the planned maintenance.
20.18.17	Screw battery cell vent plugs tight while making or breaking connections.
20.18.18	Examine the ventilation tubes of battery boxes regularly to ensure that they are free from obstruction.

20.18.19	Fasten the lids of battery boxes while they are open for servicing and secure them properly again when the work is finished.
20.18.20	Secure batteries in place to prevent shifting in rough weather.
20.18.21	Keep alkaline and lead-acid batteries in separate compartments or keep them apart using screens. Store equipment, tools and materials for the servicing of lead-acid and alkaline batteries separately to prevent contamination and/or electrolyte mixing, which can result in a dangerous chemical reaction.
20.18.22	Both acid and alkaline electrolytes are highly corrosive. Immediately wash off any accidental splashes on the person or equipment. Always wash your hands as soon as the work is finished.
20.18.23	Batteries should always be transported in an upright position to avoid electrolyte spillage. Be aware of **manual handling** techniques before moving batteries as they are heavy and this may result in painful strains or injury (see Chapter 10).

20.19 Storage batteries: lead acid

20.19.1	When preparing the electrolyte add the concentrated sulphuric acid slowly to the water. If water is added to the acid, the heat generated may cause an explosion of steam, splattering acid over the person handling it.
20.19.2	Wear goggles, rubber gloves and a protective apron when handling acid.
20.19.3	To neutralise acid on skin or clothes, use plenty of clean, fresh water.
20.19.4	An eyewash bottle should be available in the compartment for immediate use on the eyes in case of accident. This bottle should be clearly distinguishable by touch from acid or other containers so that a person who is temporarily blinded can find it easily.
20.19.5	The corrosion products that form round the terminals of batteries can cause injury to skin or eyes. Remove them by brushing away from the body. Protect the terminals with petroleum jelly.
20.19.6	An excessive charging rate causes acid mist to be carried out of the vents onto adjacent surfaces. Clean this off with diluted ammonia water or soda solution and dry the affected areas.

20.20 Storage batteries: alkaline

20.20.1	The general safety precautions with this type of battery are the same as for the lead-acid batteries, but with the following exceptions.

| 20.20.2 | The electrolyte in these batteries is alkaline but is similarly corrosive. Do not allow it to touch skin or clothing. If an accident happens wash the affected parts with plenty of clean, fresh water. Treat burns with boracic powder or saturated solution. Wash eyes out thoroughly with water, followed immediately with a solution of boracic powder (at the rate of 1 teaspoonful to 0.5 litre or 1 pint of water). This solution should always be readily accessible when electrolyte is handled. |

20.20.3	**Warning**
	Metal cases of alkaline batteries remain live at all times. Do not touch them or allow metal tools to come into contact.

20.21 Work on apparatus on extension runners or on the bench

20.21.1	Chassis on extension runners should be firmly fixed, either by self-locking devices or by use of chocks, before any work is done.
20.21.2	Get assistance where units are awkward or too heavy for one person to handle easily (see Chapter 10). Strain, rupture or a slipped disc can result from a lone effort.
20.21.3	Wedge any chassis on the bench firmly or otherwise secure it to prevent it overbalancing or moving. If a live chassis overbalances do not try to grab it.
20.21.4	Temporary connections should be soundly made. Flexible extension cables should have good insulation and adequate current-carrying capacity.

20.22 Servicing radio and associated electronic equipment

20.22.1	Strictly follow the manufacturers' recommendations to avoid exposure to microwave radiation. Operate radar sets only when the wave guide is connected.
20.22.2	Work within the safety radius of a satellite terminal antennae only when its transmitter has been isolated.
20.22.3	Direct viewing of a radar aerial and wave guide can damage the eyes. Avoid doing this while the radar is in operation or where arcing or sparking is likely to occur.

20.22.4	Exposure to dangerous levels of X-ray radiation may occur in the vicinity of faulty high-voltage valves. Take care when fault tracing in the modulator circuits of radar equipment. An open-circuited heater of such valves can lead to X-ray radiation where the anode voltage is in excess of 5000 volts.
20.22.5	Follow the manufacturers' instructions for maintenance and repair activities. Vapours of some solvents used for degreasing are toxic, particularly carbon tetrachloride, which should never be used.
20.22.6	Some dry recorder papers used in echo sounders and facsimile recorders give off toxic fumes when in use. Keep the equipment well-ventilated to avoid inhalation of the fumes.
20.22.7	Do not operate radio transmitters and radar equipment when people are working near aerials; isolate the equipment from mains supply and earth radio transmitters. When equipment has been isolated, post warning notices on transmitting and radar equipment and at the mains supply point. This is to prevent apparatus being switched on until the people doing the outside work have given clearance that they have finished.
20.22.8	Rig aerials out of reach of seafarers standing at normal deck level or mounting easily accessible parts of the superstructure. If that is impractical, put up safety screens instead.
20.22.9	Post notices warning of the danger of high voltage near radio transmitter aerials and lead-through insulators.

20.23 Additional electrical hazards from radio equipment

20.23.1	Where accumulators are used, disconnect them at source. Otherwise take precautions to prevent the short circuiting of the accumulator, with the consequent risk of burns.
20.23.2	Live chassis connected to one side of the mains are usually marked appropriately; handle them with caution. Where the mains are AC and a transformer is interposed, the chassis is usually connected to the earth side of the supply, but check this using an appropriate meter.
20.23.3	Modern equipment often consists of a master crystal enclosed in an oven. The supply to the oven comes from an independent source and is not disconnected when the transmitter is switched off and the mains switch is off. Mains voltage will be present inside the transmitter so take care.
20.23.4	Before beginning work on the extremely high-tension section of a transmitter or other high-tension apparatus, with the mains switched off, discharge all high-tension capacitors using an insulated jumper. Insert a resistor in the circuit to slow the rate of discharge. Take this precaution even where the capacitors have permanent discharge resistors fitted.

20.23.5 Replace an electrolytic capacitor that is suspect, or shows blistering, because it is liable to explode when electrical supply is on. There is a similar risk when an electrolytic capacitor is discharged by a short circuit.

20.23.6 Avoid work at or near live equipment if possible, but where it is essential for the safety of the ship or for testing take the additional precautions described in section 20.13.8.

20.24 Valves and semi-conductor devices

20.24.1 When removing valves from equipment that has recently been operating grasp them with a heat-resistant cloth. Allow large valves (such as power amplifiers, output valves and modulators, which reach a high temperature in operation) time to cool down before removing them. Severe burns can result if they touch bare skin.

20.24.2 Handle cathode ray tubes and large thermionic valves with care. Although they implode when broken there is still a risk of severe cuts from sharp-edged glass fragments. Some special-purpose devices (such as trigatrons) contain vapour or gas at high pressure but these are usually covered with a protective fibre network to contain the glass if they explode.

20.24.3 Beryllia (beryllium oxide) dust is very dangerous if inhaled or if it penetrates the skin through a cut or abrasion. Symptoms of poisoning include respiratory troubles or cyanosis (grey/blue discolouration of the skin); they may develop within a week or after a latent period of up to several years.

20.24.4 Beryllia may be present in some electronic components. Cathode ray tubes, power transistors, diodes and thyristors containing beryllia will usually be identified in the manufacturers' information provided. However, if there is no such information there is no guarantee that beryllia is not present. The heat sink washers that contain beryllia are highly polished and look like dark brass. Store these items carefully in their original packaging until required.

20.24.5 Physical damage to components containing beryllia, whether new or defective, is likely to produce dangerous dust. Avoid abrasion and do not work on components with tools. Encapsulations should be left intact. Excessive heat can be dangerous, but it is safe to do normal soldering with thermal shunt. Pack damaged or broken parts separately and securely, following the manufacturer's instructions for return or disposal.

20.24.6 Seafarers handling parts containing beryllia should wear PPE, including gloves, to prevent the substance touching the skin. Use tweezers where practicable. If the skin is contaminated with the dust they should clean the affected parts without delay, particularly any cuts, following their instructions for handling dangerous chemicals.

- Follow the manufacturer's instructions and guidance.
- Monitor the insulation of hot surfaces through a heat (thermographic) survey or using infra-red thermometers to ensure that surface temperatures do not exceed 220°C.
- Paint tank tops and bilges a light colour wherever practicable. Keep them clean and well-lit near pressure oil pipes so that leaks may be easily found.

- Work on safety-critical systems should be authorised by the master and chief engineer.
- Any stored energy should be effectively isolated and discharged by competent people and supervised by authorised people.
- Chemicals and cleaning materials can cause hazardous gases; consider this when using them in work.

21 Hazardous substances and mixtures

21.1 Introduction

21.1.1 Many substances and mixtures found on ships are capable of damaging the health and safety of those exposed to them. They include not only substances displaying hazard-warning labels (particularly those declared as dangerous goods in ships' stores) but also dusts (including hardwood dusts), fumes and fungal spores which could arise from goods, machinery or activities aboard ship.

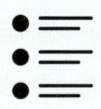

21.1.2 This chapter deals with the use of hazardous substances and mixtures (referred to in this chapter as 'hazardous substances') carried on board ships, such as in a ship's stores. Chapter 28 covers dangerous substances carried as dry cargo. Chapter 29 covers tankers and other ships carrying bulk liquid cargoes.

Key points
- Many substances and mixtures found on board ship can damage the health and safety of people exposed to them.
- Seafarers should ensure that they can recognise potentially dangerous materials, including any warning labels, and that they comply with company instructions to control or minimise risk.
- A hazard-warning label includes a pictogram, a precautionary statement, a hazard statement (eg 'Carcinogenic', 'Flammable') and, where required, a signal word (either 'Danger' or 'Warning'). Seafarers should familiarise themselves with the meaning of such labels.

Your organisation should
- based on the findings of the risk assessment and with reference to the manufacturer's safety data sheet (SDS), identify where seafarers are working in the presence of hazardous substances and put in place appropriate measures to remove, control or minimise the risk
- ensure that seafarers understand the risks arising from their work and the precautions to take
- ensure that seafarers' exposure to carcinogens and mutagens does not exceed statutory limits.

21.1.3 The company's risk assessment will identify where seafarers are working in the presence of hazardous substances and evaluate any risks from exposure. Take appropriate measures to remove, control or minimise the risk. Refer to the manufacturer's SDS before exposing seafarers to any hazardous substance, to select appropriate personal protective equipment (PPE) and working methods.

21.2 Instruction/training of seafarers

21.2.1 The company should instruct and inform seafarers so that they know and understand the risks arising from their work and the precautions to take. Employers should tell seafarers the results of any monitoring of exposure.

21.2.2 Where possible, seafarers should avoid direct contact with hazardous substances, wear appropriate gloves and, if necessary, safety glasses/goggles, and follow the manufacturer's instructions.

21.2.3 The company should instruct seafarers to take appropriate precautions and make them aware of the potentially hazardous by-products that may be produced from mixing hazardous substances together. For example, mixing chlorine-based toilet cleaner with descaler will cause a hazardous gaseous by-product which may result in an asphyxiating, explosive or other hazardous atmosphere.

21.2.4 The risk assessment will also provide information on whether health surveillance is appropriate as a result of exposure to hazardous substances (see Chapter 7).

21.2.5 To help identify hazards and assess risks from hazardous substances, check the SDS. In Europe the manufacturer is required to supply this with all hazardous substances and mixtures.

21.2.6 For more specialist advice relating to particular work activities, where appropriate refer to the series of publications by the Health and Safety Executive (HSE) under the Control of Substances Hazardous to Health (COSHH) Regulations (see Appendix 2).

21.3 Health surveillance

21.3.1 The company should take reasonable steps to ensure that any control measures are properly used and maintained. Where appropriate, monitor and record exposure levels. For some hazardous substances, seafarers must not be exposed at work beyond a statutory level. These workplace exposure limits are published by HSE in *EH40/2005 Workplace exposure limits* available on the HSE website.

21.3.2 The risk assessment will also provide information to determine whether health surveillance is appropriate as a result of exposure to hazardous substances (see Chapter 7).

21.4 Prevention or control of exposure

21.4.1 First consider how to prevent exposure by removing the substance (such as by replacing it with a less harmful one.)

21.4.2 Where removing the substance is not reasonably practicable, exposure can be prevented or controlled by any combination of the following:

- total or partial enclosure of the process and handling systems
- using plant, processes and systems of work, which minimise the generation of, or suppress and contain/prevent, spills, leaks, dust fumes and vapours of hazardous substances
- local exhaust ventilation (to remove toxic fumes and, therefore, limit exposure)
- limiting the quantities of a substance at the place of work
- keeping the number of people who might be exposed to a substance to a minimum, and reducing the period of exposure
- prohibiting eating, drinking and smoking in areas that may be contaminated by the substance
- hygiene measures, including providing adequate washing and laundering facilities, and regular cleaning of walls/bulkheads and other surfaces
- designation of those areas that may be contaminated and the use of suitable and sufficient warning signs
- safe storage, handling and disposal of hazardous substances and use of closed and clearly labelled containers
- using appropriate procedures for the measurement of hazardous substances, in particular for the early detection of abnormal exposures resulting from an unforeseeable event or an accident
- taking individual/collective protection measures
- where appropriate, drawing up plans to deal with emergencies likely to result in abnormally high exposure.

21.4.3 Apply these measures to reduce the risk to seafarers to the minimum, but where they do not adequately control the risk to health, provide PPE in addition.

21.4.4 Seafarers should comply fully with the control measures in force.

21.4.5 For certain substances (eg asbestos and benzene), very specific control measures apply. Where failure of the control measures could result in risk to health and safety, monitor the exposure of personnel and keep a record for future reference.

21.4.6 Where the adequacy or efficiency of control measures is in doubt, stop work until outside advice has been sought and action taken proportionate to the risks involved.

21.5 Carcinogens and mutagens

21.5.1 The Merchant Shipping and Fishing Vessels (Health and Safety at Work) (Carcinogens and Mutagens) Regulations 2007 (the 2007 Regulations) specifically require that the risk assessment considers the risk arising from exposure to carcinogens and mutagens. A carcinogen is a substance or mixture for which evidence exists to establish a link between exposure to it and the development of cancer. A mutagen is a substance or mixture for which evidence exists to establish a link between exposure to it and heritable genetic damage.

SI 2007/3100; MGN 624 (M+F) Amendment 2

21.5.2 Hazardous substances that are found on ships and considered carcinogens and mutagens include:

- aflatoxins
- arsenic
- asbestos (see section 21.6)
- hardwood dusts
- rubber dust and rubber fumes
- used engine oils
- welding fumes (see Chapter 24).

21.5.3 The supplier of a hazardous substance or mixture is required to:

- identify the hazards of the substance or mixture
- provide information about the hazards to their customers. This information is usually on the package itself (eg on a hazard label) and, if supplied for use at work, in an SDS
- package the chemical safely (classification of carcinogens is described in Annex 21.1).

Use the hazard information to help the company comply with the 2007 Regulations.

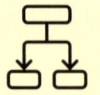

21.5.4 Where the risk assessment reveals there is a risk to seafarers' health from carcinogens and mutagens, and the measures set out in section 21.3 do not completely remove that risk, the company should ensure that the exposure never exceeds the limit values set out in the regulations.

21.5.5 All cases of cancer that can be identified as resulting from occupational exposure to a carcinogen or mutagen, and have been confirmed in a report from a doctor, must be reported to the MCA (see section 7.3).

21.6 Asbestos dust

21.6.1 The use of asbestos in ship construction has been banned internationally, but asbestos is still being discovered in non-approved parts such as gaskets and brake linings. Take care when getting spare parts because some may contain asbestos even when declared 'asbestos free'. Measures to protect seafarers' health where there is a risk of exposure to asbestos are in the Merchant Shipping and Fishing Vessels (Health and Safety at Work) (Asbestos) Regulations 2010 and associated marine guidance notes (MGNs).

SI 2010/2984; MGN 669 (M+F) Amendment 1

21.6.2 Asbestos has a fibrous structure and can produce harmful dust if the surface exposed to the air is damaged or disturbed. The danger is not immediately obvious because the fibres that can damage the lungs and cause lung cancer are too small to be seen with the naked eye. Asbestos that is in good condition is unlikely to release fibres, but where the material is damaged or deteriorating, or work is undertaken on it, airborne fibres can be released.

21.6.3 Dry asbestos is much more likely to produce dust than asbestos that is thoroughly wet or oil-soaked. Asbestos is particularly likely to occur on older vessels in insulation and panelling, but certain asbestos compounds may also be found elsewhere and on other vessels in machinery components such as gaskets and brake linings.

21.6.4 The company should advise masters of any location where asbestos is known or believed to be present on their ship. Masters and/or safety officers should keep a written record of this information. They should also note any other position where asbestos is suspected, but they should not probe or disturb any suspect substance. Warn crew members who work regularly near asbestos or a substance likely to contain it of the need for caution and to report any deterioration in its condition such as cracking or flaking.

21.6.5 The condition of old asbestos may deteriorate so where reasonably practicable consider removing it. This work should be done in port and a specialist removal contractor used to ensure adequate protective procedures. Where the port is in the UK and the work involves asbestos insulation or asbestos coating, it is usually necessary for the contractor to hold a licence issued by HSE. If the work is done outside the UK, the contractor should be of equivalent competence.

21.6.6 If it is essential to carry out emergency repairs likely to create asbestos dust while the ship is at sea, take strict precautions, including the use of the appropriate protective clothing and respiratory protective equipment, in line with the guidance in the relevant merchant shipping notice (MSN). See also the general guidance on the assessment and control of risks from hazardous substances in section 3.11.

21.6.7 Where asbestos or asbestos-containing materials are carried as a cargo, generally in shipping containers, take extreme caution to prevent exposure.

21.7 Use of chemical agents

21.7.1 Consult the relevant MGNs which give further guidance on the handling of chemicals. There is a particular emphasis on health monitoring for people exposed to chemicals (see Chapter 7).

SI 2010/330; MSN 1888 (M+F) Amendment 3

21.7.2 Never use a chemical from an unlabelled package or receptacle unless its identity has been positively established. In addition to transport labelling, packaged substances supplied in Europe may also display similar or additional labelling for supply and use for compliance with the European regulation on classification, labelling and packaging of substances and mixtures ('the CLP Regulation').

European Regulation EC 1272/2008

21.7.3 Employers should ensure workers are instructed to familiarise themselves with the accompanying SDS for any chemical agents they may use in their work. They should also be aware that potentially hazardous gaseous by-products may be produced from the reaction of a cleaner/descaling product and the object itself, or products used together, because this may result in an asphyxiating, explosive or other hazardous atmosphere.

21.7.4 Handle chemicals with the utmost care. Industrial formulations may be stronger. Eyes and skin should be protected from accidental exposure or contact.

21.7.5 Follow manufacturers' or suppliers' advice on the correct use of chemicals. Some cleaning agents (eg caustic soda and bleaches) may burn the skin even when used domestically. The product's hazard-warning label should identify where skin corrosion/serious eye damage hazards are present. Instructions on handling such chemicals safely will be made clear in the precautionary statements.

21.7.6 Do not mix chemicals unless it is known that no dangerous reaction will be caused.

21.7.7 Employers should ensure that they give any necessary training in the use of chemicals.

21.8 Safe use of pesticides

21.8.1 Read the following guidance in conjunction with MSN 1718 (M) and MGN 576 (M), which have mandatory force under the Merchant Shipping (Carriage of Cargoes) Regulations 1999.

📖 *SI 1999/336; MSN 1718 (M); MGN 576 (M)*

21.8.2 Where pesticides are used in the cargo spaces of ships or cargo units, safety procedures should be in line with the International Maritime Organization (IMO) publication MSC 1/Circ 1624. Keep a copy of this publication on board and accessible for all crew members.

📖 *MSC 1/Circ 1624*

Where pesticides are used in other spaces of ships, safety procedures should be in line with MSC.1/Circ.1358.

📖 *MSC 1/Circ 1358*

21.8.3 Where the crew are carrying out space and surface-spraying operations the master should ensure that they wear the appropriate protective clothing, gloves, respirators and eye protection.

21.8.4 The ship's personnel should not handle fumigants; only qualified operators should do this type of work. Fumigation should only be done with the authority of the ship's master. (HSG251 gives health and safety guidance on fumigation and is available from the HSE website.)

21.8.5 Display the 'fumigation warning' sign conspicuously on cargo units or spaces being fumigated. Station a member of personnel to watch and prevent access to areas of risk by unauthorised personnel.

21.8.6 In exceptional circumstances, in-transit fumigation may be permitted only after first referring to the requirements of the ship's own national administration, and seeking the approval of the administration of the state of the vessel's next destination or port of call. The master should provide safe working conditions and ensure that at least two members of the crew, including one certificated officer, have received the appropriate training. They should be familiar with the recommendations of the fumigant manufacturer concerning the methods of detection of the fumigant in air, its behaviour and hazardous properties, symptoms of poisoning, relevant first-aid treatment, special medical treatment and emergency procedures.

21.9 Biological agents

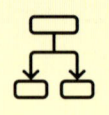

21.9.1 Read the following guidance in conjunction with MSN 1889 (M+F) Amendment 3 on biological agents. Biological agents are classified in groups 1 to 4. These groups are defined in Annex 21.1.

 SI 2010/323; MSN 1889 (M+F) Amendment 3

21.9.2 As well as following the guidance given in 21.8.1, employers must keep a list of personnel exposed to biological agents of group 3 or higher.

21.9.3 Any worker involved with the handling of, or being exposed to, biological agents should have appropriate training and advice.

21.9.4 Before any work takes place, carry out a risk assessment and put procedures in place for any potential accident to minimise its effects.

21.9.5 The most likely areas for contamination by biological agents are as follows:

- food preparation
- contact with animals and/or products of animal origin
- health care
- work with air-conditioning and water-supply systems
- work involving waste disposal and the sewage plant.

21.9.6 Solid carbon dioxide

Solid carbon dioxide (Drikold™, cardice, dry ice) can be used as an emergency refrigerant for preserving deep-frozen food supplies in their hard frozen condition.

21.9.7 Take the following precautions when using solid carbon dioxide:

- Carbon dioxide does not diffuse readily because it is heavier than air. Therefore take special care to test the atmosphere thoroughly and ventilate such compartments/enclosed spaces before entering.
- The door of the compartments/enclosed spaces should remain open while the seafarer is inside the cold compartments/enclosed spaces.
- Always wear gloves when handling solid carbon dioxide to prevent blistering of the skin.

21.9.8 Dry-cleaning operations

The principal hazard of a dry-cleaning solvent is that it is highly volatile, producing chemicals that are harmful if inhaled. Therefore provide effective mechanical ventilation in any compartment containing dry-cleaning plant. Do not allow smoking in compartments when the solvent is present.

Dry-cleaning solvent is also a potential cause of skin damage; wear suitable PPE.

Appoint a competent person to take overall responsibility for the security and operation of the dry-cleaning plant, and control access.

Chapter 15 gives guidance on entering enclosed spaces and the procedures to follow before entry and while inside.

- Be aware of hazardous chemical warning signs and their meanings.
- Carry out tasks with hazardous substances and mixtures only if appropriate training has been given.

- Always review and follow the manufacturer's instructions before working with any hazardous substances or mixtures.
- Always follow instructions based on risk assessments when working on board in areas exposed to hazardous substances and mixtures.
- Wear PPE, correctly as directed.
- If labels are missing or contents are unknown, always identify any container or package before handling it.

Annex 21.1 Classification of carcinogens and biological agents

Classification of carcinogens and mutagens

Tables 21.1 and 21.2 show the classifications of carcinogens and mutagens respectively.

In the case of mutagens, there are three similar categories with analogous descriptors, based on the strength of evidence for heritable genetic damage.

All categories should be treated as hazardous substances or mixtures.

This categorisation is available in Table 3.8.1 of EC no.1272/2008 http://data.europa.eu/eli/reg/2008/1272/oj

Table 21.1   Classification of carcinogens

Carcinogenic category 1	Substances known to cause cancer on the basis of human experience.
Carcinogenic category 2	Substances that it is assumed can cause cancer on the basis of reliable animal evidence.
Carcinogenic category 3	Substances where there is only evidence in animals and it is of doubtful relevance to human health; in other words the evidence is not good enough for categories 1 or 2.

Table 21.2   Classification of mutagens

Mutagenic category 1	Substances known to cause heritable genetic damage on the basis of human experience.
Mutagenic category 2	Substances that it is assumed can cause heritable genetic damage on the basis of reliable animal evidence.
Mutagenic category 3	Substances where there is only evidence in animals and it is of doubtful relevance to human health; in other words the evidence is not good enough for categories 1 or 2.

Classification of biological agents

Table 21.3   Hazard group definitions

The hazard group definitions in Table 21.3 are referred to in the HSE publication *The Approved List of Biological Agents* available at https://www.hse.gov.uk/pubns/misc208.pdf

Group 1	Unlikely to cause human disease.
Group 2	Can cause human disease and may be a hazard to employees; it is unlikely to spread to the community and there is usually effective prophylaxis or treatment available.
Group 3	Can cause severe human disease and may be a serious hazard to employees; it may spread to the community, but there is usually effective prophylaxis or treatment available.
Group 4	Causes severe human disease and is a serious hazard to employees; it is likely to spread to the community and there is usually no effective prophylaxis or treatment available.

22 Boarding arrangements

22.1 Introduction

22.1.1 This chapter provides a general overview of the best practices and general principles that must be followed regarding boarding arrangements.

22.1.2 You must provide safe means of access between the ship and the shore, or another ship alongside, to which the ship is secured. Providing safe access to and from a ship is an integral part of ensuring a safe working environment on board, as required by the Merchant Shipping and Fishing Vessels (Health and Safety at Work) Regulations 1997, as amended, regulation 5(2)l.

22.1.3 If you conform to the principles and guidance in this chapter you will generally demonstrate compliance with the duty to provide a safe working environment on board ship. Where alternative measures are taken to provide a safe means of access, these must be fit for purpose and provide at least an equivalent level of safety in the operating conditions at the time.

SI 1997/2962, as amended

22.2 General principles

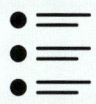

Key points
- When suitable access equipment is provided from the ship, from the shore or from another ship, any person boarding or disembarking from the ship shall use that equipment.
- Gangways and accommodation ladders are considered as lifting equipment. Test and maintain them and keep records available for verification (see Annex 22.2 for more information).
- Rigging equipment should not form a trip hazard. Ships should comply with inspection, testing and maintenance requirements.

22.2.1 To ensure a safe means of access, equipment provided must be:

- placed in position promptly
- properly rigged and deployed
- safe to use and adjusted as necessary to maintain safe conditions for use.

MSC 1/Circ 1331

Your organisation should

- Make arrangements for boarding that are fit for purpose, comply with the appropriate standards in this chapter and are properly maintained in line with section 22.6.
- Provide pilot ladders and accommodation ladders that comply with the construction and testing requirements laid out in SOLAS V, regulation 23 as amended. Annex 22.1 gives guidance on these standards.

22.2.2 The means of access should be:

- inspected by a responsible officer or **competent person** 🔍 (s) to ensure that it is safe to use after rigging
- further checked, under the supervision of a responsible officer or competent person(s), to ensure that adjustments are made when necessary due to tidal movements or change of trim and freeboard.

Keep rigging equipment such as guard ropes and chains taut at all times and keep stanchions rigidly secured.

22.2.3 When access equipment is provided from the shore, it is still the master's responsibility to ensure as far as is reasonably practicable that the equipment meets these requirements.

22.2.4 Ensure any access equipment and immediate approaches to it are adequately lit. For appropriate standards of lighting see Annex 11.2.

22.2.5 Keep the means of boarding and its immediate approaches free from obstruction and, as far as is reasonably practicable, any substances likely to cause a person to slip, trip or fall. Where this is not possible, display appropriate warning notices in the immediate area. If necessary the surfaces should have a suitable non-slip treatment.

22.2.6 Each end of a gangway or accommodation or other ladder should provide safe access to a safe place or to an auxiliary safe access.

SIP 014

22.2.7 Use a portable ladder for access to a ship only when no safer access is reasonably practicable. Use a rope ladder only between a ship with a high freeboard and one with a low freeboard, or between a ship and a boat if no safer means of access is reasonably practicable.

22.2.8 You must provide a lifebuoy with a self-activating light and also a separate buoyant safety line attached to a quoit or some similar device. This must be ready for use at the point of access aboard the ship.

📖 *MGN 533 Amendment 2*

22.3 Safety nets

22.3.1 The ship must carry enough safety nets of a suitable size and strength, or they must otherwise be readily available. Where there is a risk of a person falling from the access equipment, or from the quayside, or ship's deck adjacent to the access equipment, mount a safety net where reasonably practicable.

22.3.2 Safety nets are used to minimise the risk of injury arising from falling between the ship and the quay or falling onto the quay, deck or between two vessels. Cover the whole length of the means of access if possible. Safety nets should be securely rigged, using attachment points on the quayside where appropriate.

📖 *SI 1997/2962, as amended*

22.4 Positioning of boarding equipment

22.4.1 Always keep the angles of inclination of a gangway or accommodation ladder within the limits for which it was designed.

22.4.2 When the inboard end of the gangway rests on or is flush with the top of the bulwark, provide a bulwark ladder. Fence any gap between the bulwark ladder and the gangway adequately to a height of at least 1 metre.

22.4.3 Do not rig gangways or other access equipment on ships' rails unless the rail has been reinforced for that purpose. They should comply with the guidance in Annex 22.1.

22.4.4 Position the means of access so it is clear of the cargo working area and no suspended load passes over it. Where this is not practicable, supervise access at all times.

22.4.5 When an accommodation ladder is being rigged, keep the ladder horizontal so that seafarers working on it can be safely attached with a safety line to the deck. Secure the ladder to reduce any unnecessary movement.

22.4.6 During rigging, stay alert for your safety, fence any gaps in the railing as explained in Annex 22.1 section 2.3, and follow guidance for working at height (see also Chapter 17).

22.5 Portable and rope ladders (excluding pilot ladders)

22.5.1 Where, in exceptional cases, a portable ladder is used to access the ship, a competent person must inspect it regularly, taking account of vessel movement and tide changes.

22.5.2 When it is necessary to use a portable ladder for access, position it at an angle of 75° from the horizontal. The ladder should extend at least 1 metre above the upper landing place unless there are other suitable handholds. Secure it firmly against slipping, shifting sideways or falling and place it to afford a clearance of at least 150 mm behind the rungs.

22.5.3 When a portable ladder is resting against a bulwark or rails, there should be suitable safe access to the deck.

22.5.4 When using a rope ladder to board a vessel, follow the standards that apply to pilot ladders (see sections 22.8, 22.9 and Annex 22.1 section 4). Do not use the pilot ladder for any other purposes.

22.6 Maintenance of equipment for means of access

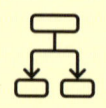

22.6.1 A competent person should inspect any equipment used for boarding or hoisting boarding equipment, including lifting wires at appropriate intervals, and they should keep it properly maintained. Renew parts in accordance with the manufacturer's instructions. Make additional checks each time the equipment is rigged, looking out for signs of distortion, cracks or corrosion. During inspections check the welding connections in particular.

SOLAS II.1/3-9; MSC 1/Circ 1331

22.6.2 Arrange to periodically inspect the underside of ladders and gangways, including the turntable. Report any defects affecting the safety of any access equipment, including access provided by a shore authority immediately to a responsible person and repair them before further use.

22.6.3 Examine aluminium equipment for corrosion and fracture following the instructions in Annex 22.2.

22.6.4 Record all inspections, maintenance work and repairs. The record should include the date of the most recent inspection, the name of the person or body doing the inspection, the due date for the next inspection and the dates for renewal of wires for supporting the equipment.

22.6.5 Test gangways, accommodation ladders and winches used for lifting or access in the same way as all other lifting appliances and keep records, including any test certificates.

22.7 Special circumstances

22.7.1 In some circumstances, it may not be practical to mount proper safe boarding arrangements by conventional means. Examples include where there is frequent movement of the ship during cargo operations, or where access is required between the ship and an offshore structure. On such occasions, supervise boarding carefully and consider providing alternative means of access.

22.7.2 Further guidance on safe access to offshore structures is given in Chapter 31.

22.7.3 Small boats or tenders used between the shore and the ship should be safe and stable for the expected conditions, suitably powered, correctly operated, properly equipped with the necessary safety equipment and, if not a ship's boat, approved for that purpose.

22.7.4 Where a vessel is moored alongside another vessel, there should be agreement between the two vessels to provide suitable and safe boarding arrangements. Generally the ship lying outboard should provide the access, but where there is a great disparity in freeboard, the ship with the higher freeboard should provide it.

22.7.5 Take care at all times, particularly at night, when boarding or disembarking from a ship, or when moving through the dock area. Avoid the edges of the docks, quays, etc. and strictly follow any sign prohibiting entry to an area. Where there are designated routes follow them exactly. This is particularly important near container terminals or other areas where rail traffic, straddle carriers or other mechanical handling equipment is operating, because the operators of such equipment have restricted visibility, placing anyone walking within the working area at risk.

22.7.6 Transfer of personnel between two unsecured ships at sea is potentially a particularly dangerous manoeuvre, so avoid it where possible. Where it is unavoidable:

- Carry out a risk assessment of the transfer arrangements and put appropriate safety measures in place to ensure the safety of those involved.
- Equip both vessels properly and/or modify them to allow the boarding to take place without unnecessary risk.
- Provide a proper embarkation point and clearly agree on the boarding procedure.
- Consider the relative movements of both vessels in any seaway and the varying sea, tide and swell conditions when deciding whether to do a transfer.
- The master responsible for the transfer operation should have full sight of the area of transfer and, with at least one designated crew member, be able to communicate at all times with the crew member making the transfer.
- Vessels doing ship-to-ship transfers while under way should carry equipment designed to aid in the rapid recovery of a casualty from the waters.

SI 2002/1473; SI 2020/0673; MGN 432 Amendment 1

22.7.7 Crew should wear working lifejackets when there is a risk of falling into the water during a transfer to a vessel or structure that is not alongside. Baggage and other items should be transferred by the crews of the vessels and not by those boarding.

22.8 Access for pilots

22.8.1 Pilot ladders and arrangements should comply with the appropriate standards for design, construction and testing.

SOLAS V/23; SI 2020/0673; MSN 1874 (M+F) Amendment 7; Resolution A.1045(27); Resolution A.1108(29); MSC 1/Circ 1428; BS ISO 799-1:2019; BS ISO 799-3:2022

22.8.2 The master must ensure the following:

- All pilot ladders used for pilot transfer are clearly identified with tags or other permanent marking. This enables identification of each appliance for the purposes of survey, inspection and record keeping. Keep a record on the ship of the date the identified ladder is placed into service and make any repairs.
- Each pilot ladder, accommodation ladder and their associated equipment are properly maintained and stowed, and regularly inspected to ensure that, so far as is reasonably practicable, each is safe to use.
- Each pilot ladder is used only for the embarkation and disembarkation of pilots and by officials and other people while a ship is arriving at or leaving a port.
- The rigging of the pilot ladder, accommodation ladder and associated equipment is supervised by a responsible officer who is in communication with the navigating bridge. This officer's duties will include arranging for the pilot to be escorted by a safe route to and from the bridge. Advice on safe rigging of such equipment is given in section 22.9.
- Personnel engaged in rigging or operating any mechanical equipment are instructed in the safe procedures to be adopted and that the equipment is tested before each use.

22.8.3 Always keep at hand and ready for use at the point of boarding a safety line and harness for seafarers rigging the pilot ladder, a lifebuoy with a self-igniting light, and a heaving line.

22.8.4 Ensure that there is adequate lighting for the pilot ladder, accommodation ladder and the position where the person embarks and disembarks on the ship.

22.8.5 It is very important that the ship offers a proper lee to the pilot boat. Site the arrangements for boarding as near the middle of the ship as possible. However, they should never be in a position that could lead to the pilot boat running the risk of passing underneath overhanging parts of the ship's hull structure. For further information see Marine Guidance Note MGN 301 (M+F) Amendment 1.

MGN 301 (M+F) Amendment 1

22.9 Safe rigging of pilot ladders

22.9.1 In addition to the general points in section 22.2, to minimise the danger to pilots when embarking on and disembarking from ships, pay particular attention to the following:

- Rig pilot ladders so that the steps are horizontal and the lower end is at a height above the water to allow ease of access to and from the attendant craft, as shown in Figure 22.1.
- The ladder should rest firmly against the side of the ship.
- When an accommodation ladder is used with a pilot ladder, the pilot ladder should extend at least 2 metres above the bottom platform.
- Provide safe, convenient and unobstructed access to anyone embarking or disembarking between the ship and the head of the pilot ladder.
- A lifebuoy with self-igniting light should always be available at the point of access to the ship.
- At night, the pilot ladder and ship's deck should be lit by a forward-shining, overside light.
- Secure the top of the pilot ladder to the approved strong fixing point on deck, not to handrails. Do not rig ladder steps or spacers in a position where they are taking the weight of the ladder.
- Ensure that steps and spacers do not become entrapped or twisted.

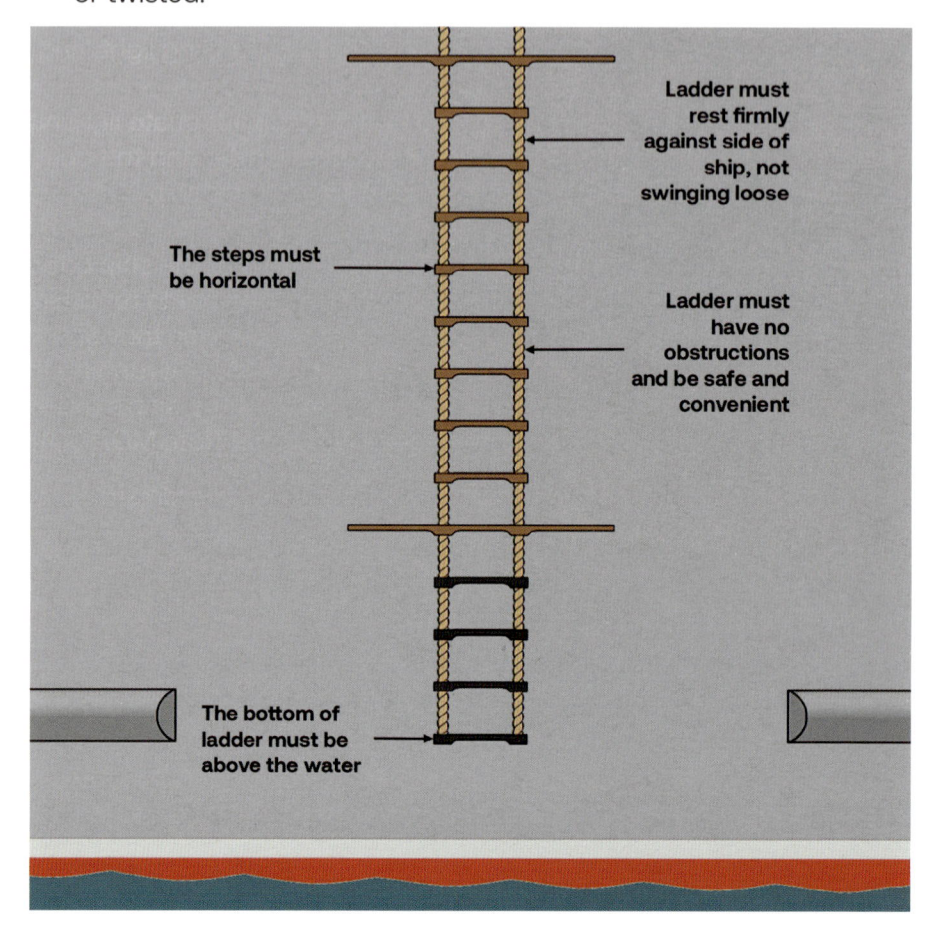

Figure 22.1 Example of safe rigging for a pilot ladder

See the 'Required boarding arrangements for pilot' diagram on the International Maritime Pilots' Association website, which is listed in Appendix 2.

22.10 Safe access to small craft

22.10.1 Ports and harbours may not have areas specifically designed for safe access to and from small vessels. When determining how to provide access consider the options below, starting with gangways before moving to the next level. Identify the most suitable means of access by risk assessment, considering which safety measures are required.

22.10.2 All these methods for gaining access to small craft can be used safely, provided you take appropriate safety measures.

22.10.3 The industry's recommended hierarchy of access arrangements for small craft, starting with the safest first, is:

- a gangway between small craft and the quay, quay steps, quay wall, pier or other vessel/small craft
- stepping directly (short step, level access) between the small craft and the quay, quay steps, quay wall, pier, other vessel/small craft or pontoon
- fixed ladder from the quay, quay wall, pier or jetty
- portable ladder between the small craft and the quay, quay wall, pier or jetty.

SIP 021

Assess ports and harbours for suitability for access to and from vessels.

Carry out a risk assessment for boarding arrangements.

Annex 22.1 Standards for means of access

📖 *BS ISO 5488:2015; ISO 7061:2015; MSC 1/Circ 1331*

1. General

1.1 Accommodation ladders and gangways should comply with appropriate international standards such as BS ISO 5488:2015 Ships and marine technology – Accommodation ladders and ISO 7061:2015 Ships and marine technology – Aluminium shore gangways for seagoing vessels.

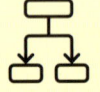

📖 *BS ISO 5488:2015*

1.2 The structure of accommodation ladders and gangways and their fittings should allow regular inspection and maintenance of all parts and, where necessary, lubrication of their pivot pin. Each accommodation ladder or gangway should be clearly marked at each end with a plate showing any restrictions on safe operation or loading, including minimum and maximum permitted design angles or inclination, design load and maximum load on the bottom end plate. Where the maximum operating load is less than the design load, that should also be shown on the marking plate.

1.3 Gangways should be carried on ships of 30 metres in length or over and accommodation ladders must be carried on ships of 120 metres in length or over, complying with the specifications in section 2 of this annex. Access equipment must be of good construction, sound material and adequate strength, free from patent defect and properly maintained. Rope ladders must comply with the requirements in section 4 of this annex.

1.4 Gangways and accommodation ladders must be clearly marked with the manufacturer's name, the model number, the maximum designed angle of use and the maximum safe loading, both by numbers of persons and by total weight.

2. Gangways

2.1 Gangways must comply with the specifications set out in standard BS MA 78:1978 or equivalent, and should be fitted with suitable fencing along their entire length.

BS MA 78:1978

2.2 Do not use gangways at an angle of more than 30° from the horizontal, unless designed and constructed for use at greater angles.

2.3 Do not fix gangways to the ship's railings unless the railings are designed for that purpose. If gangways are rigged in an open section in the ship's bulwark or railings, any remaining gaps should be adequately fenced.

3. Accommodation ladders

3.1 An accommodation ladder should be designed so that:

- it rests firmly against the side of the ship where practicable
- the angle of slope is no more than 55°. Treads and steps should provide a safe foothold at the angle at which the ladder is used
- it is fitted with suitable fencing (preferably rigid handrails) along its entire length, except that fencing at the bottom platform may allow access from the outboard side
- at a maximum inclination, the lowest platform of the ladder is no more than 600 mm above the waterline in the lightest seagoing condition, as defined in SOLAS III/3.13
- the bottom platform is horizontal, and any intermediate platforms are self-levelling
- it provides direct access between the head of the ladder and the ship's deck by a platform that is securely guarded with guardrails and adequate handholds
- it can easily be inspected and maintained
- it is rigged as close to the working area as possible but is clear of any cargo operations.

BS MA 39, Part 2: 1973; SOLAS III/3.13

3.2 After installation, test the winch and ladder operationally to confirm proper operation and condition of the winch and ladder after the test. This test should include raising and lowering the accommodation ladder at least twice (in accordance with ISO 7364:2016). Keep records, including any test certificates.

ISO 7364:2016

3.3 When a bulwark ladder is to be used, it must comply with the specifications set out in the Shipbuilding Industry Standard No. SIS 7 or BS MA 39, Part 2:1973 Specification for ships' ladders, or be of an equivalent standard. Provide adequate fittings to enable the bulwark ladder to be properly and safely secured.

4. Pilot ladders

See SOLAS V 23; SI 2020/0673; MSN 1874 (M+F) Amendment 7; IMO Resolution A.1045(27); IMO Resolution A.1108(29); MSC 1/Circ 1428; BS ISO 799-1:2019; BS ISO 799-3-2022

4.1 A rope ladder must be of adequate width and length and so constructed that it can be efficiently secured to the ship:

- The steps must provide a slip-resistant foothold of not less than 400 mm × 115 mm × 25 mm and must be so secured that they are firmly held against twist, turnover or tilt.
- The steps must be horizontal and equally spaced at intervals of 310 mm (± 5 mm).
- The side ropes, which should be a minimum of 18 mm in diameter, should be equally spaced.
- There should be no shackles, knots or splices between rungs.
- Ladders of more than 1.5 metres in length must be fitted with spreaders not less than 1.8 metres long. The lowest spreader must be on the fifth step from the bottom and the interval between spreaders must not exceed nine steps. The spreaders should not be lashed between steps.

4.2 New or replacement pilot ladders installed on or after 1 July 2012 should be certified by the manufacturer as being compliant with international standards. Merchant Shipping Notice MSN 1874 (M+F) Amendment 7 provides more information. A pilot ladder conforming to BS ISO 799-1:2019 is acceptable, provided that it meets the regulation requirements.

4.3 In addition to the standards above, every pilot ladder should be positioned and secured so that:

- it is clear of any possible discharges from the ship
- it is, where practicable, within the mid-ship half-section of the ship (but see section 22.8.4)
- it can rest firmly against the ship's side
- the person climbing it can safely and conveniently board the ship after climbing no more than 9 metres.

4.4 Where replacement steps are fitted, they should be provided by the manufacturer and secured in position by the method used in the original construction of the ladder. No pilot ladder should have more than two replacement steps secured in position by a different method. Where a replacement step is secured by means of grooves in the sides of the step, such grooves should be in the longer sides of the step.

4.5 Provide two man-ropes of not less than 28 mm in diameter, properly secured to the ship.

4.6 Where access to the ship is by a gateway in the rails or bulkhead, provide adequate handholds. Shipside doors used for this purpose should not open outwards.

4.7 Where access is by bulwark ladder, securely attach the ladder to the bulwark rail or landing platform. Provide two handhold stanchions, between 700 mm and 800 mm apart, each of which should be rigidly secured to the ship's structure at or near its base and at another higher point. The stanchions should be not less than 32 mm in diameter and extend no less than 1.2 metres above the deck to which it is fitted.

4.8 Where the freeboard of the ship is more than 9 metres, provide a combination of accommodation and pilot ladders on each side of the ship.

4.9 Such accommodation ladders should comply with the standards in paragraph 2.1 of this annex, and in addition:

- The pilot ladder should extend at least 2 metres above the accommodation ladder's bottom platform.
- The bottom platform should be in a horizontal position, at least 5 metres above sea level and secured to the ship's side when in use.
- If a trap door is fitted in the bottom platform to allow access to the pilot ladder, the opening should be no less than 750 mm square, open upwards and be secured flat on the platform or against the rails. The remainder of the bottom platform should be fenced, as should be the rest of the ladder. In this case, the pilot ladder should extend above the lower platform to the height of the handrail and remain in alignment with and against the ship's side.

Annex 22.2 Corrosion and fractures of accommodation ladders and gangways

Figure 22.2 provides a checklist of things to look out for when examining ladders and gangways. In particular be aware of the following:

- Aluminium alloys are highly susceptible to galvanic corrosion in a marine atmosphere if they are used in association with dissimilar metals. Take care when connecting mild steel fittings, whether or not they are galvanised, to accommodation ladders and gangways constructed of aluminium.
- Use plugs and joints of neoprene or other suitable material between mild steel fittings, washers, etc. and aluminium. The plugs or joints should be significantly larger than the fittings or washers.
- Make only temporary repairs using unsuitable materials such as mild steel doublers or bolts made of mild steel or brass. Make permanent repairs, or replace the means of access, at the earliest opportunity.
- The manufacturer's instructions should give guidance on examination and testing of the equipment. However, it is difficult to examine certain parts of accommodation ladders and gangways closely because of their fittings and attachments.
- Aluminium welds are susceptible to fracture. Where there are fractures repair them at the earliest opportunity.

It is essential, therefore, to remove the fittings periodically to thoroughly examine the parts most likely to be corroded. Turn over accommodation ladders and gangways to examine the underside carefully.

Pay particular attention to the immediate perimeter of the fittings. Test this area for corrosion with a wire probe or scribe. Where the corrosion appears to have reduced the thickness of the parent metal to 3 mm, fit back plates inside the stringers of the accommodation ladder or gangways.

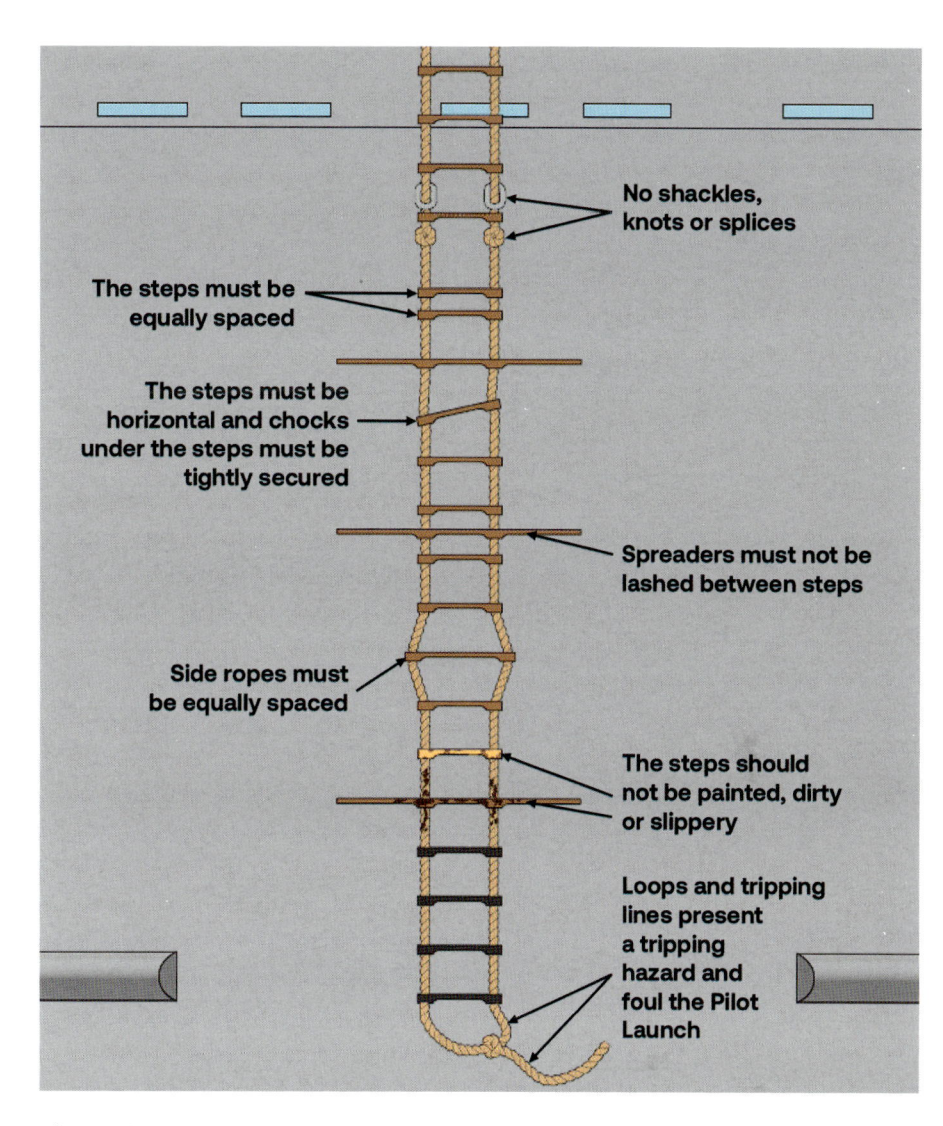

Figure 22.2 Checklist for safe means of access

23 Food preparation and handling in the catering department

23.1 Introduction

23.1.1 Catering staff should have proper training in food safety and personal hygiene, as they are responsible for ensuring that high standards of personal hygiene and cleanliness are maintained at all times while working in the galley, pantry and mess rooms. Only qualified catering staff can handle and prepare food on board ship. See MSN 1846 (M) Amendment 1 about training requirements and MSN 1845 (M) Amendment 1 for further guidance on provision of food and water.

23.1.2 If food has been prepared by an outside caterer they will have followed the equivalent shoreside precautions.

    MSN 1845 (M) Amendment 1; MSN 1846 (M) Amendment 1

For more information on topics covered in this chapter, see MCA's Wellbeing at Sea: A Guide for Organisations, section 3.8.1, and Wellbeing at Sea: A Pocket Guide for Seafarers, sections 1.1 and 1.2.

23.1.3 ### Health and hygiene

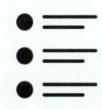

Key points
- Catering staff must be fully aware of how important their personal health and hygiene are when handling foods or liquids.
- Use the correct equipment for specific tasks when processing food and liquids and clean it according to safety requirements.
- Store, handle and process foods and liquids in line with the instructions for the type of food or liquid.
- Catering staff must wear clean protective clothing with appropriate protective gloves, if required, when handling food and preparing meals (Figure 23.1 shows some typical examples.)
- Safety and warning notices must be clearly visible and the working environment kept clean to prevent injuries from equipment, slips, trips and falls (see section 9.1).

Your organisation should

- Ensure that catering staff responsible for the preparation and processing of foods and liquids on board ships are appropriately qualified in food safety and food hygiene.
- Provide all catering staff with correct clothing and protective equipment.
- Provide the correct galley equipment and facilities for catering staff to process foods and liquids.
- Ensure safety signage in the galley is clearly visible and that appropriate fire safety equipment is accessible.
- Apply anti-slip surfaces on deck areas immediately outside the entrance to refrigerated rooms.

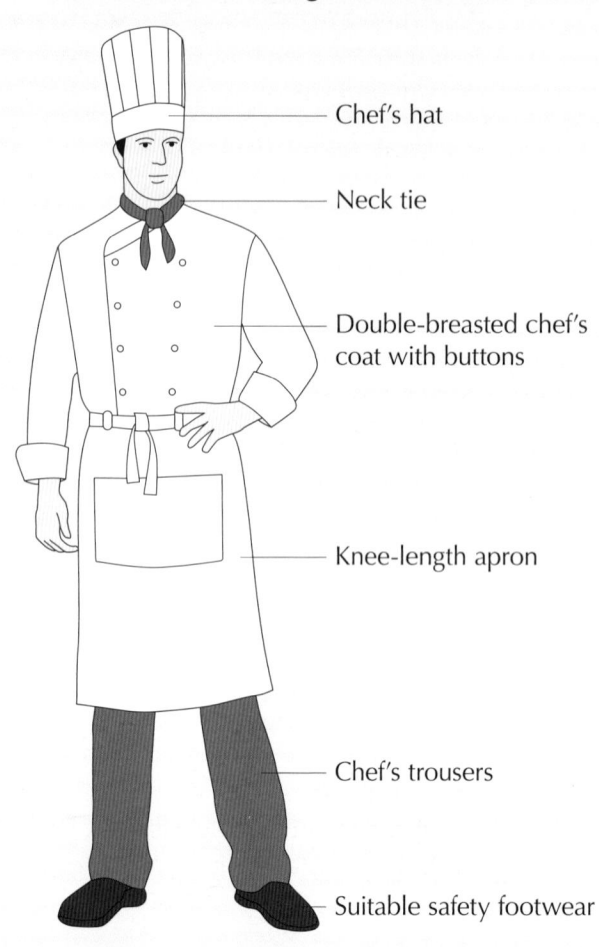

Figure 23.1 Examples of protective clothing for catering staff

23.2 Personal preparation before handling food or liquids

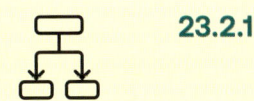

23.2.1
- Wash your hands and fingernails using a dedicated hand basin, antibacterial soap from a dispenser and disposable towels before handling food and liquids, after using the toilet, blowing your nose, or handling refuse or contaminated food (see Figure 23.2).
- Use alcohol gel only to supplement hand washing with soap and water, not as a replacement.
- Do not wear jewellery (except for a plain wedding band).
- Always wear clean protective clothing, including appropriate protective gloves, if necessary, when handling food and preparing meals.
- Do not smoke or use an e-cigarette in galleys, pantries, storerooms or other places where food is prepared or stored (see section 9.3.5). Smoke or use an e-cigarette only in designated areas.

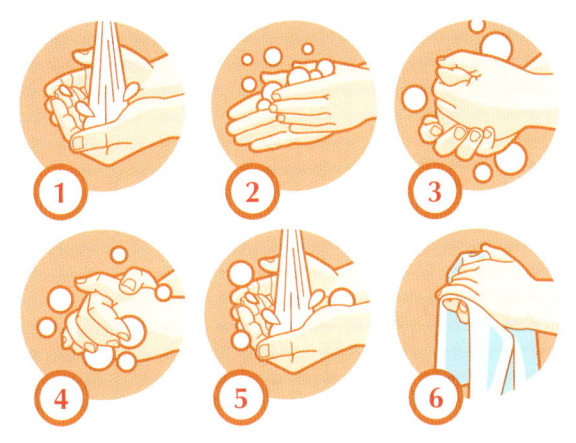

1) Wet your hands with clean running water, then apply soap.
2) Rub your hands together to lather the soap.
3 & 4) Scrub the soap around your hands, making sure you get under the nails and cover the palms and backs of your hands. This should take at least 20 seconds.
5) Rinse your hands with running water to wash the soap off.
6) Dry your hands with a clean, dry towel or dryer.

Figure 23.2 How to wash your hands

23.2.2 The cleanliness of all food, crockery, cutlery, linen, utensils, equipment and storage is vital. Do not use cracked or chipped crockery or glassware. Throw away any foodstuffs that may have come into contact with broken glass or broken crockery.

23.2.3 Personal healthcare

- Report all cuts, no matter how small, for first aid attention to prevent infection.
- Cover any cut, burn or abrasion with a blue waterproof dressing and change it regularly.
- Report any illnesses, coughs and colds, rashes or spots, however mild, immediately when the symptoms appear.
- Anyone with any type of infection, septic cut, boil, stye or anything requiring regular first aid treatment should stop working with food until it has completely healed.

23.2.4 A person suffering from diarrhoea and/or vomiting, which may be signs of food poisoning or a sickness bug, should not work in food-handling areas until medical clearance has been given.

23.3 Preparation and handling of food and liquids

- Keep all food and liquid at the correct temperature to prevent the multiplication of bacteria.
- Wash fresh fruit and salad thoroughly in fresh water before serving.
- Defrost frozen food in cool, controlled conditions, such as an area separate from other foods.
- Place food to be defrosted on grids in a container or on a shelf and do not allow it to sit in the thaw liquid. Do not refreeze food that has been defrosted.

23.3.1 Do not cross-contaminate foods or liquids.

23.3.2 Eliminate the risks of **cross-contamination** by thoroughly taking apart and cleaning the relevant parts of equipment each time different foods are used (especially raw and cooked foods). Always wash your hands after handling raw meat, fish, poultry or vegetables.

23.3.3 Keep raw food apart from cooked food or food that requires no further treatment before eating or drinking (eg milk). It is better to use separate refrigerators. If raw and cooked foods have to be stored in the same fridge always place the raw food at the bottom to avoid drips contaminating ready-prepared food. Cover food to prevent drying out, cross-contamination and absorption of odour.

23.3.4 Set aside separate work surfaces, chopping boards and utensils to prepare raw meat. Do not use them to prepare foods that will be eaten without further cooking. Colour coding is an established way of ensuring separation between the two activities.

23.3.5 Do not leave crockery or glassware in washing-up water where it may easily be broken and cause injury. Wash knives and any utensils or implements with sharp edges individually. It is better to wash crockery, glassware and utensils in a dishwasher, which will wash at higher temperatures than hand washing.

23.4 Cleaning surfaces and equipment

23.4.1 Some domestic cleaning substances contain bleach (sodium hypochlorite) or caustic soda (sodium hydroxide) while some disinfectants contain carbolic acid (phenol). These can burn the skin and they are poisonous if swallowed. Treat them with caution and do not mix them together or use them at more than the recommended strength.

23.4.2 If you accidentally come into contact with toxic chemicals or other harmful substances report this immediately so you can get appropriate medical care. Store cleaning substances and materials in a suitable locker or cupboard away from food-handling areas. Use cleaning products that are not injurious to individuals or the environment when possible.

23.5 Waste disposal

23.5.1 Food waste, empty food containers and other garbage are major sources of pollution and disease. Place them in proper covered storage facilities safely away from foodstuffs. It is prohibited to throw them into the sea (with limited exceptions; see MGN 632 (M+F).

MGN 632 (M+F) Amendment 1; SI 2020/621

23.6 Slips, falls and tripping hazards

23.6.1 Wear suitable footwear, with slip-resistant soles, at all times. Many catering staff suffer injuries because they wear sandals, plimsolls or flip-flops, which do not grip greasy decks or protect the feet from injury, burns or scalds if hot or boiling liquids are spilt.

23.6.2 **Keeping work areas clean and clear**

- Keep decks and gratings clear from grease, rubbish, ice and anything that might be on the floor, to avoid slipping.
- Clean up any spillages immediately.
- Always use a brush and dustpan to clear up broken glass or crockery; never use your hands.

23.6.3 **Moving about the galley or decks**

- Take care when you use stairs and companionways; always keep one hand free to grasp the handrail.
- Carry any items correctly so you can see sills, storm steps or any other obstructions.
- Avoid reaching for objects that are too high or too low.
- Do not stand on any unsecured objects to get something that is out of reach.

23.7 Galley stoves, steam boilers and deep fat fryers

23.7.1 If your ship has an oil-fired stove you should operate safety procedures according to the manufacturer's instructions, particularly when lighting the stove. Display the instructions clearly in the galley.

23.7.2 Catering staff should not try to repair electric or oil-fired ranges or electric microwave ovens. Always report defects so that proper repairs may be made. Keep broken equipment out of use and display a warning notice until it has been repaired.

Risk of electric shock, hot liquid spills and burns

23.7.3 It is very dangerous to use too much water when hosing down and washing equipment in the galley, particularly when there are electrical installations, as there is a danger of electric shock.

> **23.7.4 Warning**
> When washing down the galley deck switch off the power and isolate the supply to all electric equipment. Keep water away from the equipment.

23.7.5 Always use range guardrails in adverse weather. Never fill pots and pans so full that the contents spill over when the ship rolls.

23.7.6 Instruct all catering staff in how to avoid burns from hot surfaces, serving tables, bains-marie, steamers and tilting pans.

23.7.7 When handling hot pans and dishes always use dry cloths, pot holders or heatproof oven gloves that are long enough to cover the arms. Wet cloths conduct heat quickly and may scald the hands.

23.7.8 Do not stand directly in front of an oven when the door is opened – the initial heat blast can cause burns.

23.7.9 Turn off the steam supply to pressure cookers, steamers and boilers and release the pressure before opening their lids.

23.8 Liquid petroleum gas appliances

23.8.1 Gas leak detection alarms should be provided and securely fixed in the lower part of the galley, as gas is heavier than air. A gas detector should have both an audible and a visible alarm, and these should be tested frequently. A suitable notice, detailing the action to take when an alarm is given by the gas detection system, should be clearly displayed and include the following information:

- Shut off the gas supply to the appliance.
- Open doors and hatches to ventilate the area.
- Do not use switches or create flames which could spark and ignite any remaining vapour.
- If the alarm continues to sound, follow onboard fire safety procedures.

📖 *MGN 280 (M), Annex 5, section 8 Emergency Action*

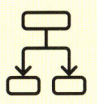

23.8.2 Equipment should be fitted, where practicable, with an automatic gas shut-off device which operates in the event of flame failure.

23.8.3 When gas burning appliances are not in use, turn off the controls. If they are not going to be used again for some time shut the main regulators close to the storage bottles. See Marine Guidance Note MGN 280 (M) Annex 5 for further information on the safe operation of liquid petroleum gas appliances.

📖 *MGN 280 (M) Annex 5*

23.8.4 Establish a safe system of working, training and supervision for lighting and operating procedures.

23.8.5 If you can smell gas check for a defect in a joint, valve or connection. Catering staff should not attempt to repair electric, oil or gas appliances.

23.9 Deep fat frying

23.9.1 **Warning**
Never pour water into hot oil; the water turns to steam, throwing the oil in multiple directions. This may cause severe burns to personnel, and possibly start a fire.

23.9.2 Oil fires

- If hot oil catches fire, smother the flames with a dedicated fire blanket or suitable extinguisher. Never use water on the fire.
- Only remove the container from the heat source if there is no risk to personal safety.
- Report the fire as you have been trained and instructed.

23.9.3 The manufacturer's safety data sheet (SDS) will tell you the flash point of the cooking medium. Use the thermostat(s) to monitor the temperature to avoid reaching the flash point.

23.9.4 Deep fat fryers should have suitable safety lids. Keep these in position when the fryers are not in use.

23.9.5 To minimise the risk of fire from failure of the control thermostat, all deep fat fryers should have both a primary and a backup thermostat, with an alarm to alert the operator if either thermostat fails. Maintain the thermostats and check them in line with the manufacturer's instructions.

23.9.6 Switch off electrically operated deep fat fryers immediately after use. Make arrangements to automatically shut off the electrical power when the fire extinguishing system is activated.

23.9.7 Establish a safe system of work for cleaning and draining fat fryers.

23.9.8 Set up a strict schedule of cleaning for galley uptakes/grills so that fat deposits cannot accumulate.

23.9.9 Display a notice clearly, explaining what to do if there is a deep fat fryer fire.

23.10 Microwave ovens

23.10.1 When using microwave ovens:

- Cook food thoroughly and evenly. This is particularly important with deep frozen foods; defrost these thoroughly before cooking.
- Follow the manufacturer's instructions carefully, along with the information on the packaging of the foodstuff.
- Check that there is no damage to the microwave oven's door, its seals or that its interlock is out of use.
- Check microwave radiation levels regularly.
- Display a notice of each microwave oven's last safety check.

23.11 Catering equipment

23.11.1 Warning
Do not use electrical equipment with wet hands.

23.11.2 No one should use catering equipment unless they are supervised by a **competent person** 🔍 or have been trained in its use and fully instructed in the precautions to take.

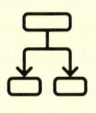

23.11.3 Guard the dangerous parts of catering machines properly and keep the guards in position when using the machine.

23.11.4 Inspect machines and equipment routinely for faults, wear and tear, damage or defective parts. Report any machine or piece of equipment that has faulty or defective parts, guards or safety devices. Take it out of service and disconnect the power until it has been repaired.

23.11.5 When cleaning a power-operated machine or removing a blockage switch it off and **isolate** 🔍 it from the power supply. Some machines will continue to run down after switching off, so make sure all dangerous parts have stopped before starting cleaning.

23.11.6 Establish a safe procedure for cleaning all machines and follow this carefully. Take every precaution where cutting edges (eg on slicing machines) are exposed because guards need to be removed for thorough cleaning. Replace the guards properly and securely as soon as the job is done.

23.11.7 Unless they are properly supervised, a seafarer under 18 years of age should not clean any power-operated or manually driven machine with dangerous parts that may move during the cleaning operation.

23.11.8 Use appropriate kitchen tools, not fingers, to feed materials into processing machines.

23.11.9 A competent person should inspect all electrical equipment regularly.

23.12 Knives, meat saws, choppers, etc.

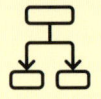

23.12.1 Handle sharp implements with care at all times and treat them with respect. Do not leave them lying around working areas where someone may accidentally cut themselves. Do not mix them with other items for washing up but clean them individually and store them in a safe place.

23.12.2 Keep knives tidily in secure racks or sheaths when not in use.

23.12.3 Securely fix the handles of knives, meat saws, choppers, etc. and keep them clean and free from grease. Keep the cutting edges clean and sharp.

23.12.4 Use proper can openers in clean condition to open cans. Using other implements is dangerous and may leave jagged edges on the can.

23.12.5 Always pay full attention when you are chopping meat. The chopping block must be firm, the cutting area of the meat well on the block and hands and your body clear of the line of strike. There must be adequate room for movement and nothing in the way of the cutting stroke. Take particular care when the vessel is moving in a seaway. Wear appropriate gloves for use when cutting meat.

23.12.6 When you are chopping a foodstuff with a knife do not feed it towards the blade with outstretched fingers. Tuck your fingertips on the free hand in towards the palm of your hand with your forefinger overlapping your thumb, as shown in Figure 23.3. Angle the knife blade away from the work, using the knuckles as a guide to keep the blade away from the fingers.

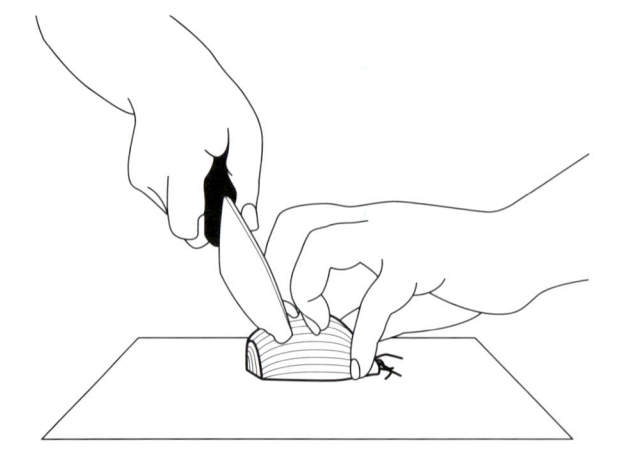

Figure 23.3   How to chop with a knife

23.12.7		Leave a falling knife to fall; do not try to catch it.
23.12.8		Guide a meat saw by putting the forefinger of your free hand over the top of the blade. The use of firm, even strokes will allow the blade to feel its way. If you force the saw it may jump, possibly causing injury.

23.13 Refrigerated rooms and store rooms

23.13.1 Fit all refrigerated room doors with a means of opening the door from both sides. It should be possible to sound an alarm from inside the room.

23.13.2 Test the alarm bell and check the door clasps and inside release regularly, at least once a week.

23.13.3 Workers using the refrigerated room should know how to release the door from the inside in darkness and how to sound the alarm.

23.13.4 All refrigerated room doors should have a strong enough means to hold the door open in a seaway. Doors should be secured open while stores are being handled. These doors are extremely heavy and can cause serious injury to a person caught between the door and the jamb.

23.13.5 Anyone going into a refrigerated room should take the padlock, if any, inside with them, and should tell another person they are going in.

23.13.6 Do not enter cold stores or refrigerated rooms if it is suspected that there has been a leakage of refrigerant. Post a warning notice to this effect outside the doors.

23.13.7 Stow all stores and crates securely so that they do not shift or move in a seaway.

23.13.8 When opening wooden boxes or crates remove any fastenings that stick out, or make them safe.

23.13.9 Stow metal meat hooks in a special container when not in use. Where you cannot easily remove the hooks keep them away from passageways or areas where people are working.

23.13.10 When entering meat and fish storage rooms use appropriate thermal personal protective equipment.

- Be aware of your surroundings and keep areas clean to prevent slips, trips and falls.
- Be familiar with warning notices and procedures so you are prepared in the event of fires and be aware of any potential hazards when using equipment.

- Wash your hands frequently and between handling different types of foods or liquids to prevent cross-contamination.
- Wear clean personal protective clothing and appropriate gloves as instructed.
- Report injuries or incidents for appropriate medical treatment.
- Follow instructions for the handling and processing of foods and liquids.
- Switch off the mains electricity supply to equipment when washing equipment down.

24 Hot work

24.1 Introduction

24.1.1 Based on the hazards and findings identified from the risk assessment, companies should put appropriate control measures in place to protect seafarers who may be affected. This chapter identifies areas to consider in respect of hot work.

Key points
- Hot work in places other than the workshop should have a permit to work (see Chapter 14).
- Operators should be trained in the process, familiar with their equipment and instructed in any special precautions.
- Before starting welding, flame cutting, angle grinding or other hot work, check that there are no combustible solids, liquids or gases at, below or adjacent to the work area that might be ignited by heat or sparks from the work.
- Never work on surfaces covered with grease, oil or other flammable or combustible materials. Move combustible materials and dunnage to a safe distance before starting work. Keep these places free of materials that could release a flammable substance if disturbed.
- Keep suitable fire extinguishers ready for use during the operation. Station a person with an extinguisher to keep watch on affected areas not visible to the seafarer doing the work.
- Because of the risk of delayed fires resulting from the use of burning or welding apparatus, check frequently for at least two hours after the work has stopped.
- When doing hot work near open hatches, put up suitable screens to prevent sparks dropping down hatchways or hold ventilators.
- Close port holes and other openings through which sparks may fall where practicable.
- When doing work close to or at bulkheads, decks or deckheads, check the far side of the divisions for materials and substances that may ignite, and for cables, pipelines or other services that may be affected by the heat.

Your organisation should

- ensure cargo tanks, fuel tanks, cargo holds, pipelines, pumps and other spaces that have contained flammable substances are certified free of flammable gases before any repair work is started
- ensure testing includes adjacent spaces, double bottoms and cofferdams according to risk assessments. Further tests should be carried out at regular intervals and before hot work is recommenced following any suspension of the work
- be aware that when doing hot work on tankers and similar ships all tanks, cargo pumps and pipelines should be thoroughly cleaned. Care should be taken with the draining and cleaning of pipelines that cannot be directly flushed using the ship's pumps
- provide adequate lighting where needed; this should be clamped or otherwise secured in position (not handheld) with leads kept clear of the working area
- ensure that hot work is properly supervised and kept under regular observation.

24.2 Gas cutting

24.2.1 Galvanising paint and other protective materials can produce harmful fumes.

24.2.2 Warning
The use of gas-cutting equipment can deplete oxygen in the atmosphere and produce noxious gases.

24.2.3 Provide adequate ventilation for welding and flame cutting. Check the effectiveness of the ventilation at intervals while the work is in progress. If appropriate consider using local exhaust ventilation (LEV) and breathing apparatus.

24.3 Welding

24.3.1 Exposure to welding fumes (including from mild steel) may cause lung and kidney cancers, and oxygen in the atmosphere can be depleted when welding. Provide LEV, even for welding in the open. If LEV fails to prevent exposure to the fumes, workers should wear adequate and suitable respiratory protection equipment (RPE).

BS EN 12941/2:1998+A2:2008

24.3.2 Annex 24.1 suggests procedures for lighting up and shutting down.

24.4 Personal protective equipment

24.4.1 The operator and their assistants must wear personal protective equipment (PPE) that complies with the relevant standard specifications or their equivalent. This is to protect them from particles of hot metal and slag, and to protect their eyes and skin from ultra-violet and heat radiation. Where workers wear RPE to prevent exposure to welding fumes, it must be compatible with any other PPE that they wear at the same time.

BS EN ISO11611:2015

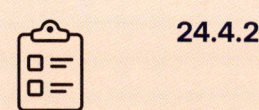

24.4.2 The operator should normally wear the PPE shown in Figure 24.1:

- welding shields or goggles with the appropriate shade of filter lens to EN 169 (goggles are recommended only for gas welding and flame cutting)
- leather gauntlets
- a leather apron (in appropriate circumstances)
- where necessary to avoid exposure to welding fumes, RPE for toxic atmospheres
- a long-sleeved natural-fibre boiler suit or other approved protective clothing.

BS EN ISO 16321-1:2022; ISO 16321-2; ISO 16321-3; BS EN 1146

Safety hat with ear protectors

Flame-retardant, long sleeved natural fibre clothing – which is free from grease, oil and other flammable substances

Leather gauntlets

Leather apron

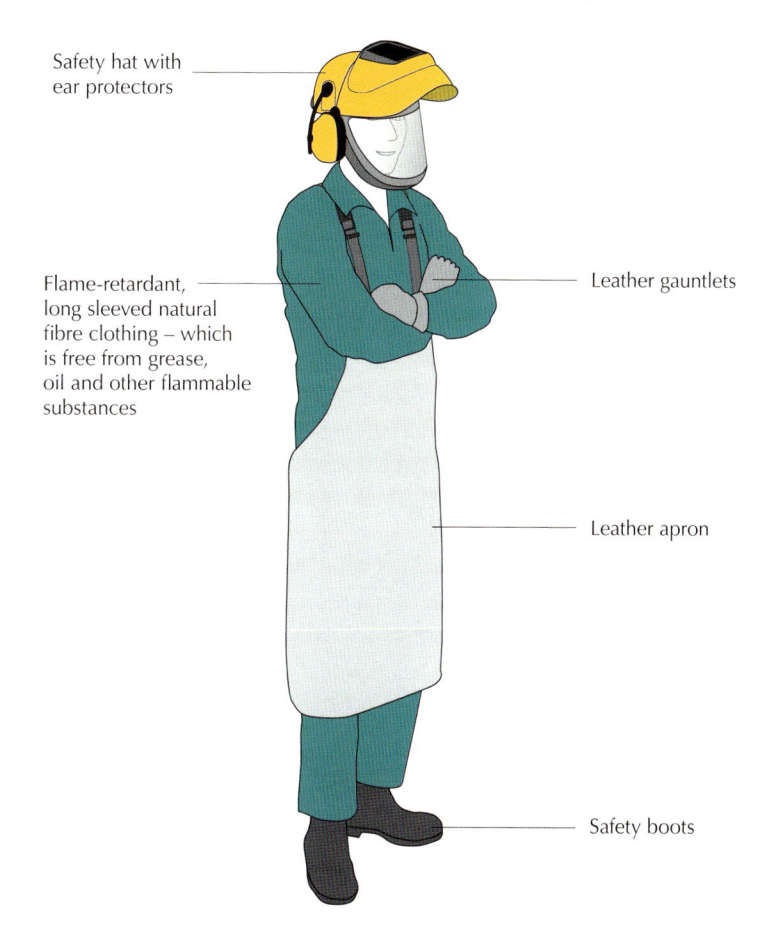

Safety boots

Figure 24.1 Examples of PPE for hot work

24.4.3		Clothing should be free of grease, oil and other flammable substances.

24.5 Pre-use equipment check

24.5.1	A **competent person** 🔍 should check that hot work equipment is in a serviceable condition before use.
24.5.2	In cold weather, moisture trapped in the equipment may freeze and not work properly; for example, valves may malfunction. Use hot water and cloths to thaw equipment; never use naked flames.

24.6 Electric welding equipment

24.6.1	To minimise personal harm from electric shock, electric welding power sources for shipboard use should have a direct current (DC) output not exceeding 70 volts, with a minimum ripple. See section 24.6.13 for further information on DC power sources.
24.6.2	When DC equipment is not available, AC output power sources may be used providing:

- they have an integral voltage-limiting device to ensure that the idling voltage (the voltage between electrode and workpiece before an arc is struck between them) does not exceed 25 V rms
- the proper function of the device (which may be affected by dust or humidity) is checked each time a welding set is used.

> 💡 **24.6.3** Some voltage-limiting devices are affected by their angle of tilt from the vertical. Therefore mount and use them in the position specified by the manufacturers. This requirement can be affected by adverse sea conditions.

24.6.4	Use a **go and return** 🔍 system using two cables from the welding set. Clamp the welding return cable firmly to the workpiece.
24.6.5	The workpiece should be **earthed** 🔍 to protect against internal insulation failure of the welding transformer, by keeping the workpiece at (or near) earth potential until the protective device (such as a fuse) operates to cut off the mains supply.
24.6.6	Where the welding circuit is not adequately insulated from the earthed referenced mains supply (it is not constructed to one of the standards listed in Annex 24.2), earth the workpiece. The 'return' cable of the welding set and each workpiece should be separately earthed to the ship's structure. Do not use a single cable with hull return. The workpiece earthing conductor should be robust enough to withstand possible mechanical damage. It should be connected to the workpiece and a suitable earth terminal by bolted lugs or secure screw clamps. The clamping point should be clean, and free of rust and paint to provide a good electrical connection.

24.6.7 If an alternative method of protecting against welding transformer insulation failure is used, avoid the hazards caused by stray welding currents by not earthing the workpiece or the welding output circuit. Self-contained engine-driven welding sets, and welding power sources that comply with the standards listed in Annex 24.2, do not need the workpiece to be earthed. Note, however, that other equipment connected to the workpiece may require earthing for safe operation (eg welding sets not constructed to one of the standards listed in Annex 24.2 or electrical pre-heating systems).

24.6.8 To avoid voltage drop in transmission, the lead and return cables should be of the minimum length practicable for the job and of an appropriate cross-section.

24.6.9 Inspect cables before use; if the insulation is impaired or conductivity reduced, do not use them.

24.6.10 Cable connectors should be fully insulated when connected, and so designed and installed that current-carrying parts are adequately recessed when disconnected.

   BS EN 60529:1992+A2:2013

24.6.11 Electrode holders should be fully insulated so that no live part of the holder is exposed to touch. Where practicable they should have guards to prevent accidental contact with live electrodes, and as protection from sparks and splashes of weld metal.

24.6.12 Provide a local switching arrangement or other suitable means to rapidly cut off current from the electrode if the operator gets into difficulties, and to isolate the holder when changing electrodes.

24.6.13 The direct current output from power sources should not exceed 70 volts open circuit. The ripple on the output from the power source should not exceed the values shown in Table 24.1. The ripple magnitudes are expressed as percentages of the DC, and the ripple peak is that with the same polarity as the DC.

Table 24.1   Suitable ripples on output from power sources

Ripple frequency (Hz)	50/60	300	1200	2400
Max. rms O/C voltage ripple (%)	5	6	8	10
Max. peak O/C voltage ripple (%)	10	12	16	20

24.6.14	The conditions in Table 24.1 are normally met by DC generators incorporating commutators and by rectifier power sources that have a three-phase bridge rectifier operating from a three-phase 50/60 Hz supply. Do not operate rectifier power sources from a power supply of less than 50 Hz.
24.6.15	If you need to use a power source with a DC output with a ripple magnitude in excess of those in Table 24.1 (eg a single-phase rectifier power source) include a voltage-limiting device in the power source to ensure that the idling voltage does not exceed 42V.

24.7 Precautions to be taken during electric arc welding

24.7.1	As well as the PPE specified in section 24.4.1, the welding operator should wear non-conducting safety footwear complying with BS EN ISO 20345:2012/BS EN 50321-1:2018.

> ⚠️ **24.7.2 Warning**
> Keep clothing as dry as possible to protect against electric shock. Gloves in particular should be dry because wet leather is a good conductor.
>
> 📖 BS EN ISO 20345:2012; BS EN 50321-1:2018

24.7.3	An assistant should attend continuously during welding operations. They should be alert to the risk of accidental shock to the welder and ready to cut off power instantly, raise the alarm and provide artificial respiration without delay. A second assistant may be needed in difficult conditions.
24.7.4	Where people other than the operator are likely to be exposed to harmful radiation or sparks from electric arc welding, protect them with screens or another effective means.
24.7.5	In restricted spaces, where the operator may be in close contact with the ship's structure or is likely to make contact during ordinary movements, protect them with dry insulating mats or boards.
24.7.6	There are increased risks of electric shock to the operator if welding is done in hot or humid conditions because body sweat and damp clothing greatly reduce body resistance. In these conditions delay the operation until an adequate level of safety can be achieved.
24.7.7	A welder should never work while standing in water or with any part of their body immersed.

24.7.8 Isolate the electrode holder from the current supply before removing a used electrode and inserting a new electrode. This precaution is necessary because some electrode coatings have extremely low resistance. Even a flux coating, which is normally insulating, can become damp from sweating hands and thus potentially dangerous.

24.7.9 When the welding operation is completed or temporarily suspended, remove the electrode from the holder.

24.7.10 Eject hot electrode ends into a suitable container; do not handle them with bare hands.

24.7.11 Keep spare electrodes dry in their container until required for use.

24.8 Compressed gas cylinders

24.8.1 Always handle compressed gas cylinders with care, whether full or empty. Secure them properly and store them in a suitable place for their intended use and the risks that an inadvertent release of gas may present. Secure the cylinders so they can be released quickly and easily (eg in case of fire). Use cylinder trolleys where appropriate to transport cylinders from one place to another.

24.8.2 If the cylinder design permits protective caps over the valve, screw them in place when the cylinders are not in use or are being moved. Where the cylinder design does not permit protective caps over the valve protect the valve system from damage (eg from impact). Close valves when cylinders are empty or not in use.

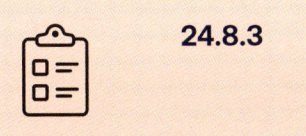

24.8.3 Store flammable gases for hot work carefully. The storage should:

- be separated according to type of gas, and empty cylinders kept separate from full ones
- be well ventilated
- not have extremes of temperatures
- not contain any sources of ignition, including electronic devices
- be prominently marked 'No smoking' and have safety signs in line with the standards in Annex 9.1.

24.8.4 For compressed gas cylinders also take the following precautions:

- Keep the cylinders' valves, controls and associated fittings free from oil, grease and paint; do not operate controls with oily hands.
- Do not take gas from such cylinders unless the correct pressure-reducing regulator has been attached to the cylinder outlet valve.
- Take cylinders with leaks that cannot be stopped by closing the outlet valve to the open deck away from any sources of heat or ignition, then slowly discharge them to the atmosphere.

24.8.5 Identifying marks on cylinders are set out in section 9.6.

24.9 Gas welding and cutting

24.9.1 Although this section deals mostly with oxygen and acetylene, other fuel gases may be used and similar precautions should be taken.

24.9.2 The pressure of oxygen used for welding should always be high enough to prevent acetylene flowing back into the oxygen line.

24.9.3 **Warning**

Do not use acetylene for welding at a pressure exceeding 1 atmosphere gauge because it could explode when under excessive pressure, even without any air.

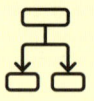

24.9.4 Fit non-return valves adjacent to the torch in the oxygen and acetylene supply lines.

24.9.5 Provide flame arrestors in the oxygen and acetylene supply lines. They will usually be fitted at the low-pressure side of regulators, although there may be other ones at the torch.

24.9.6 If a cylinder backfires shut the oxygen valve on the blowpipe to prevent internal burning. Immediately shut off the fuel gas at the blowpipe valve. Then follow stages 3–6 of the shutting-down procedure in Annex 24.1. Once the cause of the backfire is known, the fault rectified and the blowpipe cooled down, the blowpipe may be re-lit.

24.9.7 If there is a flashback into the hose and equipment, or a hose fire or explosion, a fire at the regulator connections or gas supply outlet points, isolate the oxygen and fuel gas supplies at the cylinder valves or gas supply outlet points if safe. Further action should follow in line with the vessel's fire-drill requirements.

24.9.8 Monitor acetylene cylinders to ensure that they are not becoming hot. A hot cylinder could be a sign of acetylene decomposition and there is an increased risk of explosion.

24.9.9 If an acetylene cylinder is hot, close the cylinder stop valve immediately to limit/reduce the decomposition. Take emergency action which should include:

- evacuating the area
- prolonged cooling with water (either immersion or large volume of flow)
- possibly jettisoning the cylinder overboard (but movement can cause rapid decomposition and cooling should be continued during any movement).

> **24.9.10** **Warning**
> Approach any acetylene cylinder suspected of overheating with extreme caution as any impact could result in explosion.
>
> **24.9.11** Couple acetylene cylinders only if their pressures are approximately equal.

24.9.12 In fixed installations, manifolds should be clearly marked with the gas they contain.

24.9.13 Manifold hose connections, including inlet and outlet connections, should be such that the hose cannot be interchanged between fuel gases and oxygen manifolds and headers.

24.9.14 Use only hoses specially designed for welding and cutting operations to connect any oxy-acetylene blowpipe to gas outlets.

24.9.15 Discard any length of hose in which a flashback has occurred.

24.9.16 The connections between hose and blowpipe and between hoses should be securely fixed with fittings that comply with BS EN 1256 (see Annex 24.3 for more detailed guidance on hose connections and assemblies).

📖 BS EN 1256:2006

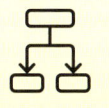

24.9.17 Arrange hoses so they are not likely to become kinked, tangled, tripped over, cut or otherwise damaged by moving objects or falling metal slag or sparks. A sudden jerk or pull on a hose could pull the blowpipe out of the operator's hands, cause a cylinder to fall or a hose connection to fail. Cover hoses in passageways to avoid them becoming a tripping hazard.

24.9.18 Use soapy water only for testing leaks in hoses. If there are leaks that cannot easily be stopped, isolate the gas supply and take the leaking components out of service, replace or repair them. If the leak is at a cylinder valve or pressure regulator ('bull-nose') connection, move the cylinder to a safe place in the open air. If it is a fuel-gas cylinder, take it well clear of any source of ignition.

24.9.19 Never use excessive force on cylinder valve spindles or hexagon nuts of regulator connections when trying to stop a leak. Do not use sealing tape and other jointing materials to prevent leaks between metal–metal surfaces that are designed to be gas tight. With an oxygen cylinder, this could result in a metal–oxygen fire.

24.9.20 Light blowpipes with a special friction igniter, stationary pilot flame or other safe means.

24.9.21 If a blowpipe-tip opening becomes clogged clean it only with purpose-designed tools.

24.9.22 When changing a blowpipe shut off the gases at the pressure-reducing regulators.

24.9.23 To prevent a build-up of dangerous concentrations of gas or fumes during a temporary stoppage or after work, securely close the supply valves on gas cylinders and gas mains and blowpipes. Remove hoses and moveable pipes to lockers that open onto the open deck.

24.9.24 Never use oxygen to ventilate, cool or blow dust off clothing.

24.10 Further information

Detailed advice on the selection and standards for equipment used in hot work is available in the HSE guidance note 'HSG139 The safe use of compressed gases in welding, flame cutting and allied processes' on the HSE website.

- Personnel must wear PPE according to company procedures and as required by the manufacturer's instructions.
- Some manufacturers may recommend earthing electrical welding equipment to reduce electrical interference. Although this is not a safety-related measure, follow the manufacturer's advice.

- Use a **permit to work**, which includes a risk assessment, for hot work outside of a workshop.
- Provide personnel with PPE that is appropriate and conforms to the necessary official standards.

Annex 24.1 Hot work: lighting up and shutting down procedures

These procedures are appropriate for oxy-fuel gas equipment and air-aspirated blowpipes with a few changes.

Pre-lighting up

1. Do a risk assessment including a survey of all adjacent spaces and get a hot-work permit to work.
2. Post fire sentries in all adjacent compartments.

Lighting up

1. Ensure that the pre-use equipment checks have been made.
2. Check that the outlets of adjustable pressure regulators are closed; in other words that the pressure-adjusting screw of the regulator is in the fully unwound (anti-clockwise) position.
3. Check that the blowpipe valves are closed.
4. Slowly open the cylinder valves (or gas supply point isolation valves) to avoid sudden pressurisation of any equipment.
5. Adjust pressure regulators to the correct outlet pressures, or check that the pressures in distribution pipework are suitable for the equipment and process.
6. Open the oxygen valve at the blowpipe and allow the flow of oxygen to purge* air out of the oxygen hose and equipment. If necessary, reset the pressure regulator to ensure the correct working oxygen pressure.
7. Close the oxygen valve at the blowpipe.
8. Open the fuel gas valve at the blowpipe and allow the gas flow to purge* air or oxygen from the fuel gas hose and equipment. If necessary, reset the pressure regulator to ensure the correct working fuel gas pressure.
9. Light the fuel gas immediately, preferably with a spark lighter.
10. Open the oxygen valve at the blowpipe. Adjust it and the fuel gas valve to give the correct flame setting.

* Purging is important. It removes flammable gas mixtures from the hoses and equipment, which could result in explosions and fires when the blowpipe is first lit. Do it in a well-ventilated area. It may take from several seconds to a minute or more depending on the length of the hose and gas flow rates.

Shutting down

1. Close the fuel gas valve at the blowpipe.
2. Immediately close the oxygen valve at the blowpipe.
3. Close the cylinder valves or gas supply point isolation valves for both oxygen and fuel gas.**
4. Open both blowpipe valves to vent the pressure in the equipment.
5. Close the outlets of adjustable pressure regulators by winding out the pressure-adjusting screws.
6. Close the blowpipe valves.

** Step 3 is not necessary when the equipment is to be used again in the immediate future.

On completion of hot work and following shutdown, visit all adjacent compartments to ensure that all is well.

Annex 24.2 Earthing of arc-welding systems' transformer casing

Earthed	Class I appliance
Not earthed	Class II appliance

Transformer secondary

Earthed

This is an obsolete type of equipment and should be taken out of service. Failure of the weld-return connection might not be noticed, and damage to other earthed metallic paths could result.

Isolated

The absence of a weld-return conductor will prevent welding being carried out. However, a failure of isolation within the welding set could cause the work item to become live. For this reason, the workpiece should be earthed.

Isolated with double or reinforced insulation

This is the most recent standard to which equipment is being built. Owing to the strengthened insulation, the workpiece need not be earthed. Furthermore, to prevent the possibility of stray weld-return currents in the supply system earth conductors, it is recommended that the workpiece is not earthed. Such welding power sources may be identified by the additional symbol if they were manufactured in line with the relevant parts of BS EN IEC 60974-1:2018+A1:2019, or they will be marked with the standards numbers BS EN IEC 60974-11:2010 and related references.

BS EN IEC 60974-1:2018+A1:2019

Annex 24.3 Hot work: hoses and connections/assemblies

Hoses

Rubber hoses complying with Standard BS EN ISO 3821:2019 are recommended for use in gas-welding and cutting processes, which are often carried out in aggressive working environments. Hoses meeting these standards are reinforced with an outer protective cover designed to be resistant to hot surfaces, molten slag or sparks, and made with linings that resist the action of hydrocarbons (for liquefied petroleum gas (LPG) hose), acetone or dimethyl formamide (for acetylene hoses) and ignition in an atmosphere of oxygen (for all services). Burst pressure is 60 bar g and maximum working pressure 20 bar g.

BS EN ISO 3821:2019

Hoses made of thermoplastic materials are not generally suitable for welding and cutting, because they do not have the same resistance to hot surfaces or particles as reinforced rubber hoses.

BS EN 16436-1:2014+A3:2020; BS EN 16436-2:2018

Connections

Hose connections (comprising hose nipples and 'bull-nose' hose connections) comply with BS EN 1256, ISO/TR 28821:2012 or equivalent. Thread sizes specified in these standards are based on Whitworth dimensions, which are generally used in this field in many countries. Right-hand threads are used for oxygen and non-combustible gases; left-hand threads are used for fuel gases, with the hexagon nuts on their union connections notched to aid identification.

BS EN 1256:2006

Hose connections may also be made with a quick-action coupling. A male probe is fitted to the end of the hose and a female connector with a self-sealing valve is usually fitted to a fixed piece of equipment or gas supply outlet point. The probe is pushed into the female fitting where it locks in position and automatically opens the internal valve. Connections of this type are simple and quick to operate, and there is no need to use a spanner to tighten any nuts.

Problems include the male probe becoming damaged (eg from being dragged along the ground or overuse) causing the coupling to leak. There is also a possibility of connecting the hose to the wrong gas outlet. Both should be avoided if couplings comply with Standard EN 561 or ISO 7289:2018. These require probes to be constructed of hard material. Their design dimensions are intended to prevent oxygen and fuel gas connections being used interchangeably.

BS EN 561:2002

Hose assemblies

Hose lengths are usually supplied in the UK as pre-assembled units complete with connection fittings crimped to the ends of the hose. Hose and hose-nipple dimensions are matched by the supplier to ensure a good fit. The recommended standard for hose assemblies is BS EN 1256:2006, which specifies requirements for leak tightness and resistance to axial loading. Worm drive or similar clips are not recommended for fastening hoses.

BS EN 1256:2006

25 Painting

25.1 Introduction

25.1.1 Based on the findings of the risk assessment, appropriate control measures should be put in place to protect those who may be affected. This chapter identifies some areas that may require attention in respect of painting.

Key points
- Check packaging information before you start.
- If you are painting inside, make sure the space is properly ventilated.
- Never use paints containing lead, mercury or other toxic compounds.

Your organisation should
- plan painting work with appropriate control measures to protect everyone who may be in danger
- warn seafarers about the risks of using paints
- provide suitable personal protective equipment (PPE).

25.2 Preparation and precautions

25.2.1 Because the origin of any paint to be removed may be unknown always take precautionary measures. Rub down painted surfaces wet to reduce dust from the old paint, which may be toxic if inhaled. Where the dust is known to contain lead, use other dust-treating methods. Appropriate respiratory protective equipment should be worn as protection against other dusts.

25.2.2 If the surface to be rubbed down is known to contain lead, use methods that do not create dust. It is safer to avoid or minimise dust creation than to try to clean up the dust afterwards. Avoid sanding or abrasive blasting. Never burn off lead-based paint because fumes will contain metallic lead in a readily absorbed form.

25.2.3 Rust removers are acids so avoid contact with unprotected skin. Eye/face protection should be worn against splashes (see Chapter 8). If painting at height or otherwise near ropes, take care to avoid splashes on ropes, safety harness, lines, etc. (see sections 18.28.16 and 18.28.17 on the effect of such contamination on ropes).

25.2.4 Keep interior and enclosed spaces well ventilated, both while painting is in progress and until the paint has dried. See Chapter 15 on how a space can become hazardous.

25.2.5 There should be no smoking or use of naked lights during painting or until the paint has dried hard. Some vapours, even in low concentrations, may decompose into more harmful substances when passing through burning tobacco.

25.2.6 When painting near machinery, turn off the power supply and immobilise the machinery in such a way that it cannot be moved or started up inadvertently. Appropriate warning notices should be posted. Close-fitting clothing should be worn, as shown in Figure 25.1.

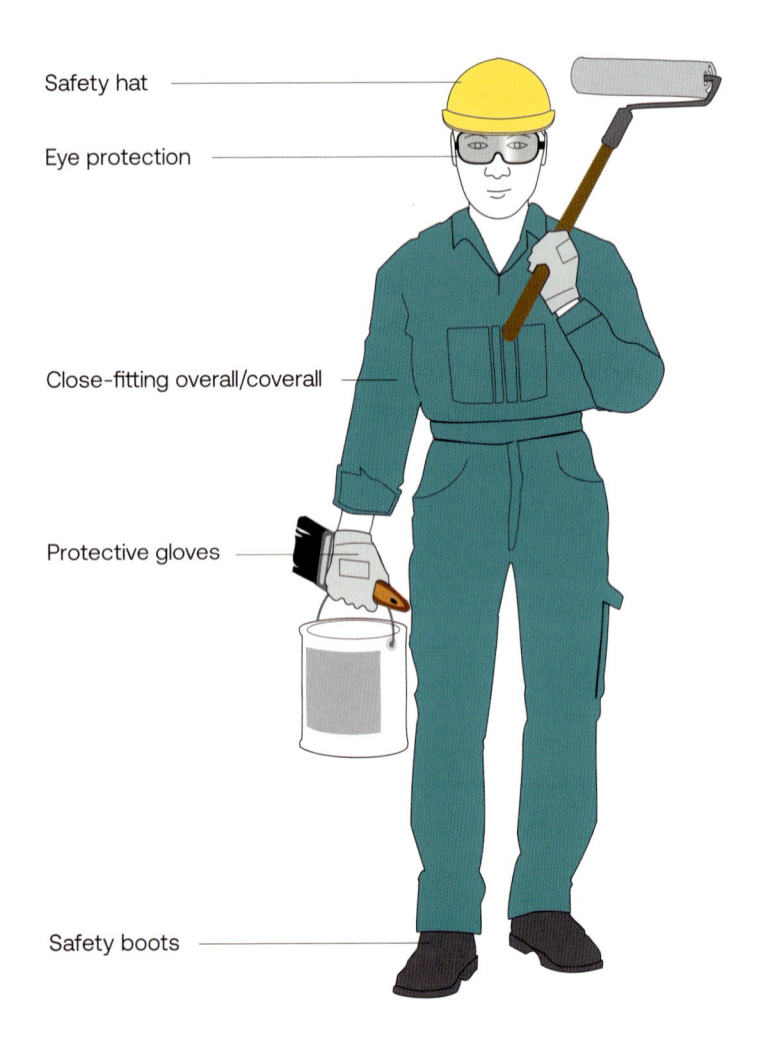

Figure 25.1 PPE for general painting

25.3 Application of new paint

25.3.1 Paints are hazardous substances and mixtures, and may present risks that require precautions to be taken. Packaging is required to be marked with warning signs, which will give the first indication of any risks. Carry out a risk assessment using the safety data sheet provided with the product. Warn seafarers using such paints of the particular risks arising from their use.

25.4 Use of paint-spraying equipment

25.4.1 Because there are many different types of paint-spraying equipment in use, seafarers should comply with the manufacturer's instructions for use.

25.4.2 Airless spray-painting equipment is particularly hazardous because the paint is ejected at a very high pressure and can penetrate the skin or cause serious eye injuries. Do not allow spray to come into contact with the face or unprotected skin.

25.4.3 Suitable protective clothing such as a combination suit, gloves, cloth hood and eye protection should be worn during spraying, as shown in Figure 25.2.

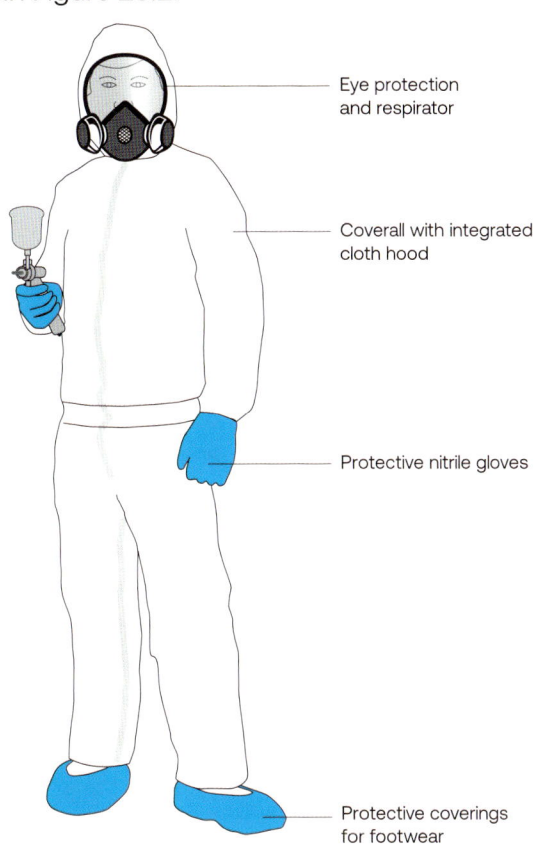

Figure 25.2 PPE for spray-painting

25.4.4 Do not use paints containing lead, mercury or similarly toxic compounds.

25.4.5 A respirator suitable for the nature of the paint being sprayed should be worn. In exceptional circumstances it may be necessary to use specialist breathing apparatus (see section 8.7).

25.4.6 The pressure in the system should not exceed the recommended working pressure of the hose. Inspect the system regularly for defects.

25.4.7 As an additional precaution against the hazards of a hose bursting, a loose sleeve (eg a length of 2 to 3 metres (6 to 10 feet) of old air hose) may be slipped over the portion of the line that is adjacent to the gun and paint container.

How to unblock a spray nozzle

25.4.8 If a spray nozzle becomes blocked:

- Lock the trigger of the gun in a closed position.
- Before you remove the nozzle or try to dismantle the equipment, relieve pressure from the system as explained above.
- If the nozzle is reversible, blow through it to remove the blockage. Keep yourself clear of the nozzle mouth.

- Check the packaging information before you start.
- Rub old paint down wet.
- Do not burn off lead-based paint.

- Wear suitable PPE.
- Do not smoke around wet paint.

26 Anchoring, mooring and towing operations

26.1 Introduction

26.1.1 All seafarers involved in anchoring, mooring and towing operations should have additional instruction on the specific equipment and mooring configurations used on the vessel. This should include (but may not be limited to):

- the types of winches and windlass and their operation
- the location of emergency stop buttons
- the types of ropes and/or wires used
- the location and use of rollers, dollies and leads.

26.1.2 Records of instruction should be maintained.

26.1.3 Based on the risk assessment, appropriate control measures should be put in place. The risk assessment must consider the consequences of failure of any equipment. This chapter identifies some areas that require attention when anchoring, mooring or conducting towing operations. The risk assessment and control measures should be reviewed for each new mooring operation, taking account of the expected mooring configuration, particularly to the risk of snap-back.

26.1.4 When anchoring, mooring or towing operations are taking place, all seafarers should be adequately briefed on the mooring configurations, including any limitations of the mooring equipment, and wear appropriate personal protective equipment.

Key points
- On joining a vessel ensure instruction is given on specific equipment and mooring configurations
- When making fast or casting off the **monkey's fist** must only be made of rope and not weighted, to prevent injury to those receiving the lines.
- Equipment must be regularly inspected for defects.
- When mooring lines are under strain all seafarers in the vicinity should remain in safe positions.
- Maintain communications during anchoring, mooring and towing operations.

Your organisation should
- carry out a risk assessment of anchoring, mooring and towing operations
- arrange shoreside communications when self-mooring
- be aware of tidal and weather conditions during anchoring, mooring and towing operations.

26.2 Anchoring and weighing anchor

26.2.1 Before using an anchor, a competent seafarer must check that:

- the brake is securely on, then clear all securing devices
- a responsible person is in charge of the anchoring party, with a suitable means of communication with the vessel's bridge
- the anchoring party wear protective clothing, including safety helmet, safety shoes, gloves and goggles, to protect them from injury by rust particles and debris that may be thrown off the cable during the operation. Where the noise levels generated may be harmful, they should consider hearing protection; however, take into account the length of exposure and the greater risk from impaired communication
- during anchoring the anchoring party stand aft of, or at a safe distance from, the windlass/capstan and be aware of the risk of anchor cable failure.

Intertanko (2019); OCIMF (2010a); OCIMF (2010b); The Swedish Club (2016)

26.2.2 Where communication between bridge and anchoring party is via portable radio, the identification of the ship should be clear to prevent confusion caused by other users on the same frequency.

26.2.3 Before anchoring the vessel, the anchor should be 'walked out' clear of the hawsepipe, as shown in Figure 26.1. This is to confirm it can be released, before anchoring the vessel.

Check the following:

- that there are no small craft near the vessel that will be impacted by the operation, such as the vessel's bow, the main engines and thrusters, if used
- what the anticipated movement of the vessel will be while anchoring.

For very large ships with heavy anchors and cables, walk the anchor out either at intervals or all the way to avoid excessive strain on the brakes (and on the bitter end, if the brakes fail to stop the anchor and cable).

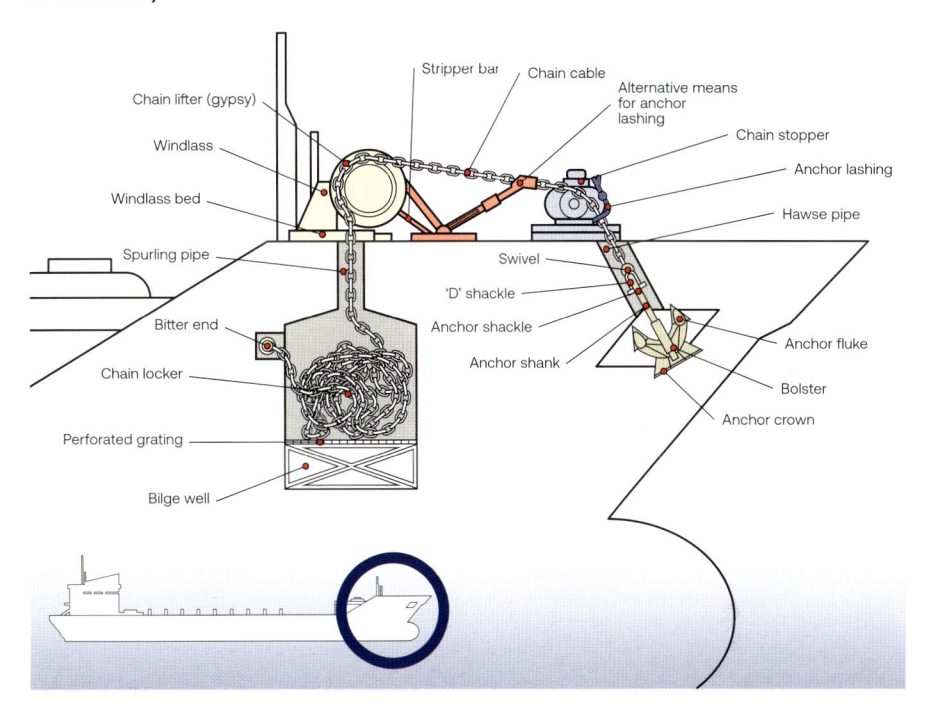

Figure 26.1   Anchoring

26.2.4   In situations where the cable has to be walked out after the anchor has been landed onto the seabed, the speed of the vessel over the ground should not exceed the design hoisting speed of the windlass.

26.2.5   Seafarers should also be aware of the following limitations for the design of anchoring equipment, which are referenced by the International Association of Classification Societies (IACS) Unified Requirements (UR):

- The windlass design load lift for vessels without a deep water anchoring (DWA) notation is three shackles (82.5 metres) and the weight of the anchor when hanging vertically.
- The maximum environmental loads for sheltered waters that are exposed to no waves is a wind velocity of 25 m/s and a current velocity of 2.5 m/s.
- The maximum environmental loads for outside sheltered locations that may be exposed to a significant wave height of up to 2 metres is a wind velocity of 11 m/s and a current velocity of 1.5 m/s.

IACS UR A2 Rev 5 (Sept 2020), IACS Recommendation No 10 Rev 4 (Sept 2020)

26.2.6 When the anchor is let go from the stowed position, if, on release of the brake, the anchor does not run, seafarers should not attempt to shake the cable. Reapply the brake, place the windlass in gear and walk the anchor out clear before release.

26.2.7 The cable should stow automatically. If for any reason it is necessary for seafarers to enter the cable locker, they must first take proper precautions for entering an enclosed space. They should stand in a protected position and be in communication with the windlass/capstan operator.

26.2.8 When weighing anchors, make every effort to wash cables, as far as safely practicable, as they are heaved on board. Mud or debris left on the cable will dry and could become a hazard to the anchor party when next letting go the anchor.

26.2.9 Secure properly anchors that are housed and not required to prevent accidental release.

26.3 Making fast and casting off

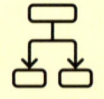

26.3.1 During mooring and unmooring operations, enough seafarers should always be available both forward and aft of the vessel to ensure a safe operation. At the same time minimise the exposure of seafarers to risk on mooring decks when lines are under tension.

When making fast and casting off:

- A responsible person should be in charge of each of the mooring parties and should have a clear line of sight to all lines and members of the mooring parties.
- There should be a suitable means of communication between the responsible people and the vessel's bridge team. If this means using portable radios the ship should be clearly identified by name to prevent confusion with other users.
- All seafarers involved must wear protective clothing, including safety helmet, safety shoes and gloves, and be fully briefed on the berthing plan.

For more information, see Effective Mooring (2019, 4th edition) from the Oil Companies International Marine Forum (OCIMF).

26.3.2 Due to the design of mooring decks, consider the entire area dangerous in the event of snap-back or the sudden release of a line under tension. All crew working on a mooring deck should be aware of this, with clear visible signage.

| 26.3.3 | Seafarers, other than the mooring parties, should keep clear of the mooring decks during mooring and unmooring operations. |

26.3.4 Avoid painting **snap-back zones** 🔍 on mooring decks because they may give a false sense of security. Annex 26.1 shows the potential snap-back zone for a complex mooring system, where there is a single point of failure in the line. However, this is only illustrative; a line or lines may break at any point, and may part at more than one point, increasing the area of risk from snap-back. See section 18.28 for guidance on the use of different types of ropes and wires.

26.3.5 Working on enclosed mooring decks adds hazards so take extra care. In particular ensure there is adequate lighting.

26.3.6 To prevent injury to seafarers receiving heaving lines, the monkey's fist should be made with rope only and must not contain added weighting material. Safe alternatives include a small high-visibility soft pouch, filled with fast-draining pea shingle or similar, with a weight of not more than 0.5 kg. Under no circumstances is a line to be weighted by items such as shackles, bolts or nuts, or twist locks.

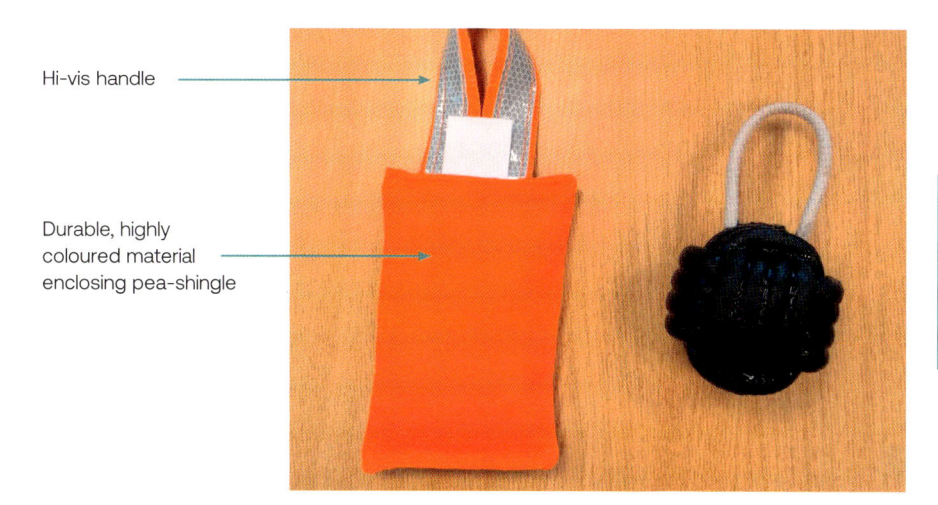

Figure 26.2 Monkey's fist and weighted bag

26.3.7 Keep areas where mooring operations take place tidy and clutter-free. Stow all mooring ropes properly, coil heaving lines and stoppers away and clean up any oil and grease immediately. Decks should have anti-slip surfaces provided by fixed treads or anti-slip paint coating, and the whole working area should be adequately lit for operations undertaken during periods of darkness.

26.3.8 Inspect equipment used in mooring operations regularly for defects. Record and correct any defects. Pay attention to oil leaks from winches. The surfaces of fairleads, bollards, bitts and drum ends should be clean and in good condition, and drum ends should not be painted. Rollers and fairleads should turn smoothly; do a visual check that corrosion has not weakened them.

Pedestal roller fairleads, lead bollards, mooring bitts, etc. should be:

- properly designed for the task
- able to meet all foreseeable operational loads and conditions
- correctly sited
- fixed to a part of the ship's structure that is suitably strengthened.

Personnel should be aware of the safe working load of the equipment.

26.3.9 Ropes and wires that are stowed on reels should not be used directly from stowage, but should be run off and flaked out on deck in a clear and safe manner, ensuring sufficient slack to cover all contingencies. If you are unsure how much to use, run off the complete reel.

26.3.10 The following principles will ensure that the ship's equipment is used to best effect:

- **Breast lines** provide most of the **athwartships restraint.**
- **Spring lines** provide the largest proportion of the **longitudinal restraint.**
- Avoid very short lengths of line when possible because they will take a greater proportion of the total load when the ship moves.
- Compensate for very short lengths by running the line on the bight.

26.3.11 Consider the layout of moorings carefully:

- to use the leads that are most suited, without creating sharp or multiple angles. Either of these increases the wear on lines and the potential snap-back area if the line parts
- not to feed ropes and wires through the same leads or bollards
- to plan operations and do a risk assessment, especially where unusual or non-standard mooring arrangements are used. The risk assessment and control measures should consider the shoreside mooring equipment at the berth, bearing in mind the consequences of a line snagging on a shoreside structure, such as a fender, and then suddenly releasing and coming under tension.

26.3.12 There is a risk of a snagged mooring line suddenly releasing under tension without breaking. When this happens the trajectory of the line could be in either the horizontal or vertical plane, or a combination of both, and may involve 'snaking'. Do a risk assessment to account for the possibility that the sudden release of tension of a snagged line or the recoil of a parting line may have a vertical component.

26.3.13 Seafarers should never stand in a bight of rope or wire. Winches should be operated by competent seafarers to ensure that excessive loads do not arise on moorings. Warping a vessel along a berth involves risk; do a risk assessment before using mooring lines to warp a vessel along a berth.

26.3.14 The risk assessment should take account of:

- the environmental conditions
- the vessel's size
- the type and length of rope used
- the capability of the shoreside mooring arrangements.

Seafarers doing this must be competent and there should be a toolbox talk before operations start. All participants should maintain good communications with each other throughout.

26.3.15 When moorings lines are under strain, seafarers nearby should remain in safe positions, avoiding potential snap-back zones. Seafarers should always stay well clear of an entrapped line due to the risk of it suddenly releasing under tension. A bird's-eye view of the mooring deck arrangement should be produced to identify danger areas. Seafarers should also be aware of other dangerous areas – the whole mooring deck is a danger zone. Risk exists in any area where lines could come under tension or snap-back, including side decks.

26.3.16 Warning
Take immediate action to reduce the load if any part of the system appears to be under excessive strain.

26.3.17 Ensure that ropes or wires will not jam when they come under strain, so they can be slackened off quickly if necessary.

26.3.18 Where a mooring line is led around a pedestal roller fairlead, the snap-back zone will change and increase in area. Where possible do not lead lines round pedestals, except when mooring the ship. Thereafter, lines should be made up on bitts, clear of pedestals if possible.

26.3.19 When heaving moorings on a drum end, the winch operator must have a full view of all activity. The 'fleet angle' or lead angle of the rope onto the drum should be no more than a few degrees. One seafarer should be at the drum end, backed up by a second seafarer standing at least a metre away, backing and coiling down the slack. Three turns on the drum end are usually sufficient to do this successfully and avoid riding turns.

26.3.20 Never use a wire on a drum end as a check wire. Never surge a synthetic rope on the drum end. After hauling a rope tight, use a stopper to allow the rope to be removed from the warping drum and then placed on a bollard or bitts by using either single turns or figures of eight. For wire rope secure at least the top three lays of the figure of eight by a fibre rope to prevent jumping. Use like for like stopper material (natural for natural ropes, and chain for wire ropes).

26.3.21 Never lead a wire across a fibre rope on a bollard. Keep wires and ropes in separate fairleads or bollards.

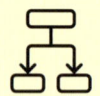

26.3.22 When stoppering off moorings:

- Stopper natural fibre rope with natural fibre.
- Stopper synthetic fibre rope with synthetic fibre stopper (but not polyamide).
- The double-tail 'West Country' stopper is preferable for ropes; see Figure 18.1.
- Stopper wire moorings with chain, using two half-hitches in the form of a cow hitch, suitably spaced with the tail backed up against the lay of wire. This ensures that the chain does not jam or open up the lay of the wire.

26.4 Mooring to a buoy

26.4.1 Where seafarers are mooring to a buoy from a ship's launch or boat they must wear a working lifejacket (personal flotation device). A lifebuoy with an attached lifeline should be available in the boat.

26.4.2 Provide a means to recover a person overboard. If a boarding ladder with flexible sides is used, it should be weighted so that the lower rungs remain below the surface.

26.4.3 Where mooring to a buoy is undertaken from the ship, a lifebuoy with an attached line of sufficient length should be available for immediate use.

| 26.4.4 | When using slip wires for mooring to buoys or dolphins, never put the eyes of the wires over the bitts because at the time of unmooring it may not be possible to release the load sufficiently to lift the eye clear. To prevent accidental slippage of the wire eye(s) over the bitts or other obstruction seize the eyes, partially closing them. |

26.5 Mooring arrangements with an upward lead angle

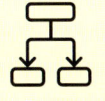

| 26.5.1 | There is a risk of loss of control or containment of mooring lines when using open fairleads and lines have an upward lead angle. This situation can arise due to changes in water level, as well as during transfer of cargo between moored ships due to changes in relative vessel freeboard. It is better to use closed fairleads when transferring cargo in this way, and in any other situation where lines have an upward lead angle. |

26.6 Towing

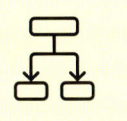

| 26.6.1 | There have been accidents when making fast and releasing a tow. The gear can become taut without warning, causing the messenger to part and strike anyone within the snap-back zone, resulting in serious injury. Poorly controlled towing operations are also a significant hazard to tug crews. |

| 26.6.2 | Maintain equipment used for towing adequately and inspect it before use. During towing operations, excessive loads may be applied to ropes, wires, fairleads, bitts and connections. If the quality of the towline is in doubt reject it and use an alternative line. |

| 26.6.3 | Before starting towing operations the master (and pilot) should establish a suitable means of communication, exchange relevant information (eg speed of vessel) and agree a plan for the tow with the tug master. |

26.6.4 Seafarers involved in towing operations must:

- understand their duties, the operation and the safety precautions to take
- have PPE including safety helmets, safety shoes and gloves
- take care during hours of darkness to ensure that floodlighting will not dazzle and destroy the night vision of the tug master.

26.6.5 On instruction from the bridge, throw the heaving line over to the tug from the shoulder (when taking a tow forward) of the vessel and not from the position of the Panama/Suez fairlead. The position in front of the vessel's (bulbous) bow is the most dangerous for the tug. The tug will then attach a messenger, which is placed on a winch and used to heave the tug's main towline on board. Use only enough turns of the messenger on the drum end to heave in the towline (see section 26.3.16). Use a stopper while the eye is placed around the bollard. On tankers, do not place the towline's eye over the same bollard that the fire wire was made fast to. Take off the fire wire if there is no bollard available. Do the whole operation efficiently to allow the tug to withdraw to a safe position without undue delay.

26.6.6 Once the tow is connected, seafarers should keep clear of the operational area. If anyone needs to remain in this area or to attend to towing gear during the towing operation they should keep clear of bights of wire or rope and the snap-back zone at all times.

26.6.7 During operations, communications should be maintained between:

- the towing vessel and both the bridge team and the foredeck of the vessel under tow
- the tow party and the bridge team.

All parties should identify themselves clearly to avoid misunderstandings. The tug master should be kept informed of engine movements and proposed use of thrusters. Seafarers in charge of the mooring party should monitor the towline to warn the crew if the towline becomes taut for whatever reason.

26.6.8 When letting go, do not try to heave in the messenger to release the towline before making positive communications with the tug. The vessel's master or pilot should do this when the tug has indicated that it is ready to receive the towline back. The instruction to release must come from the vessel's master.

	26.6.9	Use the tug's messenger to heave in the towline and then stopper it off before taking the eye off the bollard. Use turns of the messenger around the bollard to control the speed at which the towline goes out and is retrieved on board the tug. Take care where the towline goes into the water as it may foul the tug's propellers. If the towline is allowed to run out uncontrolled, it could whiplash and strike a crew member, causing severe injuries. Never try to handle towlines that have weight on them.
	26.6.10	Further recommendations on towing are available in the relevant merchant shipping notices (MSNs).
		📖 MGN 592 (M+F) Amendment 1

26.7 Safe mooring of domestic passenger craft and ships' launches to quays

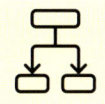

	26.7.1	The recognised and safe method for securing vessels and launches alongside a quay or wharf in a good seafarer-like manner is by using all the following ropes: • a fore spring • a back spring • a head rope • a stern rope.
	26.7.2	Do a risk assessment for the full mooring arrangement, including a diagram.
	26.7.3	Annex 26.2 shows the full and safe mooring arrangement for domestic passenger craft and ships' launches.
	26.7.4	Reduced mooring arrangements may be used in exceptional circumstances, only after taking into account the weather and sea conditions, tidal state and tidal flow. Do a risk assessment for all arrangements that are different from the full safe arrangement in section 26.7.1.
	26.7.5	Keep passengers and unnecessary seafarers clear of all mooring and unmooring operations. All seafarers should keep clear of any area that could be reached by mooring equipment or ropes.

26.7.6 Where mid-ships mooring is the only means of making fast, **breast lines** 🔍 may be run from mid-ships in addition to spring lines from the bow and stern.

26.7.7 Avoid single-point mooring and steaming on a spring.

26.7.8 Providing safe access to a vessel during an operation, in accordance with section 22.1, essential to ensure a safe working environment on board. Always ensure safe access irrespective of the form of mooring operation.

26.7.9 Moor a vessel securely considering the guidance in this chapter. Ensure that the mooring arrangement is effective in restricting movement of the vessel for the foreseeable weather and tidal conditions.

26.8 Safe self-mooring operations

26.8.1 Have appropriately trained shoreside personnel available to assist with mooring operations where practicable.

26.8.2 However, the act of mooring or unmooring a vessel with the exclusive use of the vessel's crew (self-mooring) is a common occurrence in the small vessel sector. It may be reasonably practicable provided that the unique hazards have been mitigated.

26.8.3 Provide a safe system of work for the activity, incorporating a risk assessment and method statement. Such documents should consider that access to some quays, jetties, berths and terminals may give rise to additional risks:

- working at height
- water safety
- restricted working areas
- unguarded edges and vertical ladders.

Additionally, consider the size and type of vessel in relation to the berth and/or mooring buoy, the potential dangers posed to personnel from the prevailing environmental conditions, communication between those involved and appropriate supervision by a **competent person** 🔍.

26.8.4 As many berths may not have been designed for self-mooring, do risk assessments in collaboration with the berth operator/owner to address shared risks.

26.8.5 Seafarers and other workers engaged in the operation are responsible for the health and safety of themselves or any other person on board the vessel who may be affected by the conduct of the operation. The master or coxswain of a vessel remains responsible for the safety of their crew during self-mooring operations and implementation of the control measures in the company risk assessment, including all foreseeable hazards such as unintentional entry into the water and if the craft is moving in a way that poses a risk of crushing or other injuries.

26.8.6 Consider posting a member of the crew, with good communication with the helmsman, to monitor that those leaving and returning to the vessel do so safely.

26.8.7 Where the vessel side is guarded, seafarers should not climb over bulwarks or transit along a rubbing band. Further guidance is provided in Chapter 11 (especially section 11.4 on lighting), Chapter 22, MGN 533 (M) Amendment 2, MGN 591 (M+F) Amendment 1, Port Skills and Safety SIP 014 Guidance on safe access and egress, SIP 021 Guidance on safe access to fishing vessels and small craft in ports and SIP005 Guidance on mooring operations.

MGN 533 (M) Amendment 2; MGN 591 (M+F) Amendment 1; SIP 014; SIP 021; SIP 005

Recommended hierarchy of means of self-mooring

26.8.8 Notwithstanding the points above, and ensuring that the vessel is tight alongside, the following self-mooring operations may be permissible, in order of hierarchy:

1. A means of self-mooring that can be conducted safely within the confines of the vessel using pre-rigged lines or lassoing bollards from the vessel. The vessel should be fully secured before opening the bulwark gate and/or transiting to the quay/berth.

2. A safe and effective means of partially self-mooring within the confines of the vessel using not less than two lines, before opening the bulwark gate and/or transiting to or from the quay/berth to complete the operation. In such circumstances, additional control measures, as appropriate to the size and capabilities of the vessel, may be necessary to address hazards regarding unrestrained movement of the vessel.

3. Only when it is not possible to achieve full or partial mooring of the vessel prior to embarkation or disembarkation, as detailed above, should crew transit to or from the quay or berth while the vessel is unsecured. The risk assessment should account for the manoeuvrability of the vessel, its handling characteristics, the stability of the platform, the vessel's ability to hold position alongside in the prevailing circumstances and the conditions for the transit of personnel to conduct the mooring operation.

26.8.9 Annex 26.3 includes examples of mooring arrangements, showing the use of fore and aft lines doubled back onto the vessel to allow for safe departure without leaving the vessel.

26.8.10 The use of appropriately trained shoreside personnel, such as linesmen, shall be considered as a reasonably practicable measure in all circumstances where additional hazards have been identified or existing control measures have been assessed as impractical or ineffective given a change of condition or defect.

- Equipment used should be adequately maintained and inspected prior to use.
- Seafarers should be adequately briefed before anchoring, mooring or towing operations.
- Seafarers should wear appropriate PPE as instructed.

- Clear communication is needed between parties on the ship and shoreside personnel during anchoring, mooring or towing operations.
- Personnel should always stay well clear of an entrapped line due to the risk of it suddenly releasing under tension.
- There is a higher risk to seafarers carrying out self-mooring operations and this requires a risk assessment.

Annex 26.1 Complex mooring system, illustrating the snap-back zone

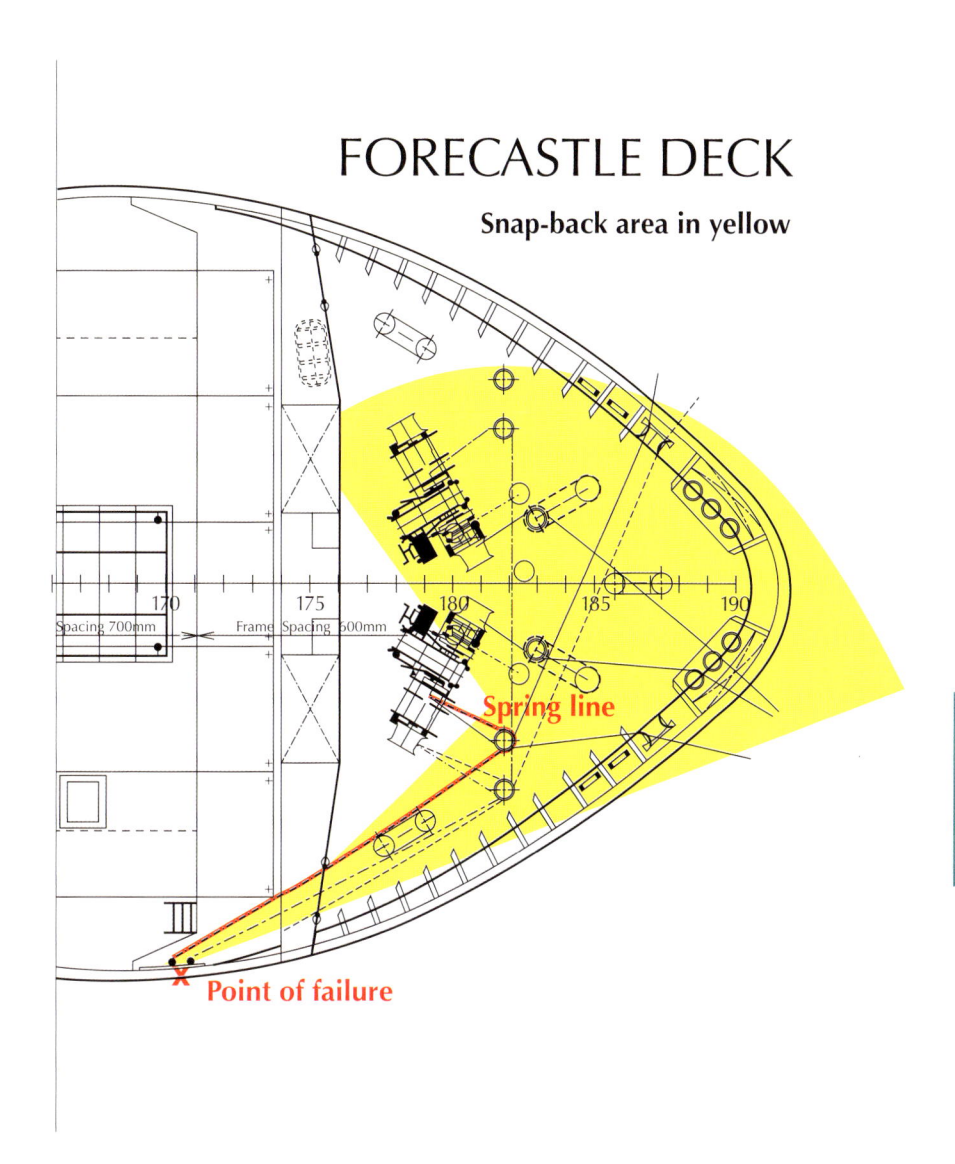

FORECASTLE DECK
Snap-back area in yellow

Spring line

X Point of failure

The diagram shows the potential area of danger (snap-back zone) when the spring line parts at the spring line fairlead.
The snap-back zone would be increased if both pedestal fairleads were used.

(Swedish Accident Investigation Authority Report S-95/11 Morraborg)

Annex 26.2 The full and safe mooring arrangement for small, domestic passenger craft and launches

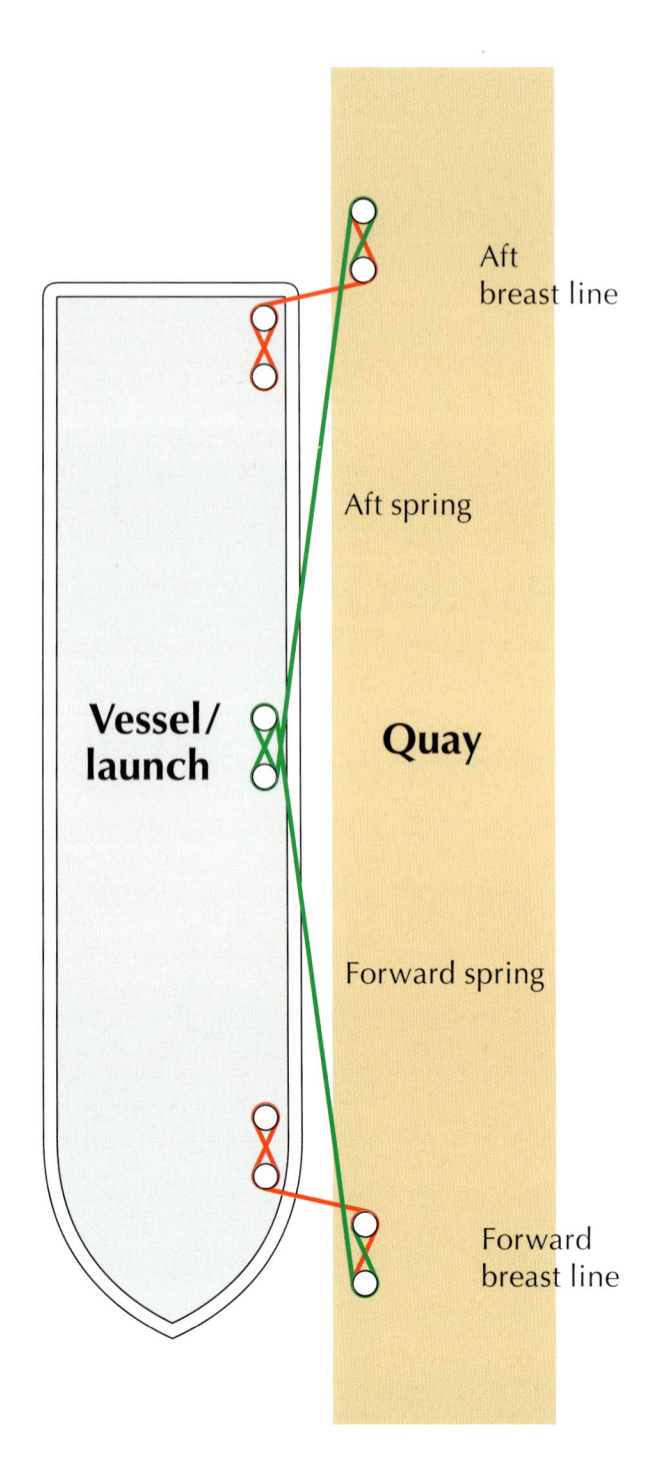

Annex 26.3 Examples of mooring arrangements

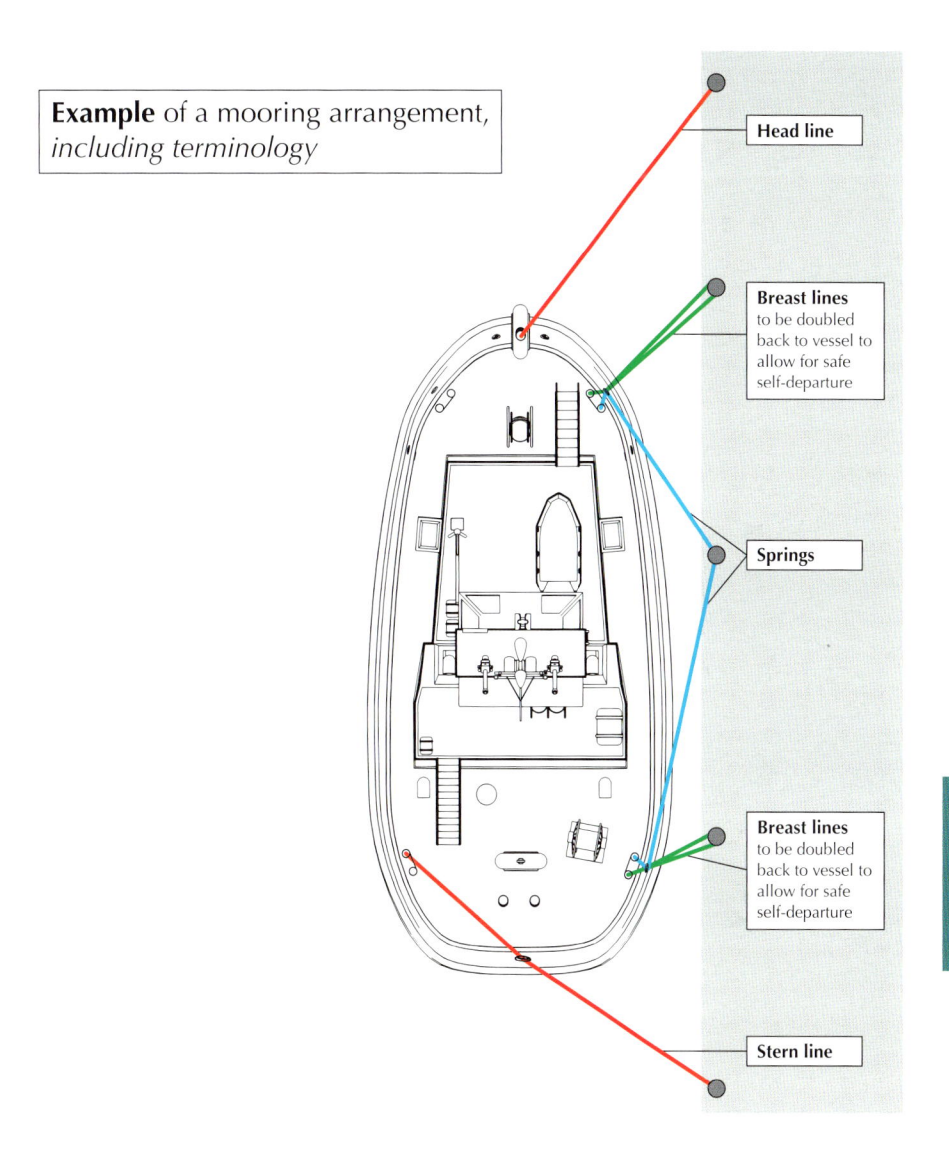

Example of a mooring arrangement, *including terminology*

- **Head line**
- **Breast lines** to be doubled back to vessel to allow for safe self-departure
- **Springs**
- **Breast lines** to be doubled back to vessel to allow for safe self-departure
- **Stern line**

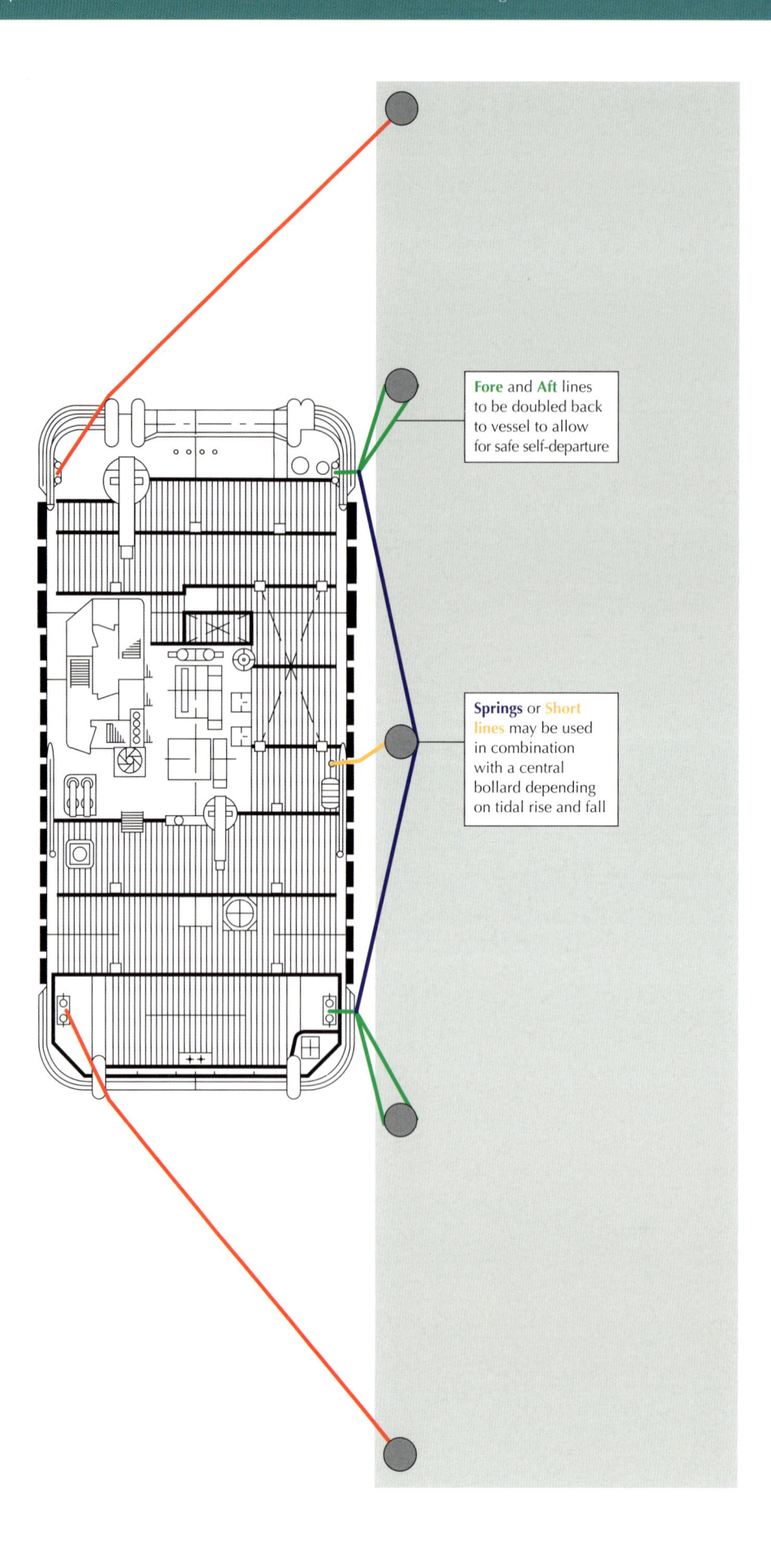

410 Anchoring, mooring and towing operations   Code of Safe Working Practices for Merchant Seafarers

Fore and **Aft** lines to be doubled back to vessel to allow for safe self-departure

Springs or **Short lines** may be used in combination with a central bollard depending on tidal rise and fall

27 Roll-on/roll-off ferries

27.1 Introduction

27.1.1 This chapter gives general advice for the safety of personnel working on the vehicle decks of roll-on/roll-off (ro-ro) ferries. Where other documents or chapters of this code apply, these are cross-referenced and should be read with this chapter.

📖 *MGN 341 (M)*

Key points
- A responsible ship's officer should supervise the movement, stowage and securing of vehicles on vehicle decks and ramps, assisted by at least one **competent person** 🔍.
- Do not permit smoking or naked flames on any vehicle decks. Display clear 'no smoking' or 'no smoking/naked lights' signs.
- Do not allow unauthorised people onto vehicle decks at any time, including when the vessel is at sea.
- For their own personal safety, passengers and drivers should not remain on vehicle decks without the express authority of the responsible ship's officer. Minimise the period before disembarkation when you ask passengers and drivers to return to their vehicles.
- Closed-circuit television (CCTV) cameras should have an uninterrupted view of the vehicle deck. Even if you use CCTV for continuous watch car-deck patrols might still be necessary; for example, coupled with fire patrols of passenger accommodation.

Your organisation should

- ensure that only competent people authorised by a responsible ship's officer operate ships' ramps, car platforms, retractable car decks and similar equipment, in line with the company's work instructions
- provide safe systems of work so as not to put the health and safety of crew or passengers at risk
- operate ramps only when the deck and ramp are clear of people. Stop the operation immediately if any person appears on the deck while the ramp is moving. Fit such ramps and decks with audio and visual alarms where possible
- provide training in the use of equipment that combines theoretical instruction (enabling the trainee to appreciate the factors affecting the safe operation of the plant) with supervised practical work
- not expose personnel working on vehicle decks to the equivalent of 85 dB(A) or greater when averaged over an eight-hour day
- provide hearing protection for when the noise level is equivalent to or exceeds 80 dB(A) averaged over an eight-hour day.

For further guidance on noise levels see Chapter 12 and MCA's Code of Practice for Controlling Risks due to Noise on Ships (revised 2009).

27.2 Ventilation

27.2.1 Vehicle decks should have adequate ventilation at all times, with special regard to hazardous substances.

SI 1998/1011; SI 1998/1012

27.2.2 On passenger vessels, run ventilation fans in closed ro-ro spaces continuously whenever vehicles are on board. More frequent air changes may be required when vehicles are being loaded or unloaded, or where flammable gases or liquids are stowed in a closed ro-ro space. Merchant shipping regulations specify the special requirements for cargo space ventilation.

27.2.3 To reduce the accumulation of fumes:

- Instruct drivers to stop their engines as soon as practicable after embarking.
- Drivers should not start their engines before departure until instructed to do so.
- During loading and discharging, ventilation may be improved by keeping both bow and stern doors open, as long as there is adequate freeboard at these openings.

27.2.4 When there is doubt about the freshness of the atmosphere, arrange for testing to ensure that you maintain 20% oxygen and a carbon monoxide content below 30 ppm in the atmosphere of the space.

27.3 Fire safety and prevention

27.3.1 Switch on fire detection systems whenever vehicle decks are unattended. Train deck and engine crew to use and operate the drencher systems. Ensure there is continuous monitoring of vehicle decks by CCTV or regular fire patrols.

27.3.2 Keep all fire doors closed on vehicle decks when the vessel is at sea.

27.3.3 Plan for the inherent specific risk involving vehicle battery fires by providing suitable equipment for crew who fight electrical battery fires to contain and extinguish the fire safely and without personal risk. Both shore and seagoing personnel responsible for the safety of the vessel should know how to do this. Guidance is available in MGN 653 (M) Amendment 1 Electric vehicles onboard passenger roll-on/roll-off (ro-ro) ferries.

📖 *MGN 653 (M) Amendment 1*

27.4 Safe movement

27.4.1 Warn pedestrians of vehicle movements when entering or crossing car or vehicle decks and to keep to walkways when moving about the ship.

Figure 27.1   Example of a well mapped out car deck with pedestrian walkway

27.4.2   Separate vehicle routes from pedestrian passageways as far as possible, as shown in Figure 27.1. Vehicle ramps should not be used for pedestrian access unless there is suitable segregation of vehicles from pedestrians. Segregation can be achieved by means of a suitably protected walkway, or by ensuring that pedestrians and vehicles do not use the ramp at the same time (see the Code of Practice on the Stowage and Securing of Vehicles on Roll-on/Roll-off Ships, section 2.6).

27.4.3   Crew members should take great care when supervising the driving, marshalling and stowing of vehicles to ensure that no one is put at risk. Take the following precautions, as shown in Figure 27.2:

- Crew should be easily identifiable by passengers. Personnel who need to be on the vehicle decks should wear appropriate personal protective equipment, including high-visibility clothing.
- Deck officers and ratings should communicate clearly and concisely with each other to keep passengers and vehicles safe.
- There should be suitable traffic-control arrangements, including speed limits and, where appropriate, signallers. Crew may need to collaborate with shoreside management where they also control vehicle movements on board ship.
- Loading supervisors and personnel directing vehicles should use unambiguous hand signals.
- Ensure there is adequate lighting.

- Vehicle deck personnel must maintain situational awareness at all times when vehicles are moving. They must never be distracted by the use of any personal hand-held communication devices such as mobile phones.
- The risk assessment procedures must be followed and periodically reviewed to identify any areas where personnel could become trapped by reversing vehicles during loading operations. These areas should be marked and not used by personnel, unless appropriate mitigating measures have been put in place (see section 1.2.6 on risk awareness and risk assessment and MGN 621 (M+F) Roll-on/roll-off ships – guidance for the stowage and securing of vehicles, section 2.4 risk assessment).
- Personnel should keep out of the way of moving vehicles, particularly those that are reversing, by standing to the side. Where possible they should remain within the driver's line of sight. They should take extra care at the 'ends' of the deck where vehicles may converge from both sides of the ship.
- Crew members should be aware that vehicles may lose control on ramps and sloping decks, especially when wet, and that vehicles on ramps with steep inclines may be susceptible to damage. Ramps should have a suitable slip-resistant surface.
- Where audible alarms are fitted to vehicles, drivers should sound them when reversing.
- Provide safe systems of work to ensure that competent people direct all vehicle movements.

MGN 621 (M+F)

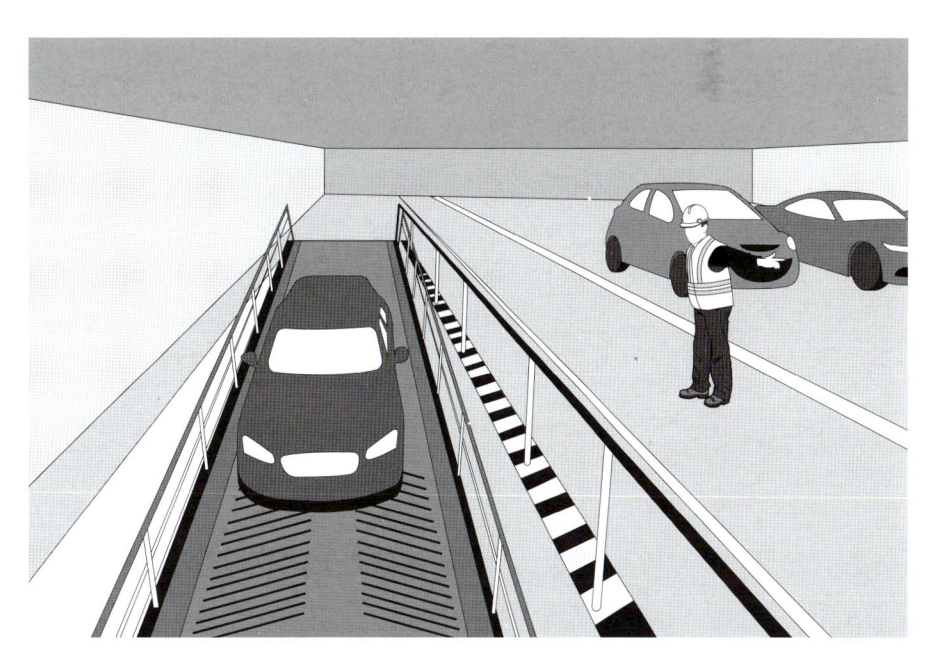

Figure 27.2 Loading a car deck

For further information see the Guidelines To Shipping Companies On Vehicle Deck Safety (produced in conjunction with the Chamber of Shipping National Maritime Occupational Health and Safety Committee).

27.5 Use of work equipment

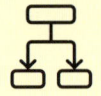

27.5.1 Keep moveable deck ramps clear of passengers when being raised or lowered. When cars are lowered on the ramps of moveable decks, they should be suitably **chocked** 🔍 as shown in Figure 27.3. If the operator cannot clearly see the whole operation from the control station, post a lookout to ensure ramp and landing areas remain clear throughout the operation.

Figure 27.3 Examples of vehicle wheel chocks

27.5.2 Ramps, retractable car decks or lifting appliances should not lift people, unless the equipment has been designed or especially adapted for that purpose.

27.5.3 Lock retractable car decks and lifting appliances securely when in the stowed position.

27.5.4 After loading all vehicles isolate the car-deck hydraulics so that they cannot be accidentally activated during the voyage, and inform the bridge.

27.5.5 Secure the ship's mobile handling equipment, which is not fixed to the ship, in its stowage position before the ship proceeds to sea.

27.6 Inspection of vehicles

27.6.1 Before accepting a freight vehicle for shipment a competent and responsible person or persons should inspect it externally to check that it is in a satisfactory condition for shipment; for example, that:

- it is suitable for securing to the ship in accordance with the approved cargo-securing manual (see also section 28.2.3)
- where practicable, the load is secured to the vehicle
- the deck or doorway is high enough for vehicles to pass through and vehicles have adequate clearance for ramps with steep inclines
- any labels, placards and marks that would indicate the carriage of dangerous goods are properly displayed.

27.6.2 Ensure, as far as is reasonably practicable, that on each vehicle the fuel tank is not so full as to create a possibility of spillage. Do not load any vehicle showing visual signs of an overfilled tank.

　　MGN 341 (M)

Seafarers should be aware of hazardous units as detailed on the stowage plan and indicated by labels, placards and marks. They should be on guard against the carriage of undeclared dangerous goods.

27.7 Stowage

27.7.1 Observe shippers' special advice or guidelines regarding handling and stowage of individual vehicles.

27.7.2 Vehicles should:

- as far as possible, be aligned in a fore and aft direction
- be closely stowed athwartships so that, if the securing arrangements fail, or from any other cause, the vehicles' transverse movement is restricted. However, provide enough distance between vehicles to allow safe access for the crew and for passengers getting into and out of vehicles, and going to and from accesses serving vehicle spaces
- be loaded so that there are no excessive lists or trims likely to cause damage to the vessel or shore structures.

27.7.3 Vehicles should not:

- be parked on permanent walkways
- be parked so as to obstruct the operating controls of bow and stern doors, entrances to accommodation spaces, ladders, stairways, companionways or access hatches, firefighting equipment, controls to deck scupper valves or controls to fire dampers in ventilation trunks
- be stowed across water spray fire curtains, if these are installed.

27.7.4 Maintain safe means of access to securing arrangements, safety equipment and operational controls. Clearly mark stairways and escape routes from spaces below the vehicle deck with yellow paint and keep them free from obstruction at all times.

27.7.5 The parking brakes of each vehicle or each element of a vehicle, where provided, should be applied and the vehicle should, where possible, be left in gear.

27.7.6 Semi-trailers should not be supported on their landing legs during sea transport unless the legs are specially designed and marked for that purpose, and the deck plating is strong enough for the point loadings.

27.7.7 Uncoupled semi-trailers should be supported by trestles or similar devices placed in the immediate area of the drawplates so as not to restrict the connection of the fifth wheel to the kingpin.

27.7.8 Drums, canisters and similar thin-walled packaging are susceptible to damage if vehicles break adrift in adverse weather. They should not be stowed on the vehicle deck without adequate protection.

27.7.9 Depending on the area of operation, the predominant weather conditions and the characteristics of the ship, stow freight vehicles so that the chassis are kept as static as possible by not allowing free play in the suspension. To do this, secure the vehicles to the deck as tightly as the lashing tensioning device will permit. Take care not to over-tighten the lashings. Use only designed tensioning arrangements; do not use additional extensions to increase the tightening force. Alternatively, jack up the freight vehicle chassis before securing it.

27.7.10 Because compressed air suspension systems may lose air, try to prevent lashings slackening off as a result of air leakage during the voyage. Arrangements may include jacking up a vehicle or releasing air from the suspension system where this facility is provided.

27.8 Securing of cargo

27.8.1 Complete cargo securing operations before the ship proceeds to sea.

27.8.2 Within the constraints laid down in the approved cargo securing manual, the master has the authority to decide on the application of securings and lashings and the suitability of the vehicles to be carried. In making this decision, due regard shall be given to the principles of good seamanship, experience in stowage, good practice and the International Maritime Organization (IMO) Code for Cargo Stowage and Securing (CSS Code).

27.8.3 Seafarers appointed to carry out the task of securing vehicles should be trained in the use of the equipment and in the most effective methods for securing different types of vehicles.

27.8.4 Securing operations should be supervised by competent persons who are conversant with the contents of the cargo-securing manual. Secure freight vehicles of more than 3.5 tonnes whenever the expected conditions for the voyage are such that the vehicles could move relative to the ship.

27.8.5 During the voyage, inspect the lashings regularly to ensure that vehicles remain safely secured. Seafarers inspecting vehicle spaces during a voyage should take care to avoid being injured by moving or swaying vehicles. Always notify the officer of the watch when inspecting the vehicle deck of the decision to alter the ship's course to reduce movement or dangerous sway when lashings are being adjusted.

27.8.6 When using wheel chocks to restrain a semi-trailer, keep them in place until the semi-trailer is properly secured to the semi-trailer towing vehicle.

27.8.7 Do not attempt to secure a vehicle until it is parked, the brakes (where applicable) have been applied and the engine has been switched off.

27.8.8 When stowing vehicles on an inclined deck:

- Chock the wheels before lashing commences. The tug driver should not leave the cab to disconnect or connect the trailer brake lines; a second person should do this.
- The parking brake on the tug should be engaged and in good working condition.
- As well as wheel chocks, leave at least two lashings holding the unit against the incline in place until the trailer's braking system is charged and operating correctly.

27.8.9	Where seafarers are working in shadow areas or have to go under vehicles to secure lashings, provide hand lamps and torches.
27.8.10	Seafarers securing vehicles should take care to avoid injury from projections on the underside of the vehicles. The driver and the lashing crew should agree on a method of signalling, preferably using a whistle or other distinct sound signal.
27.8.11	Wherever possible, attach lashings to specially designed securing points on vehicles. Attach only one lashing to any one aperture, loop or lashing ring at each securing point.
27.8.12	When tightening lashings, take care to attach them securely to the deck and to the securing points of the vehicle.
27.8.13	Apply hooks, and other devices used for attaching a lashing to a securing point, in a manner that prevents them from detaching if the lashing slackens during the voyage.
27.8.14	Lashings should be attached so that, provided there is safe access, it is possible to tighten them if they become slack.
27.8.15	All the lashings on a vehicle should be under equal tension.
27.8.16	Where practicable, have the same arrangement of lashings on both sides of a vehicle. Angle them to provide some fore and aft restraint, with an equal number pulling forward as are pulling aft.
27.8.17	The lashings are most effective on a vehicle when they make an angle with the deck of between 30° and 60°. When you cannot achieve these optimum angles use additional lashings.
27.8.18	Avoid using crossed lashings for securing freight vehicles if possible because this arrangement provides no restraint against tipping over at moderate angles of roll of the ship. Lashings should pass from a securing point on the vehicle to a deck-securing point adjacent to the same side of the vehicle. Where there is concern about the possibility of low coefficients of friction on vehicles such as solid-wheeled trailers, use additional crossed lashings to restrain sliding. Consider using rubber mats.
27.8.19	Do not release lashings for unloading before the ship is secured at the berth without the master's express permission.
27.8.20	Seafarers should release lashings with care to reduce the risk of injury when the tension is released.
27.8.21	To avoid damage during loading and unloading, keep all unused securing equipment clear of moving vehicles on the vehicle deck.

27.8.22 A competent person should inspect securing equipment to ensure that it is in sound condition at least once every six months and whenever it is suspected that lashings have experienced loads above those predicted for the voyage. Take defective equipment out of service immediately and either dispose of it or place it where it cannot be used inadvertently. Stow any unused lashing equipment away from the vehicle deck.

27.9 Dangerous goods

27.9.1 Read this section in conjunction with Chapter 21. For guidance on dealing with emergencies involving dangerous goods, see Chapter 4 and the International Maritime Dangerous Goods (IMDG) Code.

27.9.2 Before loading, examine freight vehicles carrying dangerous goods externally for damage and signs of leakage or shifting of contents. Do not accept for shipment any vehicle that is damaged or leaking or that has shifting contents. If a freight vehicle is leaking after loading inform a ship's officer. All personnel should keep well clear until it is ascertained that no danger persists.

27.9.3 Always secure freight vehicles carrying dangerous goods, and adjacent vehicles.

27.9.4 Give special attention to tank vehicles and tank containers on flat-bed trailers containing products declared as dangerous goods. Pre-voyage booking procedures should ascertain that tanks have been approved for the carriage of their contents by sea.

27.10 Specialised vehicles

27.10.1 Secure gas cylinders used for the operation and business of vehicles such as caravans adequately against movement of the ship and cut off the gas supply throughout the voyage. Refuse to ship any leaking or inadequately secured or connected cylinders.

MGN 341 (M); MGN 545 (M+F); MGN 552 (M)

27.10.2 Give special consideration to the following vehicles, trailers and loads:

- Tank vehicles or tank containers containing liquids not classified as dangerous goods. These may be sensitive to penetration damage and may act as a lubricant. Always secure these vehicles.
- Tracked vehicles and other loads making metal-to-metal contact with the deck; where possible, use rubber mats or dunnage.
- Loads on flat-bed trailers.
- Vehicles with hanging loads, such as chilled meat or floated glass.
- Partially filled tank vehicles.

> Consider the guidance in MGN 653 (M) Amendment 1, which includes the means for identifying electric vehicles (eg by wing mirror cards), fire detection, specialised fire suppression, and charging.

27.10.3 Carry small electric vehicles, such as e-bikes and e-scooters, on vehicle, special category and ro-ro spaces or on the weather deck or a cargo space. Secure them appropriately to avoid movement during transit (see SOLAS II-2 Regulation 20 for carriage requirements).

> *SOLAS II-2 Reg 20*

27.10.4 Freight vehicles carrying livestock require special attention to ensure that they are properly secured, adequately ventilated and stowed to allow access to the animals. Further guidance is available in the Department of the Environment, Food and Rural Affairs (Defra) Regulation on the Welfare of Animals During Transport: New Rules for Transporting Animals (see Appendix 2).

27.10.5 Where vehicles are connected to electrical plug-in facilities, personnel should take the precautions described in Chapter 18 for working with any electrical equipment.

27.10.6 Plug in electric vehicles for charging only at dedicated charging stations. Do not charge vehicles with modified batteries or any visible damage.

27.10.7 Do not carry damaged electric vehicles when the damage indicates that the battery may have been affected. A competent person should inspect vehicles thoroughly before they are allowed to be transported on board. The inspection should assess the risk of fire and, subsequently, the risk to the vessel. Ship crews are not likely to be suitably trained to identify these hazards so get declarations from suitably qualified persons before carriage. Where vehicles are being towed or carried by a car transporter, consider disconnecting the battery pack due to the uncertainty around the battery's performance. A suitably qualified person should do this.

27.10.8 Competent persons may include those recognised by the Institute of the Motor Industry (IMI) with TechSafe accreditation or similar. See MGN 653 (M) Amendment 1 for further details.

MGN 653 (M) Amendment 1

27.11 Housekeeping

27.11.1 Keep all walkways clear.

27.11.2 Keep all vehicle decks, ships' ramps and lifting appliances, as far as is reasonably practicable, free of water, oil, grease, any liquid that might cause a person to slip or that might act as a lubricant to a shifting load. Clean up any spillage of such liquid quickly. Provide sand boxes, drip trays and mopping-up equipment for use on each vehicle deck.

27.11.3 Keep all vehicle decks, ships' ramps and lifting appliances free of obstructions and loose items such as stores and refuse.

27.11.4 Seafarers should be careful to avoid electrical points and fittings when washing down vehicle decks.

27.11.5 Keep all scuppers clear of lashing equipment, dunnage, etc.

- Inspect vehicles carrying dangerous goods before loading.
- There should be no unauthorised personnel on vehicle decks.
- Provide adequate lighting on vehicle decks.
- Display clear signage for walkways.

- Keep all fire doors closed on vehicle decks when the vessel is at sea.
- Follow regulations and instructions to ensure there is adequate vehicle deck ventilation and test the atmosphere as required.

28 Dry cargo

28.1 Introduction

28.1.1 This chapter covers both packaged and dry bulk cargoes, except cargoes carried in roll-on/roll-off (ro-ro) ships which are covered in Chapter 27.

Chapters 10, 16 and 19 are also relevant to work on dry cargo ships.

28.2 Stowage of cargo

28.2.1 The safe stowage and securing of cargo depends upon proper planning, execution and supervision by properly qualified and experienced personnel.

28.2.2 In the case of dry bulk cargo (excluding grain), follow the *International Maritime Organization (IMO) Code of Practice for the Safe Loading and Unloading of Bulk Carriers*, with the associated IMO Ship/Shore Safety Checklist. For grain, detailed guidance is provided in the *International Code for the Safe Carriage of Grain in Bulk*.

28.2.3 Cargo other than bulk cargo should be loaded, stowed and secured in line with the ship's approved cargo-securing manual. Handling and safety instructions for securing devices are available in sections 3.1 and/or 4.1 of the manual. *The IMO Code of Practice for Cargo Stowage and Securing (IMO Resolution A.714(17))* provides further guidance. Secure all cargo before the ship proceeds to sea.

28.2.4 Timber cargo decks must be loaded, stowed and secured throughout the voyage following the *Code of Safe Practice for Ships Carrying Timber Deck Cargoes 2011* (the 2011 TDC Code).

Your organisation should

Ensure that the 2011 TDC Code details on how to load, stow and secure timber deck cargoes to prevent damage or hazard to the ship and people on board or loss overboard throughout the voyage are always complied with.

Key points
- Stow and secure all cargoes in a manner that will avoid exposing the ship and people on board to unnecessary risk.
- Agree the planned procedures for the handling of cargo with berth or terminal operators before loading or unloading.
- Arrange a system of work to limit the need to work on container tops.
- Stow all cargo taking into account the order of discharge at a port or number of ports.

28.2.5 The 2011 TDC Code provides:

- practices for safe transportation
- methodologies for safe stowage and securing
- design principles for securing systems
- guidance for developing procedures and instructions to be included in ships' cargo-securing manuals on safe stowage and securing
- sample checklists for safe stowage and securing.

SI 1999/336; MGN 107 (M); IMO Resolution A.714(17)

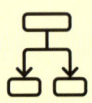

28.2.6 When planning the position of cargo and the order of loading and unloading, consider the effects that these operations will have upon access and the safety of personnel:

- Record all the cargo information, including gross mass of the cargo or cargo units and any special properties detailed on board or in the shipping documents, and use it in planning.
- Where more than one port is involved for loading or unloading, where possible load the cargo in layers rather than in tiers to avoid building high vertical walls of cargo.
- Take care not to overstow lighter cargoes with heavier cargoes, which may lead to a collapse of the stow.
- Wherever practicable, stow cargo so as to leave safe clearance behind the rungs of hold ladders and to allow safe access as may be necessary at sea.
- Minimise the need to walk across or climb onto the deck cargo where this may involve an approach to an unprotected edge with a risk of falling.
- Take care to avoid large gaps next to cargo where it is stacked against corrugated bulkheads.

28.2.7 Safe access

- Stow deck cargo in line with statutory requirements and keep it clear of hatch coamings to allow safe access.
- Always allow clear access to safety equipment, firefighting equipment (particularly fire hydrants) and sounding pipes.
- Make any obstructions in the access way, such as lashings or securing points, clearly visible by painting them white or another contrasting colour. Where this is impracticable and cargo is stowed against the ship's rails or hatch coamings to such a height that the rails or coamings do not protect personnel from falling overboard or into the open hold, provide temporary fencing (see section 11.5).

SI 1998/2241

28.2.8
Rig suitable safety nets or temporary fencing where personnel are at risk of falling because they have to walk or climb across built-up cargo.

28.2.9
When deck cargo is stowed against and above ship's rails or bulwarks, provide a wire rope pendant or a chain, extending from the ring bolts or other anchorage on the decks to the full height of the deck cargo. This is to avoid personnel having to go overside to attach derrick guys and preventers directly to the anchorages on the deck.

28.2.10
Where beams and hatch covers have to be removed at intermediate ports before unloading surrounding deck cargo, leave an access space at least 1 metre wide adjacent to any part of the hatch or hatchway that is to be opened. If this is impracticable on deck use fencing or lifelines to enable seafarers to remove and replace beams and hatch coverings safely (see section 11.5).

28.2.11
In the tween decks, the guidelines should be painted around the tween deck hatchways at a distance of 1 metre from the coamings.

28.3 Dangerous goods and substances

28.3.1
Packaged dangerous goods are marked, labelled or placarded to indicate the contents and their hazardous or polluting properties. Use this information to assess any risk to seafarers and put in place any necessary safety measures. Merchant shipping regulations lay down requirements for the carriage of dangerous substances. Follow the provisions of the International Maritime Dangerous Goods (IMDG) Code together with those in the relevant merchant shipping notices. The IMDG Code contains details of classification, documentation, marking and labelling, and packaging and advice to meet the requirements of the regulations. In particular it lists and gives details of many dangerous substances.

SI 1997/2367; MGN 340 (M)

28.3.2 The general introduction and the introductions to individual classes of dangerous goods in the IMDG Code contain many provisions to ensure the safe handling and carriage of dangerous goods, including requirements for electrical equipment and wiring, firefighting equipment, ventilation, smoking, repair work, provision and availability of special equipment. Some of these are general for all classes and others specific to certain classes. Refer to this information before handling dangerous goods. Some of the requirements are highlighted below. If in doubt consult the Maritime and Coastguard Agency (MCA) or other competent authority.

Chapter 7.8 of the IMDG Code provides advice on special requirements in the event of an incident and fire precautions involving dangerous goods. Follow this, in line with the ship's safety management system, in the event of spillage or other incident.

28.3.3 A responsible, competent officer should supervise the loading and unloading of dangerous goods. If applicable they should do this in accordance with the ship's document of compliance for the carriage of dangerous goods. Take suitable precautions, such as providing special lifting gear as appropriate, to prevent damage to receptacles containing dangerous goods.

28.3.4 Load, stow and carry dangerous substances in bulk in line with Appendix 1 of the *International Maritime Solid Bulk Cargoes Code* (IMSBC) published by IMO.

28.3.5 Establish emergency response procedures for the substances carried in line with the IMO's *Emergency Procedures for Ships Carrying Dangerous Goods* (EmS Guide). Check this to ensure that the ship carries the appropriate emergency equipment. The application of such measures is under the control of the master of the ship; it will depend on the circumstances of the incident and the location of the ship. Keep the equipment necessary for the execution of the emergency response immediately available and ensure the crew are trained and practised in its use. The *Medical First Aid Guide for Use in Accidents Involving Dangerous Goods* (MFAG – MSC 1/Circ 857) should also be available.

MSN 1706 Amendment 1

28.3.6 Emergency response procedures should include cases of accidental spillage or exposure (see section 28.3.8) and the possibility of fire.

28.3.7 Personnel who handle consignments containing dangerous substances should be able to identify dangerous goods from the labelling and placarding. They should have, and wear, personal protective equipment (PPE) including breathing apparatus, where necessary, appropriate to the hazard. More information on PPE when handling dangerous substances is available in Chapter 8.

MSN 1870 (M+F) Amendment 5

28.3.8 **Spillage of dangerous substances**

- Seafarers should promptly report any leakage, spillage or any other incident that occurs and involves exposure to dangerous substances. In the event of accidental exposure, refer to the MFAG published by IMO.
- Personnel who are required to deal with spillages or to remove defective packages should have, and wear, suitable breathing apparatus and protective clothing as the circumstances dictate. Suitable rescue and resuscitation equipment should be readily available in case of an emergency (see Chapter 8).
- Take the appropriate measures promptly to control any spillage of dangerous substances.
- Take particular care when the ship carries dangerous substances in refrigerated spaces where any spillage may be absorbed by the insulating material. Inspect insulation affected in this way and renew it if necessary.

28.3.9 Where there is leakage or escape of dangerous gases or vapours from cargo, personnel should leave the danger area and it should be treated as a dangerous (enclosed) space (see Chapter 15).

28.3.10 Further guidance on the handling and stowage of dangerous goods is available in the *Recommendations on the Safe Transport of Dangerous Cargoes and Related Activities in Port Areas* published by IMO.

28.4 Carriage of containers

28.4.1 Containers are simply packages of pre-stowed cargo, and sections of Chapters 16 and 19 may be relevant to their safe working. Guidance is also published by the UK's Port Skills and Safety organisation in its Health and Safety in Ports series, *SIP Leaflet 008 – Guidance on the storage of dry bulk cargo*.

28.4.2 Where a container holds dangerous goods, follow the relevant guidance in section 28.3. For guidance on control of substances hazardous to health see Chapter 21.

28.4.3 Freight containers should comply with the International Convention for Safe Containers 1972 (CSC), under which they must carry a safety approval plate (CSC plate). Report any defective containers or containers on which the CSC plate is missing so they can be taken out of service.

28.4.4 Containers should not be loaded beyond the maximum net weight indicated on the CSC plate, and should be in a safe condition for handling and carriage.

28.4.5
- Use equipment for lifting a container that is suitable for the load and safely attached to the container.
- The container should be free to be lifted. Lift it slowly to guard against it swinging or some part of the lifting appliances failing. This is in case there are any unsecured contents which may be unevenly loaded and poorly distributed, or in case the weight of contents incorrectly declared.
- When loading and securing goods into a container follow the IMO/ILO/UN/ECE Guidelines for Packing of Cargo Transport Units (CTUs).
- Take special care when lifting a container with a mobile centre of gravity, such as a tank container, bulk container or a container with contents that are hanging.

28.4.6 Provide safe means of access to the top of a container to release lifting gear and to fix lashings. Protect personnel from falling, where appropriate, by providing a properly secured safety harness or other suitable means.

28.4.7 Lash all containers individually and get a **competent person** to check them. Where containers are stacked, consider the appropriate strength features of the lashing and stacking-induced stress.

28.4.8 On ships that are not specially constructed or adapted for the carriage of containers, wherever possible stow the containers fore and aft and securely lashed. Do not stow containers on decks or hatches unless it is known that the decks or hatches are of adequate overall and point load-bearing strength. Use adequate dunnage.

28.4.9 Where the design for securing containers and checking lashing makes access onto container tops necessary, use the ship's superstructure, a purpose-designed access platform, or personnel cages which include a suitable adapted lifting appliance. If this is not possible, an alternative safe system of work should be in place.

28.4.10 To allow access to the tops of overheight, soft-top or tank containers, where necessary, for securing or cargo-handling operations, stow solid top or 'closed containers' between them whenever practicable.

28.4.11 Where the ship's electrical supply is used for refrigerated containers, provide the supply cables with proper connections for the power circuits and for earthing the container. Before use, inspect the supply cables and connections, repair any defects and get a competent person to test them. Handle supply cables only when the power has been switched off. Where there is a need to monitor and repair refrigeration units during the voyage, consider whether to provide safe access in a seaway when stowing these containers.

28.4.12 Personnel should be aware that containers may have been fumigated at other points in the transport chain, and there may be a residual hazard from the substances used.

28.5 Working cargo

28.5.1 For regulations and guidance on lifting equipment and lifting operations, including examination and testing requirements, see Chapter 19.

28.5.2 Safety arrangements made before working cargo should ensure that adequate and suitable lifting equipment is available, in accordance with the register of lifting appliances and cargo gear, and that all plant and equipment and any special gear necessary is available and used. Check cargo gear regularly throughout the cargo operation for damage or malfunction.

28.5.3 Do not do repair or maintenance work, such as chipping, spray painting, shot blasting or welding, in a space where cargo operations are in progress.

28.5.4 When loads are lowered or hoisted they should not pass or remain over any person working in the cargo space area, or over means of access. Personnel should take care when using access ladders in hatch squares while cargo operations are in progress.

28.5.5 Cargo information for goods should always show the gross mass of the cargo or of the cargo units. Where loads of significant gross mass are not marked with their weight, check-weigh the loads unless accurate information is available, as provided by the shipper or packer of the goods.

28.5.6	There should always be a signaller at a hatchway when cargo is being worked, unless the crane driver or winch operator has a complete, unrestricted view of the load or total working area. The signaller should be in a position that gives them a total view of the operation; where this is not possible additional signallers should assist. Guidance for signallers is given in sections 19.10.5 to 19.10.9 and Annex 19.3.
28.5.7	Before giving a signal to hoist, the signaller should have clearance from the person making up the load that it is secure. They should be sure that no-one else would be endangered by the hoist. Before giving the signal to lower, the signaller should warn personnel in the way and ensure they are clear.
28.5.8	Raise and lower loads smoothly, avoiding sudden jerks or 'snatching'. When a load does not ride properly after being hoisted, the signaller should immediately give warning of danger and the load should be lowered and adjusted as necessary.
28.5.9	Do not load hooks, slings or other lifting gear beyond their safe working loads. Strops and slings should be of sufficient size and length for safe use. Pull them tight enough to prevent the load, or any part of it, from slipping and falling. Put loads (sets) together and sling them properly before hoisting or lowering them.
28.5.10	Before swinging any heavy load give it a trial lift to test the effectiveness of the slinging.
28.5.11	Except to break out or make up slings, do not attach lifting hooks to: • the bands, strops or other fastenings of packages of cargo, unless these fastenings have been specifically provided for lifting • the rims (chines) of barrels or drums for lifting, unless the construction or condition of the barrels or drums enables safe lifting with properly designed and constructed can hooks.
28.5.12	Take suitable precautions, such as using packing or chafing pieces, to prevent chains, wire and fibre ropes from being damaged by the sharp edges of loads.
28.5.13	When using slings with barrel hooks or other similar holding devices where the weight of the load holds the hooks in place, lead the sling down through the egg or eye link and through the eye of each hook in turn so that the horizontal part of the sling draws the hooks together.
28.5.14	The angle between the legs of the slings should not normally exceed 90° because this reduces the safe working load of the sling. Where this is not reasonably practicable, you may increase the angle up to 120° as long as the slings have been designed to work at the greater angles. However, note that at 120° each sling leg is taking stress equivalent to the whole mass of the load.

28.5.15		Load trays and pallets (unit loads) using a pallet loader where available. If using slings hoist the trays and pallets with four-legged slings. Where necessary, use nets and other means to prevent any part of the load falling.
28.5.16		Sling bundles of long metal goods, such as tubes, pipes and rails, with two slings or strops and, where necessary, a spreader. Double-wrap and secure slings or strops to prevent the sling coming loose. Also attach a suitable lanyard where necessary.
28.5.17		Load or discharge logs using wire-rope slings of adequate size; do not use tongs except to break out loads.
28.5.18		Fit cargo buckets, tubs and similar appliances carefully so there is no risk of the contents falling out. They should be securely attached to the hoist (eg by a shackle) to prevent tipping and displacement during hoisting and lowering.
28.5.19		Use shackles for slinging thick sheet metal if there are suitable holes in the material; otherwise use suitable clamps on an endless sling.
28.5.20		Put loose goods such as small parcels, carboys and small drums in suitable boxes or pallets with sufficiently high sides for loading or discharge, and lift them using four-legged slings.
28.5.21		When returning slings or chains to the loading position hook them securely on the cargo hook before the signaller gives the signal to hoist. Attach hooks or claws to the egg link or shackle of the cargo hook; do not allow them to hang loose. Keep the cargo hook high enough so that slings or chains are clear of personnel and obstructions.
28.5.22		Do not take 'one-trip slings' (slings that have not been used previously for lifting and are fitted to the load before loading) back on board ship after the load has been discharged at the end of the voyage. Leave them on shore for disposal.
28.5.23		When work is interrupted or has ceased temporarily leave the hatch in a safe condition with either guardrails or hatch covers in position.

28.6 Lighting in cargo spaces

28.6.1 During cargo operations, cargo spaces should be adequately lit. Avoid strong contrasts of light and shadow or dazzle (see section 11.4). Do not use open or naked lights. Portable lights should be adequately guarded, suitable for the task, and firmly secured in such a manner that they cannot be accidentally damaged. Never lower portable lights or suspend them by their electrical leads. Run leads so they are clear of loads, running gear and moving equipment.

28.7 General precautions for personnel

28.7.1 Where crew are working alongside shore-based personnel in cargo operations, provide the same level of safety to both shore- and ship-side personnel. Each should be aware of the other's risk assessment and procedures to ensure there is common understanding.

28.7.2 Personnel working in cargo spaces should move with caution over uneven surfaces or loose dunnage, and be alert to protrusions such as nails.

28.7.3 Where vessels have been built with corrugated bulkheads, erect precautions such as suitable rails, grids or nets to prevent cargo handlers or other personnel from falling into the space between the rear of the corrugation and the stowed cargo.

28.7.4 Where personnel are working on or near the cargo 'face' the face should be secured against collapse, especially where bagged cargo may be bleeding from damage. Where it is necessary to mount a face, use a portable ladder. This should be properly secured against slipping or shifting sideways or held in position by other personnel. When personnel are working in areas where there is a risk of falling, put up safety net(s). Do not secure the nets to hatch covers.

28.7.5 Personnel should be aware that cargoes may have been fumigated at other points in the transport chain, and there is a risk that toxic fumes may build up in enclosed spaces. An enclosed space entry permit may be required to work in or around these areas. The designated competent person(s) on board the ship must do this risk assessment.

28.8 Moveable bulkheads in cargo holds

28.8.1 Moveable bulkheads are fitted in some small, multi-purpose vessels to allow more flexibility in the types of dry cargo carried from one voyage to the next.

28.8.2 There have been several serious and even fatal accidents when seafarers were moving or carrying out maintenance on these types of bulkheads. The procedures for the operation and maintenance of moveable bulkheads should be documented within the ship's safety management system.

28.8.3 Personnel doing work that involves moving the bulkhead or doing maintenance and hold cleaning should, before starting these duties, follow the risk assessment for these specific operations with these bulkheads.

28.8.4 Personnel doing these duties must be fully trained and competent in the moving operations associated with these bulkheads and, where required, with the jacking up of these bulkheads for hold cleaning. An officer or other supervisor who is familiar with these types of bulkhead and competent to oversee such operations must supervise these personnel at all times.

28.8.5 Seafarers must be trained before they are given duties associated with these bulkheads.

28.8.6 Because the operations involved with these bulkheads are so dangerous, consider issuing a **permit to work** 🔍 for any duties associated with them.

28.8.7 In the operation of certain designs of moveable bulkhead, also consider, when jacking up these bulkheads for hold-cleaning purposes, or for inspection and maintenance purposes, using additional temporary holding supports at the upper end, when the 'swing-over' wheel system for moving these bulkheads cannot be engaged.

- Load, stow and carry dangerous substances in bulk in line with IMSBC.
- When carrying dangerous goods establish the emergency response procedures within the ship's safety management system.
- Load, stow and secure cargo other than bulk cargo in line with the ship's approved cargo-securing manual.
- Load, stow and secure timber cargo decks throughout the voyage in line with the 2011 TDC Code.

- When planning the stowage of cargo, good communications are essential to ensure it is placed correctly on board and seafarers are aware of its contents.
- Seafarers must wear PPE as instructed.

29 Tankers and other ships carrying bulk liquid cargoes

29.1 Introduction

29.1.1 Seafarers appointed to work on tankers or similar vessels must meet the special training and qualifications requirements specified under Chapter V of the International Conventions on Standards of Training, Certification and Watchkeeping for Seafarers (STCW Convention), 1978, as amended.

> SI 2022/1342; MSN 1866 (M) Amendment 1

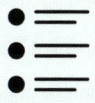

Key points

- There are specific risks involved in the carriage of bulk liquid cargoes which vary according to ship and cargo type. Cargoes may present a risk of fire or explosion; produce toxic vapours; be harmful in contact with the skin; or have other harmful characteristics.
- Seafarers should ensure that they fully understand the company's operational and emergency procedures and best practice industry guidelines relating to the carriage of bulk liquid cargoes.

> International Safety Guide for Oil Tankers and Terminals (ISGOTT) and Tanker Safety Guides (Chemicals and Liquefied Gas) published by the International Chamber of Shipping (ICS)

Your organisation should

- ensure that seafarers working on oil, chemical and gas tankers or similar vessels have had the special training and hold the relevant qualifications specified in regulation V/1-1 and V/1-2 of the International Conventions on Standards of Training, Certification and Watchkeeping for Seafarers (STCW convention), 1978 as amended
- assess the risks arising from the carriage of bulk liquid cargoes, using any information available, including relevant best practice industry guidelines
- ensure that crew have regular training in emergency procedures and special emergency equipment.

29.1.2 Companies should assess the risks arising from bulk liquid cargoes, using any information available, especially the chemical data sheets in the Tanker Safety Guides (gas and chemical) from the International Chamber of Shipping.

29.1.3 All seafarers should be trained to use the relevant personal protective equipment and in emergency procedures, including use of special emergency equipment. Guidance on general operational procedures and precautions is given in the relevant guidance documents for the ship types; details are provided in this chapter.

29.1.4 Companies are required under the International Safety Management (ISM) Code to have their own documented safety management procedures. Relevant statutory publications, best-practice industry guidelines and detailed safety management system procedures should be available on board for crew.

29.1.5 Crew members should have regular training in emergency procedures and the use of any special emergency equipment, as appropriate. Instruction should include personal first-aid measures for dealing with accidental contact with harmful substances in the cargo being carried and with inhalation of dangerous gases and fumes.

> *International Maritime Organization (IMO) EmS Guide; IMO Medical First Aid Guide (MFAG)*

29.1.6 Owing to the risks of ill-effects arising from contamination by certain liquid cargoes, especially those carried in chemical tankers and gas carriers, seafarers should maintain high standards of personal hygiene, particularly when they have been engaged in cargo handling and tank cleaning. To prevent exposure to hazardous chemicals, consider use of decontamination showers and wearing appropriate PPE such as full body chemical suits with breathing and eye protection. The primary aim must be to prevent the various routes for chemical exposure to one's body: inhalation, ingestion, injection, and eye and skin contact.

29.1.7 Seafarers on board responsible for the safe loading and carriage of the cargo should have all the relevant information about its nature and character before it is loaded. They need information about the precautions to take during the voyage, and they should also be trained in handling procedures. Other seafarers should be advised of any precautions that they too should take. The relevant material safety data sheets for the cargo being carried should be available on board before the cargo is loaded.

29.1.8 Rules restricting smoking and the carriage of matches or cigarette lighters and electronic devices should be strictly observed. Any equipment used in hazardous cargo areas should be certified as intrinsically safe.

29.1.9 Clear up any spillages and leakages of cargo promptly. Do not discard oil-soaked rags carelessly where they may be a fire hazard or ignite spontaneously. Do not allow other combustible rubbish to accumulate. Follow the procedures under the approved shipboard oil pollution emergency plan or shipboard marine pollution emergency to deal with any spills.

MARPOL Annex I/37 and Annex II/17

29.1.10 Keep cargo-handling equipment, testing instruments, automatic and other alarm systems in good working condition. Equipment should be calibrated and serviced at the manufacturer's recommended intervals. Electrical equipment for use in the cargo area should be of an approved design and certified safe. The safety of this equipment depends on its being properly maintained by competent people. Unauthorised people should not interfere with this equipment. Report immediately any faults, such as loose or missing fastenings or covers, corrosion, and cracked or broken lamp glasses.

29.1.11 Work that might cause sparking or that involves heat should not be done unless authorised by the responsible person. This work should be authorised only after the work area has been tested and found gas-free, or its safety is otherwise assured by evaluating all the risks involved.

29.1.12 Where people have to enter any dangerous (enclosed) space, the precautions given in Chapter 15 should be strictly observed. Dangerous gases may be released or leak from adjoining spaces while work is in progress so the atmosphere should be tested frequently. Follow the guidance in Chapter 14 on **permit to work** systems.

SI 2022/0096

29.2 Oil and bulk ore/oil carriers

29.2.1 Tankers and other ships carrying petroleum or petroleum products in bulk are at risk from fire or explosion arising from ignition of vapours from the cargo, which may in some circumstances enter into any part of the ship.

29.2.2 Additionally, vapours may be toxic, some even in low concentrations, and some liquid products are harmful on contact with the skin.

29.2.3 Guidance on the general precautions to take is given in ISGOTT, published by ICS, Oil Companies International Marine Forum and the International Association of Ports and Harbours. Companies are also required, under the ISM Code, to have their own documented safety management procedures. These publications and detailed procedures should be available on board and the guidance diligently followed.

29.2.4 Consider all hazards associated with the generation of static electricity during loading and discharging of cargo and during tank cleaning, dipping, **ullaging** and sampling.

29.2.5 Where an inert gas system is fitted under the regulations, the oxygen content in the inert gas delivered to cargo tanks should not exceed 5% by volume. This will help ensure that the oxygen content of the atmosphere in any part of the cargo tank(s) does not exceed 8% by volume. In this way it prevents a flammable atmosphere developing in the cargo tanks.

29.2.6 Take additional precautions in line with ISGOTT when handling static accumulator cargoes and those containing hydrogen sulphide (H_2S) and benzene.

 SI 2014/1512; FSS Code

29.3 Liquefied gas carriers

29.3.1 Tankers and other ships carrying liquified gas in bulk are at risk from vapour cloud explosion and boiling liquid expanding vapour explosion. These can be particularly dangerous to everyone on board and if in port, the local area, especially if safe systems of work are not followed.

> *Guidance on the general precautions to take on these vessels is given in the Tanker Safety Guide (Liquefied Gas) published by the ICS. The International Code for Construction and Equipment of Ships Carrying Liquefied Gases in Bulk (IGC) contains guidance on operational procedures for safe transportation of liquefied gases in bulk.*

 SI 1994/2464

29.3.2 Note that cargo pipes, valves and connections, and any point of leakage of the gas cargo, may be very cold. Accidental contact with these may cause severe cold burns.

29.3.3 Reduce pressure carefully and drain the liquid cargo from any point of the cargo transfer system, including discharge lines and cargo manifolds. When necessary purge with nitrogen before disconnection.

29.3.4 Some cargoes, such as ammonia, have a very pungent, suffocating odour. Even very small quantities may cause eye irritation and disorientation together with chemical burns. Seafarers should take this into account when moving about the vessel, and especially when climbing ladders and gangways. The means of access to the vessel should allow the cargo to be closely supervised and should be sited as far away from the manifold area as possible. Seafarers should know where to find eyewash equipment and safety showers.

29.4 Chemical carriers

29.4.1 The cargoes carried in bulk on chemical tankers range from the so-called non-hazardous to those that are extremely flammable, toxic or corrosive, or have a combination of these properties, or that have other hazardous characteristics.

29.4.2 A chemical tanker may carry just one or a few products. Alternatively it may have many cargo tanks carrying numerous products which may be completely segregated, if required, to avoid cross-contamination.

29.4.3 IMO has developed codes (International Code for the Construction and Equipment of Ships Carrying Dangerous Chemicals in Bulk (IBC Code) and the Code for the Construction and Equipment of Ships Carrying Dangerous Chemicals in Bulk (BCH Code)). The codes are statutory under merchant shipping regulations and apply to chemical tankers based on their date of construction. Ships carrying cargoes in bulk that are listed in the IBC Code must display for the information of all on board any data necessary for the safe carriage of the cargo. This includes what to do in the event of spills and leaks, countermeasures against accidental personal contact, firefighting procedures and firefighting media. Further guidance on general operational procedures and precautions to follow on chemical tankers is available in the *Tanker Safety Guide (Chemicals)* published by the ICS.

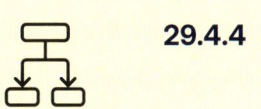

29.4.4 The ship arrangements and the equipment for cargo handling may be complex. They might require a high standard of maintenance and special instrumentation, protective clothing and breathing apparatus for entry into dangerous spaces.

SI 1996/3010

29.4.5 Many products carried on chemical tankers are loosely referred to as alcohols. Drinking these could lead to serious injury and death, and strict controls should be exercised when carrying such cargoes to prevent pilfering.

- Follow safety signs (see Chapter 9).
- Always wear appropriate PPE.
- Read the material safety data sheets for cargoes being carried.
- Use tested and calibrated gas meters, appropriate to the cargoes being carried.
- Maintain good communications throughout cargo handling operations.
- Report spillages and leakages of cargo immediately so they can be managed and cleared safely.
- Always use intrinsically safe equipment within gas-hazardous and other hazardous areas.
- Do not use any electrical or electronic equipment that has not been approved for use in gas-hazardous areas.
- Maintain positive pressure inside accommodation to prevent entry of hazardous vapours from cargo areas.

- Follow the ship's SMS and guidance on handling dangerous cargos.
- Do risk assessments to ensure the safe storage and handling of dangerous cargos.
- Always follow instructions for entry into an enclosed space, with permits to work (see Annex 14.1 and Chapter 15).

30 Port towage industry

30.1 Introduction

30.1.1 This chapter applies to seafarers working on tugs involved in towage operations within port/harbour limits. It highlights the potential hazards of towage and provides general guidance on the management of risk. This chapter refers to other relevant documents or sections of this Code so read them with it.

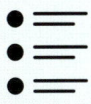

Key points
- Understand the risks of conducting towage operations.
- Take precautions during connecting/disconnecting and handling tow lines.
- Inspect and maintain towing equipment effectively.
- Maintain good communications throughout the operation.
- Move clear of hazards as soon as possible.

Your organisation should
- Based on the findings of a risk assessment, plan towage operations with appropriate control measures to protect everyone who may be at risk.
- Follow accepted best practice and the requirements and guidance of equipment manufacturers.

30.1.2 Before starting towage, prepare a comprehensive plan in liaison with the towed vessel, taking into account all relevant factors:

- the weather and sea state
- the visibility
- the condition of the equipment, and whether it is in place and fixed correctly
- which personnel to allocate to the task
- the best communication method to use
- the position of tug points, bollard pull and the limits of bitts and bollards
- the connection and disconnection points, and the destination
- the contingencies in case of emergency.

Communicate this plan to everyone involved in the operation on board.

30.2 Watertight integrity

30.2.1 Maintain the watertight integrity of a tug at all times. When the tug is engaged on any towage operation, keep all watertight openings securely fastened. The tug crew should avoid working below the waterline at this time.

30.2.2 Display signs to advise of a watertight opening that conform with Chapter 9.

30.3 Testing and inspection of towing equipment

30.3.1 Inspect all towing hooks and alarm bells daily.

30.3.2 Test the emergency-release mechanisms on towing hooks and winches both locally and, where fitted, remotely, before each operation.

30.3.3 Inspect all towing equipment in use for damage both before and after a tow. Refer to the manufacturer's instructions and company procedure for inspecting and maintaining towing equipment and when determining the safety factor and life expectancy of rope or wire.

30.3.4 Read the guidance in section 18.28 (Ropes and wires) in conjunction with this section.

📖 *MGN 592 (M+F) Amendment 1 anchoring*

30.4 Connecting and disconnecting the towing gear

30.4.1 Before beginning a tow, the master should determine which towing gear is suitable for the operation and instruct the crew accordingly.

30.4.2 When receiving heavy lines, the tug crew should be aware of the risk of injury through being struck by a **monkey's fist** 🔍 or other weighted object attached to a line. Report any dangerously weighted heaving lines.

⚠️ **30.4.3** **Warning**
Always stand clear of the area onto which the heaving line is to be thrown down. Remember that the heaving line may not land exactly where you may have indicated.

30.4.4 When connecting the tug to the assisted vessel, the tug crew should ensure that the towing gear is clear of any obstructions, able to run freely and is run out from the tug in a controlled manner.

30.4.5 During disconnection, seafarers on deck should be aware of the risk of injury if the towing gear is released by the assisted ship in an uncontrolled manner, and avoid standing directly below. They should also be aware that any towing gear that has been released and is still outboard may 'foul' on the tug's propeller(s), steelworks or fendering, causing it to come tight unexpectedly.

30.5 Use of a bridle or gog rope during towing operations

30.5.1 Use a suitable bridle, **gog rope/gob rope** 🔍 or wire in circumstances where the towline is likely to reach such an angle that a **girting** 🔍 situation may arise.

30.6 Seafarer safety during towing operations

30.6.1 The deck crew should indicate to the master once the towing gear is connected and then clear the open deck. If seafarers need to remain on deck they should stand in a safe position well clear of the winch/towline. If seafarers need to attend the towing equipment during the towing operation, they should be in contact with the bridge via radio and should keep the length of time they are exposed to the absolute minimum.

30.6.2 During towage operations, the towing gear, equipment and personnel should be continuously monitored. Personnel should relay any change in circumstances immediately to the master. This is particularly important on tugs where the master has a restricted view of those areas and personnel.

30.6.3 Where a tug is made fast, the crew of the assisted vessel should be made aware that the towing gear may have to be released in an emergency, possibly without warning.

30.6.4 Tug crews should wear appropriate personal protective equipment (see Chapter 8).

30.7 Communications

30.7.1 Before towing, the tug and the master (or pilot) of the assisted ship should agree an effective means of communication and exchange all relevant information (e.g. speed during connection). They should also agree on a secondary/alternative means of communication where possible.

30.7.2 Internal communications are equally important, and the tug master should make the crew aware of the intended operation, including any special circumstances or instructions. The master and crew should agree an effective means of communication before the operation begins and should maintain this throughout.

30.8 Interaction

30.8.1 Interaction and its effects on a tug and its handling are well known and appreciated in port/harbour towage. Masters and crew should remember that these effects increase with speed.

30.8.2 In areas where interaction exists, and when manoeuvring alongside a ship, the master should be aware of the possibility of underwater obstructions, such as bulbous bows and stabiliser fins, and areas of the ship's sides to avoid, such as pilot doors. Bow/stern thrusters and azimuth propulsion systems by the ship may present a hazard to the tug.

30.8.3 When the tug is close to or coming alongside an assisted ship, the crew should be aware of interaction and the effect it may have on the tug. There may be sudden movement or contact which could cause loss of balance or movement of equipment and other objects.

📖 *Marine Guidance Note MGN 199 (M)*

30.9 Escorting

30.9.1 Escorting as a regular operation is common within the port towage industry. It should only be done, however, after investigating whether the tug is suitable, the crew are competent and a plan can be agreed. This type of operation is carried out in both passive and active modes: passive escort means running free in close attendance and active escort means fast to the tow.

30.9.2 If active escort is being undertaken, the form of towage can be direct or indirect, depending on the speed of the tow. When the escort is made fast to the tow, masters should be aware that increased loads can be applied to towing gear, especially when operating in the indirect mode.

To maintain watertight integrity:

- Mark all watertight openings with a sign warning that they must remain closed during towage operations.
- If any openings are used while crew are moving about the tug during a towage operation they should be re-secured immediately after use.

Communicate effectively with all involved during the towing by:

- Agreeing on both the primary and secondary communication methods before starting towage operations.
- Agreeing in advance the call signs to use and the format in which towage instructions from the assisted vessel/pilot will be given.
- Being clear, concise, complete, correct and courteous, and always acknowledging requests.

- Towage operations are hazardous.
- Be prepared to respond in an emergency.
- Establish a plan for every towing operation.
- The master should conduct a towage brief with all personnel on board.
- Test and maintain equipment.

31 Ships serving offshore oil and gas installations

31.1 Introduction

31.1.1 The offshore industry has changed much over the years, affecting the way the industry works. The safe working practices set out in this chapter reflect those changes.

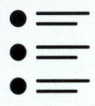

Key points
- All personnel are responsible for their own safety and the safety of those they work with. They must always act to prevent accidents and may terminate cargo operations on safety grounds at any time. All operations on deck must be risk assessed, discussed and agreed with all involved via a toolbox talk prior to the work commencing.
- When carrying out risk assessments, discussion and communications need to take place with all parties involved.
- When at sea, only seafarers involved in the cargo operation should be on the cargo deck. Other seafarers should stay clear of the work area.

Your organisation should
- ensure that appropriate personal protective equipment (PPE) is provided for all weather conditions
- ensure that arrangements are in place for crew members to have adequate breaks, particularly if working in adverse weather conditions
- assess the working areas of working decks to ensure that the noise levels, generated from both the vessel and the neighbouring platform, are within acceptable levels
- if necessary, suitable hearing protection should be provided that does not restrict or inhibit communication on the installation or between the installation and the bridge.

31.1.2 There are many different types of vessels now in regular use other than the standard platform supply vessel or anchor handling tug supply. These range from emergency response and rescue vessels (ERRVs) and their daughter crafts to more flexible types of multi-role vessels that cover the ERRV role, inter-field transfers and general cargo activities.

31.1.3 In addition, there are a wide range of specialist vessels involved in drilling, construction, platform maintenance, accommodation, diving support and other functions. This chapter provides general guidance for offshore operations, and in particular for supply vessels and anchor handling. Sources of guidance for other specialist vehicles are listed in Appendix 2.

31.2 Responsibilities

- The master has the responsibility to stop any operations that threaten the safety of the vessel or crew or the installation's integrity.
- Other pressures, whether work-related or commercial, must not interfere with the master's professional judgement. The master must inform the relevant parties of any serious conflict of interests arising from instructions or activities of other parties.
- The offshore installation manager (OIM) controls the entry of all vessels into the 500-metre zone around the installation and can modify or terminate any support vessel activity that they regard as hazardous to the installation or persons on it. However, the master of the vessel has the final responsibility for ensuring the safety of the vessel and the crew.
- The OIM may delegate operational tasks to other competent installation personnel.

31.3 General precautions

31.3.1 Seafarers working in cold and wet conditions should wear waterproof garments over warm clothing. Arrangements should be made for relief at suitable intervals to avoid undue exhaustion and hands and limbs becoming cold and numb. Consideration should be given to breaks for seafarers if operations are to continue for several hours.

31.3.2 If working on deck cannot be avoided during adverse weather, consideration should be given to adjusting the ship's heading and speed to provide as safe a platform as possible. Lifelines should be rigged on the working deck to facilitate safe movement. Decks should, as far as practicable, be kept free from ice, slush, algae and any substance or loose material likely to cause slips and falls. A lookout should be kept to give warning of imminent oncoming, quartering or following seas, or the operation suspended until the risk of shipping seas is over.

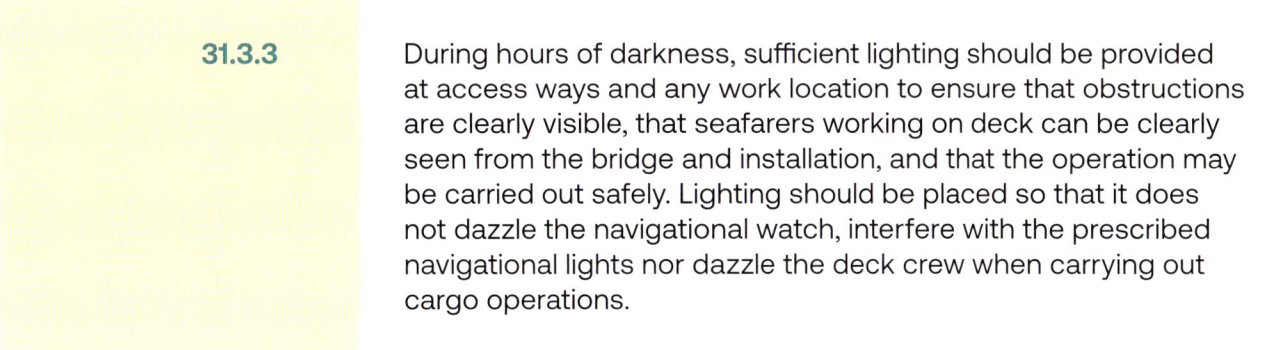

31.3.3		During hours of darkness, sufficient lighting should be provided at access ways and any work location to ensure that obstructions are clearly visible, that seafarers working on deck can be clearly seen from the bridge and installation, and that the operation may be carried out safely. Lighting should be placed so that it does not dazzle the navigational watch, interfere with the prescribed navigational lights nor dazzle the deck crew when carrying out cargo operations.
31.3.4		Owing to the unpredictable movement of vessels, especially in regard to the rise and fall, the use of tag lines should be considered only in exceptional circumstances and after a thorough risk assessment.

31.4 Personal protective equipment

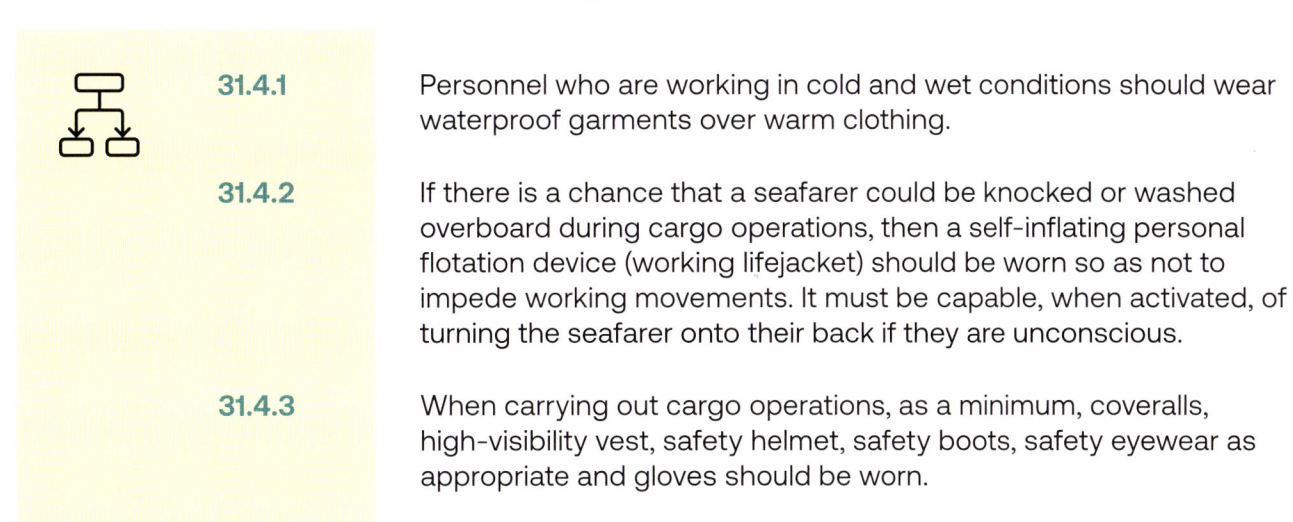

31.4.1		Personnel who are working in cold and wet conditions should wear waterproof garments over warm clothing.
31.4.2		If there is a chance that a seafarer could be knocked or washed overboard during cargo operations, then a self-inflating personal flotation device (working lifejacket) should be worn so as not to impede working movements. It must be capable, when activated, of turning the seafarer onto their back if they are unconscious.
31.4.3		When carrying out cargo operations, as a minimum, coveralls, high-visibility vest, safety helmet, safety boots, safety eyewear as appropriate and gloves should be worn.

31.5 Offshore support vessels: communications

31.5.1		Where practical, and when using very high frequency (VHF), communications between the vessel and the platform should be conducted on a different channel from the one used for general in-field communications, because this allows for better and less interrupted communications.
31.5.2		At all times when work is being done on the deck, there should be an efficient means of radio communication between the bridge, crane and seafarers involved. A back-up system should be available between the bridge and seafarers involved and this can either be a public address system or an additional radio.
31.5.3		A proper radio watch must be maintained on the bridge. This includes the appropriate emergency and calling channels as well as the current working channels.

31.6 Carriage of cargo

📖 *This section should be used in conjunction with the Oil & Gas UK publication Best Practice for the Safe Packing and Handling of Cargo to and from Offshore Locations, Guidelines for Offshore Marine Operations (G-OMO) and local supplement, and the International Maritime Dangerous Goods (IMDG) Code where applicable. Please see Appendix 2 for further information.*

31.6.1 The master is responsible for the safe and correct loading of their vessel, and should give due consideration to any known discharge priorities or order of discharge for the cargo when deciding how and where it will be loaded. They should ensure that the cargo is stowed in such a way as to allow access for the seafarers to lifts without the need to climb over cargo and allows for proper and effective lashing arrangements. This will help avoid the temptation to 'cherry pick' a specific container.

31.6.2 All oncoming cargo should be checked against the manifest to ensure that only the cargo listed is loaded. If there are any discrepancies, then loading should be stopped until they can be resolved. If necessary, the cargo should not be loaded. A cargo plan should be produced so that locations of all items are known. There should be pre-notification of any dangerous goods.

31.6.3 Careful attention must be paid to the positioning of dangerous cargo, with segregation as necessary, and the loading plan must include the locations of all dangerous goods.

31.6.4 Before securing, all containers should be given a visual check to ensure there are no defects, the container test is in date and there are no trapped strops or potential dropped objects. All containers should have been inspected prior to loading, so the check that is carried out by the deck crew will be to ensure that nothing has been damaged in transit and nothing obvious has been missed.

31.6.5 When stowing cargo, attention should be given to potential snagging hazards. These include, but are not limited to, stacking points and pad-eye protectors, which may be larger than usual, tie-down hooks, door handles, crash barriers or even entrances to safe havens.

31.6.6 When loading half-height containers, consideration should be given to whether the lifting strops may get caught on the containers' contents when discharging. A suitable material should be used to cover the equipment inside and prevent the potential for snagging hazards. This may include nets, tarpaulins, wood battens, roof bars, cord lashing and crating of equipment.

31.6.7 Crane operators should be instructed to take up the weight slowly on lifting strops in case of any snagging, and crew members must exercise particular caution and stand clear.

31.6.8		Boat-shaped skips should not be used.
31.6.9		Wherever possible, scaffolding tubes and/or boards should be pre-slung into an appropriate cargo-carrying unit designed for four-point lifting.
31.6.10		Cargo operations can continue for several hours. In such cases, careful consideration should be given to ensuring that all involved remain alert. All seafarers must ensure that they follow the hours of rest requirements at all times. This is particularly important if the vessel is sailing into port immediately after cargo work or sailing directly prior to it.
31.6.11		Any personnel working on cargo operations are entitled to stop the operation on safety grounds, until the activity has been reassessed and cleared as safe to resume.
31.6.12		Areas of the deck that are not to be used for cargo stowage should be clearly marked or otherwise indicated.
31.6.13		The safe securing of all deck cargoes should be checked by a **competent person** before the vessel proceeds on passage. To enable unloading at sea to be carried out safely, independent cargo units should, as far as practicable, be individually lashed. Where it is not practical to lash individual pieces of cargo, groups of lifts intended for the same delivery location should be secured together. Lashings should, where practicable, be of a type that can be easily released and maintained.
31.6.14		All lashings should be checked at least once during each watch whilst at sea. Seafarers engaged in the operation should be closely supervised from the bridge, particularly in adverse weather conditions. At night in adverse weather, a searchlight should be used to aid remote checking of lashings to avoid placing personnel at risk.
31.6.15		Where fitted, pipe posts should be used to restrain the movement of tubulars.

31.7 Bulk cargo operations

31.7.1   Cargoes carried in bulk range from dry-powdered products such as cement or barites to liquid products such as water, fuel oil, brine and oil-based muds.

31.7.2 Discharging bulks pose a significant risk to the environment as well as to personnel. As such, before undertaking any bulk cargo operation, check the following:

- Agreement should be attained prior to the vessel entering the 500-metre safety zone as to what product is required and how much product will be discharged, or received by, the installation.
- Pressure ratings of all equipment should be checked to ensure that they are suitable for the operation.
- Prior to commencement, the pumping rate and density of the product should be agreed, as should the proposed sequence of events. Only once these are agreed amongst all involved parties should the operation commence.
- The pumping rate should start off very slowly, to check that all connections are secure and the product is going into the agreed tank(s). Once this has been confirmed and all checks have been made, pumping can be increased to the agreed rate. Further checks of connections should be made once the final pumping rate has been reached.
- Each party should give sufficient warning if tanks need to be changed over. Confirmation should be given once this has been done.
- The vessel and the platform should regularly confirm the amount discharged or loaded. If there are any discrepancies, then the operation should stop until the error can be ascertained.
- If communications are lost at any time, then the operation should be stopped.
- Appropriate deck personnel should be available and nearby during the entire operation.
- The master and/or officer of the watch should be able to see the bulk hoses at all times.
- When discharging liquids, appropriate save-alls should be fitted and adequate spill equipment should be ready for immediate use.
- Valves shall not be closed against the cargo pump.
- Unregulated compressed air should not be used to clear bulk hoses because this can damage tanks.
- Compressed air should not be used to clear hoses that have been used for hydrocarbons because this increases the risk of explosions.
- All hoses should have sufficient flotation collars fitted.
- The hose used should be the correct type for the task.

31.8 Approaching installation and cargo-handling operations

31.8.1 At no time is an installation's exact position to be used as a global positioning system (GPS) waypoint. Waypoints should always be offset from the installation and outside the 500-metre safety zone.

31.8.2 At no time should a vessel enter the installation's 500-metre zone in autopilot. The vessel should be in hand steering.

31.8.3 Prior to entering a 500-metre safety zone, an appropriate checklist should be carried out. Normally these are company or installation specific. A typical example of such a checklist can be found in the G-OMO publication. An entry should also be made in the vessel's logbook once these checks have been completed.

31.8.4 The approximate working position needed for the planned operation should be determined and confirmed with the platform prior to entry to the 500-metre safety zone.

31.8.5 After entry into the 500-metre safety zone, the vessel should proceed to a 'set-up' position that will be at least 1.5 ship's lengths from the installation, in a drift-off situation, or 2.5 ship's lengths in a drift-on situation. This set-up period will be carried out for a minimum of ten minutes so as to allow for an accurate assessment of the prevailing weather conditions and their effect on the vessel.

31.8.6 The current industry weather working guidelines should be followed. These can be found in the G-OMO publication.

31.8.7 Cargo operations should be stopped if the vessel requires the use of more than 45% power on its engines and/or thrusters.

31.8.8 At all times, personnel should be alert to the danger of being hit or crushed should items of cargo swing during a lift or become dislodged through sudden movement of the ship. All seafarers should approach a lift only when it is safely on the deck and the weight is off the wire.

31.8.9 Once a lift is connected, the seafarers should retreat to an appropriate safe haven before it is lifted.

31.8.10 If any back-loading has to take place from the installation during the off-loading of cargo from the vessel, care should be taken to ensure that the cargo taken on board is immediately secured against movement until it can be properly stowed.

31.9 Transfer of personnel by ship to/from installation

31.9.1 Circumstances may make it necessary to transfer personnel to or from a vessel. There are a number of ways of achieving this:

- All personnel to be transferred should be briefed by a responsible person.
- Personnel to be transferred should wear working lifejackets and other PPE suitable for the environmental conditions.
- Throughout the operation, a lifebuoy, boathook and heaving line should be kept immediately available on board the vessel for use in the case of emergencies.
- The arrangements for rescue and recovery of persons near the installation, which are set out in the installation's emergency response plan, should be in place.
- Personnel transfer is to commence only if all identified parties have confirmed readiness.
- All personnel transfers should only take place after a thorough risk assessment has been completed and a toolbox talk carried out with all personnel involved.
- Further advice and guidance on personnel transfers can be found in the G-OMO publication.

31.10 Transfer by specialist craft

31.10.1 When the weather is suitable, transfers can be carried out by specialised small craft subject to the vessel having enough trained personnel to carry out such a task safely.

31.10.2 The master of the ship providing the boat should be responsible for the operation. Due consideration should be given to the effect of prevailing weather conditions on the safety of the transfer.

31.10.3 As guidance, typically, such operation should not take place if the prevailing weather conditions include one of the following:

- The significant wave height exceeds 2.5 metres.
- There are hazardous amounts of ice or snow on any of the landing areas to be used. These include access and egress routes.
- The visibility drops below 500 metres.
- The wind speed exceeds 25 knots.

31.10.4 Personnel transfers by craft should not routinely take place during the hours of darkness. However, if in exceptional circumstances this cannot be avoided, the following precautions should be implemented:

- All transfer areas should be illuminated adequately.
- All lifejackets should be fitted with a high-intensity strobe light and/or a personal locator beacon.
- Checks should be made to ensure that retro-reflective tape on jackets, coveralls, etc. is not obscured.

31.10.5 Boarding and disembarkation should be carried out in an orderly manner under the coxswain's direction.

31.10.6 The boat's coxswain should ensure an even and safe distribution of passengers. Passengers should not stand up or change their positions during the passage between ships save under instructions from the coxswain.

31.10.7 The mother ship should establish communication with the receiving vessel prior to the commencement of the operation and should maintain continuous visual contact with the boat concerned throughout the transfer. Any boat used for personnel transfers should have at least two means of radio communications.

31.10.8 If the transfer of personnel involves a standby vessel, the master should bear in mind that their vessel must, at all times, be ready to fulfil its standby vessel duties.

31.10.9 Transfers from one vessel to another shall not take place within the 500-metre zone of any installation without the explicit permission of the OIM.

31.10.10 Radio communications should be set up between the mother ship, standby vessel (if it is not carrying out the transfer) and receiving vessel.

31.10.11 The boat should be crewed by no fewer than two experienced persons, at least one of whom should be experienced in handling it. Lifejackets and, if necessary, suitable protective clothing should be worn by all personnel.

31.11 Transfer by personnel carrier

31.11.1 Transfers from ship to installations are sometimes carried out by some type of personnel carrier that is lifted by the platform's crane. These vary in design and redundancy, and can range from simple rope netting to more elaborate systems where individuals are strapped in. However, the dangers are similar and must be mitigated against. Once again, further advice and guidance on personnel transfers using a carrier system can be found in the G-OMO publication.

31.12 Transfer of personnel by ship to installation by transfer capsule

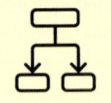

31.12.1 Figure 31.1 shows an example of a typical transfer capsule. The transfer capsule must be on deck and stable before personnel approach it.

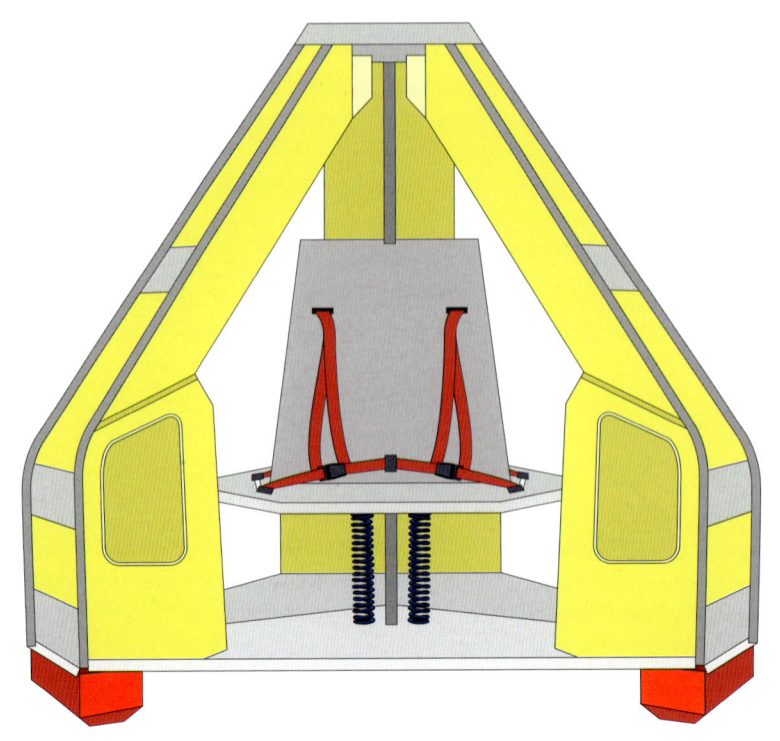

Figure 31.1 Example of a transfer capsule

31.12.2 Personnel should be escorted to the landing/loading area and approach the capsule one at a time.

31.12.3 Personnel should be secured in the transfer capsule in accordance with the manufacturer's user guidance.

31.12.4 The capacity of the capsule must not be exceeded and it is recommended that, in any case, the load should be no more than five personnel who should be evenly distributed to ensure maximum stability.

31.12.5 No baggage should be taken into the capsule. Baggage should be transferred in a separate baggage container.

31.12.6 Before lifting commences, all personnel (OIM, vessel master and crane operator) must be in agreement that they are in readiness for the transfer. Adequate radio communications should be maintained throughout the transfer.

31.12.7 The capsule should be lifted clear of the vessel and swung up and out as smoothly as possible. Once over the sea, the capsule should be lifted to the installation.

31.12.8 Once over the installation, the capsule should be lowered to the lifting/landing area. Tag lines should be cleared before the capsule is finally lowered to the landing area.

31.12.9 Transit personnel should remain seated and secured until the transfer capsule is stable on the deck and the installation personnel have removed securing and provided an escort to the reception on the installation.

31.13 Transfer by personnel carrier

31.13.1 The following procedures should be observed for the transfer of personnel from ship to installation by a personnel carrier:

- The equipment should be steadied when it is lowered to the deck. Tag lines may be used and the risk assessment must cover these. Tags lines should never be wrapped around the hands.
- Luggage should be secured within the appropriate space in the carrier or taken up separately.
- Personnel to be transferred should wear lifejackets and other PPE suitable for the water and sea conditions.
- Personnel being transferred should be evenly distributed around the carrier to ensure maximum stability.
- If using a basket type of carrier, personnel should stand outside the basket with feet apart on the board and the basket securely gripped with both arms looped through.
- When the officer in charge is satisfied that all are ready and at an appropriate moment having regard to the movement of the ship in a seaway, the basket should be lifted clear of the vessel and then swung up and out as quickly as possible before being carefully hoisted up to the installation.
- Radio communications should be set up between ship, standby vessel (if it is not carrying out the transfer) and installation.

31.14 Transfer by gangway

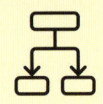

31.14.1 The master of the transfer vessel, installation OIM and ERRV master must discuss the prevailing weather conditions before deciding whether it is safe enough for the transfer to proceed. Operations should take place only in the hours of daylight.

31.14.2 Transit personnel should be escorted to the gangway access area and must use the gangway only under the direction of the gangway operator. Figure 31.2 shows an example of a vessel with a gangway in place.

Figure 31.2 Example of a vessel with gangway in place

31.14.3 Once on the installation, transit personnel are to be escorted to the reception area.

31.14.4 The capacity of the gangway should not be exceeded.

31.14.5 Personnel baggage should not be carried on the gangway. Baggage should be transferred through the use of a separate baggage container.

31.15 Anchor handling

31.15.1 Anchor handling is generally carried out by vessels commonly known as anchor handling towing supply (AHTS) vessels. As the name suggests, they are multi-purpose vessels that can carry out a number of important roles. However, generally, their primary purpose is anchor handling. This guidance should be used in conjunction with the anchor handling section of G-OMO.

31.15.2 **Warning**
All anchor-handling jobs should be risk assessed and the findings disseminated to all those involved via a toolbox talk or similar.

	31.15.3	If the AHTS vessel is engaged in cargo activities, then the safety precautions and procedures for supply vessels should be followed.
	31.15.4	Owing to AHTS vessels having a stern roller, if general supply work is undertaken by an AHTS vessel, some form of barrier is needed to prevent cargo from going over the stern. This may be something simple such as cargo chains or it can be something purpose built, such as moveable bulwarks or railings.
	31.15.5	Care should be taken on the metal decks of these vessels because they can increase the chances of slips, trips and falls. They should be regularly cleaned to prevent any build-up of algae or other residues.
	31.15.6	During adverse weather, lifelines should be rigged on the working deck to facilitate safe movement. Decks should, as far as practicable, be kept free from ice, slush, algae and any substance or loose material likely to cause slips and falls. This is particularly important for the metal section of the deck.

31.15.7 Warning

Anyone working on the deck should wear a working lifejacket at all times because of the open stern. Any lifejacket or flotation device used must turn the casualty onto their back if they are unconscious.

13.15.8

Many items used in anchor handling are large and heavy. Care should be taken when **manual handling** 🔍 any equipment and, if necessary, two persons should be used.

- Whenever an anchor is being lowered over the stern or retrieved, all seafarers should be off the working deck and within a safe area.
- Before seafarers go back on deck, the chain should be secured in the **shark jaws** 🔍 or similar securing device.
- If anchors have been retrieved from deep water, there will be a lot of tension stored within the chain or pennant. When the pin is removed, this tension will cause the pennant or chain to spin and fly into the air. It is important that all seafarers are in an appropriately safe position.

31.15.9 Warning

Never walk near or over a 'live' wire on the deck. A live wire is one that is in use, under tension or has the potential to come under tension.

	31.15.10	To reduce the likelihood of seafarers walking over a live wire, duplicate tools should be positioned on both sides of the working deck. This allows seafarers to remain on one side of the wire at all times.
	31.15.11	All equipment used is to be maintained and operated in accordance with manufacturers' instructions.
	31.15.12	There should be oxy-acetylene (or similar) cutting gear, with adequate gas, available for immediate use if needed.
	31.15.13	Seafarers should ensure that the stowage of anchors and equipment is secured in line with the planned operation, and be aware of the risk of such items moving when unsecured.
	31.15.14	Certain types of anchors are unstable and may not sit well on a flat deck. This should be considered during the initial risk-assessment stage so that adequate securing arrangements can be provided.

Further advice and guidance on offshore support vessel operations may be found in the G-OMO publication and the various Oil & Gas UK publications.

- Lifelines must be rigged on the working deck during adverse weather.
- Personnel must wear lifejackets when working on deck.
- Safety procedures to transfer personnel must be followed with communications maintained throughout the process.

Ensure that personnel are trained and aware of safety procedures for working with offshore installations.

32 Ships serving offshore renewables installations

32.1 Introduction

32.1.1 This chapter considers good practice on vessels supporting the construction, operation and maintenance of offshore renewable energy installations (OREIs). Vessels are used for survey work, transporting components and materials, transfer of personnel, construction work, dive support and accommodation.

32.1.2 Guidance on operation of vessels transiting in the vicinity of OREIs is published in MGN 372 (M+F) Amendment 1.

    MGN 372 (M+F) Amendment 1

32.1.3 Safety for diving operations is subject to HSE regulation and to the Merchant Shipping (Diving Safety) Regulations 2002 and MSN 1762 (M+F) Amendment 1.

    SI 2002/1587; MSN 1762 (M+F) Amendment 1

Key points
- Ensure coordination of communications with all parties, seafarers, workers and organisations involved in the development and ongoing activities of the offshore renewable energy installations (OREIs).
- Provide safe access to and from OREIs day and night.
- All seafarers must be fully trained, instructed and supervised for their own health and safety responsibilities and for the health and safety of others.
- Workers employed in the development, construction and maintenance of offshore windfarms may not have much experience of working in a maritime environment. Their employer is responsible for ensuring that they have the information, instruction, training and supervision necessary to safeguard their health and safety.
- The master of the vessel should ensure that the personnel carried are familiar with emergency procedures on board, and give appropriate instructions and guidance to ensure that they are aware of the vessel's working practices that affect them.
- When planning work activities that involve more than one vessel, or a vessel and an installation, known as **simultaneous operations (SIMOPS)**, it is important to identify any differences in their safety procedures, carry out a risk assessment and agree actions in advance that are clearly understood by all.

> **Your organisation should**
> - carry out risk assessments involving ongoing operations on all OREI activities and when a number of vessels may be operating together or close by
> - have a **document of compliance (DoCDG)** 🔍 (if required) for the transit of dangerous goods
> - ensure vessel transfer arrangements are risk assessed and maintain good communications between vessels throughout the process
> - maintain vessels and ensure they are properly equipped
> - arrange an emergency response coordination plan to ensure that HM Coastguard (HMCG) and search and rescue (SAR) resources have information about the fundamental details of an OREI.

Responsibility for offshore renewables personnel

32.1.4 Although the vessel provider may be a contractor with duties under Construction (Design and Management) Regulations 2015), this does not compromise the vessel master's duty to ensure the safety of the vessel, crew and passengers.

📖 *SI 2015/51*

32.2 Coordination

32.2.1 It is likely that many organisations will be involved during both the construction and ongoing operation of OREIs. Coordination is therefore key. Each OREI should have arrangements in place for the:

- provision of vessel traffic information and advice to masters
- management and coordination of all site work/activities
- emergency response (see section 32.5).

32.2.2 Any marine operations within the area should be approved through the marine coordination arrangements that are already in place. Establish clear lines of responsibility and reporting.

32.2.3 In addition, when planning work activities that involve more than one vessel, or a vessel and an installation, it is important to identify any differences in their safety procedures, carry out a risk assessment and agree actions in advance that are clearly understood by all.

32.2.4 Vessels often work near turbines, other structures and other vessels. Even where activities do not directly involve working with other vessels/installations, do a risk assessment to consider the impact of each vessel's activities on others. Where necessary, agree a sequence of actions and safe procedures before the work starts.

32.3 Safe means of access to installations

32.3.1 Chapter 22 provides guidance on safe means of access, with guidance for special circumstances in section 22.9.

32.3.2 Where passengers, industrial personnel or crew are accessing or leaving installations from a vessel, carry out a risk assessment of the transfer arrangements and put appropriate safety measures in place to ensure the safety of those involved. Take additional safety precautions during the hours of darkness. The arrangements during transfer must be compatible with the specific offshore installation and the operating company's safety management system (SMS) and comply with the statutory standards for work at height regulations. The vessel should be properly equipped and/or modified (taking into account the design of the access point on the installation) to allow the transfer to be undertaken without unnecessary risk. Provide a proper embarkation point and establish a clearly agreed boarding procedure.

IMCA M 202; SI 2005/735; SI 2007/114

Making the transfer

32.3.3 The relative movements of the vessels in varying sea, tide and swell conditions make the judgement of when to make a transfer crucial:

- The master responsible for the transfer operation should have full and direct sight of the area of transfer.
- The master and at least one designated crew member should be able to communicate at all times with the person making the transfer.
- Vessels undertaking ship-to-ship transfers while under way should carry equipment designed to aid in the rapid recovery of a casualty from the waters.
- Workers transferring and working on exposed decks during transfer should wear a personal flotation device. Consider whether they need to wear an immersion/survival suit, particularly in cold conditions.
- Baggage and other items should be transferred by the crews of the vessels and not by the people transferring.

Further guidance on the transfer of personnel to and from offshore vessels and structures is available from the International Marine Contractors Association (IMCA) (see section 31.9 and Chapter 17).

32.4 Carriage and transfer of dangerous cargoes

32.4.1 Where a workboat carries more than 30 kg or 30 litres net total quantity of dangerous goods, whether used on board for its own purposes or by the industrial personnel for their own work, the vessel generally requires a document of compliance to carry dangerous goods (DoCDG). This is issued by the MCA. The master and people ashore responsible for allocating stores and equipment to be carried should be trained in the requirements of the IMDG code.

MGN 497 (M+F) gives guidance on the storage of dangerous cargoes on board. For detailed requirements that should be complied with see MGN 280 (M); the Workboat Code, Industry Working Group Technical Standard; or the Workboat Code, Edition 2.

32.5 Emergency response plans

32.5.1 OREI operators should have in place an emergency response cooperation plan agreed with MCA SAR Operations for the construction, operation and decommissioning phases of any OREI. These plans are designed to ensure that HMCG and SAR resources have information about the fundamental details of an OREI and that all parties have access to emergency contact numbers allowing rapid contact, information sharing and effective cooperation during an emergency situation. This will ensure the effective management of incidents arising on the site. Workers operating vessels in the area may be required to take part in testing of the arrangements. The master should ensure that all seafarers on the vessel are familiar with the plan, and comply where appropriate with the arrangements set out.

32.6 Other sources of information

32.6.1 Further industry guidance is available in Appendix 2.

All personnel must:

- follow instructions and maintain communications during OREI operations and work activities
- be aware of their health and safety responsibilities.

- All seafarers should be aware of the emergency response cooperation plan for the vessel and OREI.
- Vessel transfer operations have a greater risk when vessels are underway; maintain rapid recovery equipment and keep it to hand as instructed in case of an emergency.
- Seafarers on deck during vessel transfer operations need to wear personal flotation devices.
- Follow the SMS plan and working at height regulations during transfer operations.

33 Ergonomics

33.1 Introduction

33.1.1 Ergonomics deals with the interaction between humans and work, and covers three principal areas of work:

Design and environment

- Workplace design: layout, controls, displays, temperature, light, noise, smell, vibration
- Workload and fatigue
- Safe working posture.

Work processes

- Mental workload, fatigue and work-related stress
- Human reliability, errors and violations
- Competence, capability and training.

Organisation

- Communication and teamwork
- Policies, procedures and work instructions
- Quality management and assurance.

SI 1997/2962; MGN 636 (M) Amendment 2

Key points
- Personnel must raise concerns about poor ergonomics or procedures at every opportunity.
- Use the opportunity to consider ergonomics as part of routine safety meetings.
- Be assertive in advising the shoreside organisation of the need for any changes.
- Be active in procedural or work instruction review.
- Assess how the layout could affect seafarers' situational awareness and working practices during routine and emergency operations.

Your organisation should

- ensure that everyone is familiar with the on-board equipment and knows how to use it
- ensure that seafarers have adequate individual training in the use and capabilities of display screen equipment (DSE)
- ensure that seafarers using DSE for long periods of time have the same considerations as those working ashore
- have organisational policies and practices that support the front line in all its needs; in other words, by setting people up to succeed
- have recruitment and selection practices that ensure that all personnel serving on board are fully competent for their duties, including operating all equipment on board
- provide immediate and effective on-board familiarisation training
- have working practices that ensure that vessels are fully and correctly maintained and have ready access to all the stores, tools and supplies needed, wherever they are in the world.

For more information, see 'HSG48 Reducing error and influencing behaviour' on the HSE website.

33.1.2 The quality of shipboard ergonomics plays a significant role in safety as well as efficient operational performance. Ergonomically designed ships are generally easier to operate, more efficient, less stressful, safer and more resilient.

33.1.3 Similarly, the quality of procedures, operating instructions, work instructions and maintenance instructions can play a significant role in operational performance and safety. Procedures and instructions that are clear, logical, consistent, easily understood by all users and fit for purpose will reduce violations and lead to safer operations.

33.1.4 More information about applying ergonomic principles is provided in Annex 33.1.

33.1.5 However, often seafarers will need to work in less than ideal circumstances. The following checklist helps them to work safely and efficiently in such circumstances.

33.1.6 The following guidance applies to seafarers:

- Familiarise yourself with the layout of your working areas.
- Be aware that risk factors do not operate in isolation – they combine and multiply.
- Remain vigilant and maintain situational awareness at all times.
- Help others – effective teamwork and communication are essential.
- Do not allow yourself, or others, to take shortcuts or violate procedures.
- Be aware of the effects of frustration, fatigue and stress on behaviour and performance.

33.2 Work with display screen equipment

33.2.1 In this chapter DSE includes devices or equipment that have an alphanumeric or graphic display, such as display screens, laptops, touch screens and other similar devices. There are no specific regulations governing health and safety in the use of DSE that apply to UK-registered ships. This section therefore gives guidance only, reflecting best practice ashore.

33.2.2 DSE training should include any risks from DSE work and the controls in place and, where possible:

- how to adjust equipment settings
- how to adjust furniture for correct posture as shown in Figure 33.1
- how to organise the workplace to avoid awkward or frequently repeated stretching movements
- who to contact for help and to report problems or symptoms.

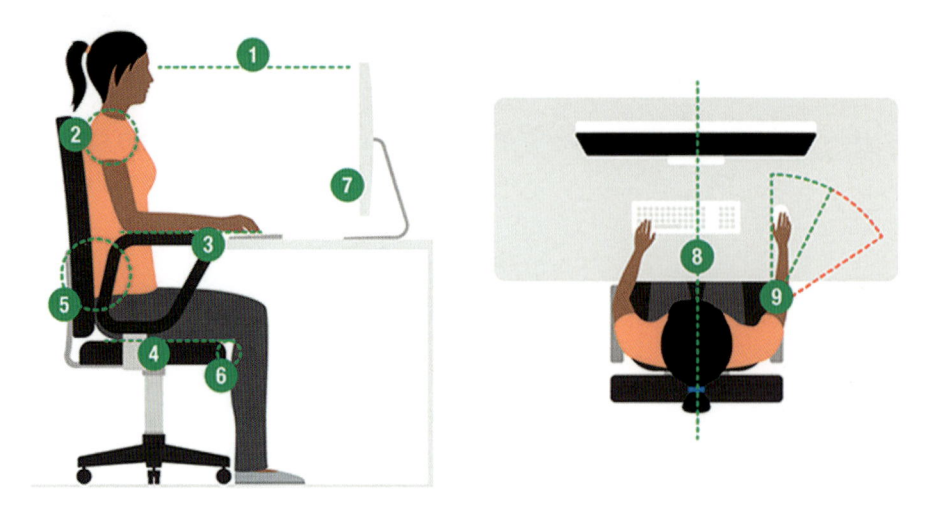

1. Top of screen level with eyes, about an arm's length away
2. Relax your shoulders – try to position yourself high enough so you don't need to shrug your shoulders
3. Keyboard just below elbow height
4. Seat height equally supports front and back of thighs (or use cushion to raise seated position)
5. Back of the seat provides good lower back support (or use a cushion, to provide additional back support)
6. Gap of 2–3 cm between front of seat bottom and back of knee
7. Computer and screen directly in front of you on desk or other surface
8. Screen and keyboard central – don't twist your back
9. Mouse in line with elbow.

Figure 33.1 Correct sitting posture for display screen equipment

33.2.3 Although the relevant regulations do not apply on ships, it is recommended that any seafarer using DSE as part of their work for continuous periods of an hour or more a day be provided with an eye test by a qualified person on request and at no cost to the seafarer.

33.2.4 Lighting should be adequate for the task, with minimum glare and reflection, and the display on screen should be clear and easy to read. The operator should adjust the brightness and contrast to suit the lighting. When appropriate, the operator should be given short rest periods away from the equipment.

33.2.5 Certain forms of medication may impair working efficiency on DSE. Personnel should be aware of this and seek medical advice if necessary.

> *Further guidance on the safe use of DSE in an office environment is available from the HSE website including 'Working with display screen equipment (DSE)'.*

> *For additional guidance on determining fatigue and musculoskeletal disorders see Chapter 1 of Wellbeing at Sea – A Pocket Guide for Seafarers and Chapter 3 of Wellbeing at Sea – A Guide for Organisations.*

- Minimise risks of musculoskeletal injuries by ensuring that seafarers follow the correct working methods and procedures.
- Review the working environment periodically and highlight to the safety representative on board any ergonomic concerns, proposing effective solutions, modifications or changes to procedures.

- Poor ergonomic design can increase stress and fatigue and encourage people to take shortcuts or procedural violations.
- When people are working with DSE for long periods regular eye tests are recommended.

Annex 33.1 Ergonomics

The underpinning principle of effective ergonomics is to make machines, equipment, processes and organisational policies fit the actual needs of people who use them. This is known as user-centred design.

In an ideal world, effective user-centred design would be the norm. In reality, many people have to adjust as best they can to the working environment they are given. This presents a number of challenges for working safely and efficiently, and seafarers need to be extra vigilant and mindful of their tasks. The challenges of poor ergonomics are as follows.

Design and construction

The design of ships, layout of workspaces and arrangement of controls and displays is not always ideal. Important or frequently used controls are not readily at hand, and controls and displays are not arranged in a logical sequence, or are difficult to see, identify, distinguish and read.

Working and living environments may be uncomfortable due to heat, cold, noise, vibration, smell or poor lighting. Communication may be difficult, making any existing language difficulties worse. Access may be inadequate; spaces may be cramped, making it difficult to operate tools and equipment.

A seafarer can encounter physical hazards as part of the ship's design, including slips, trips and falls.

Work processes

Manual work (eg cargo handling, maintenance and repair work) can be physically demanding and strain the seafarer's mind and body. The most suitable tools and equipment may not be available and the working space may not be adequate to allow them to do a job safely. This can lead to shortcuts and procedural violations.

Working at a poorly adjusted workstation for long periods can harm body posture and cause long-term health and musculoskeletal injuries.

Long hours and demanding work cause fatigue and can lead to stress. Poorly designed ships, equipment, tools and work processes, where seafarers continually have to adjust and find workarounds, increase the physical and mental workload. This will greatly increase the likelihood of errors and accidents.

Procedures and instructions must make life easier for crews. Poorly written procedures and instructions may affect the safe and successful operation of ships and equipment. Crews may interpret the instructions in different ways, leading to inconsistencies and errors. Unclear instructions may lead to procedural violations as crews struggle to find a workaround.

Risks

Ergonomic deficiencies can cause operational distractions that will affect situational awareness and efficiency. They can make errors more likely; not only in stressful or emergency situations but also during routine operations, unless crew are extra vigilant.

Safe operations depend upon well-trained seafarers who know their ship and how to use the equipment. Lower levels of competence can lead to increased workload, fatigue, stress and error rate. Poor ergonomic design of ships, equipment and procedures will increase any effect of lack of competence.

Organisational failures, including those that affect design and procedural ergonomics, can lead to operational errors through equipment breakdown, unavailability of tools and equipment, additional stress on personnel and poor resourcing.

Environmental issues beyond anyone's control (eg adverse weather; sea state) can add to any risk. Poor ergonomics will become even more difficult to manage.

Mitigation of poor ergonomics

Ship owners and operators should be proactive and:

- Consider effective ergonomic principles when commissioning ships, equipment or designing work procedures, and follow user-centred design principles.
- Consider modifying existing ships, equipment and work processes to become more user-centred.
- Encourage ships and their crews to report ergonomic issues on board.
- Ensure that procedures and work instructions are in a consistent format and are easily understood by everyone who uses them.
- Ensure that seafarers using procedures and work instructions are actively involved in developing and reviewing them.

Design and construction

Well-designed, user-centred equipment and work processes will support seafarers in their work and have a positive effect. They will reduce fatigue and stress and make work more satisfying, efficient and safe.

Procedural/work instruction ergonomics

Procedures and work instructions need to:

- reflect how tasks are actually done
- be accurate and complete
- be clear and concise but with enough detail
- be current and up to date
- be supported by training (where appropriate)
- identify hazards
- state the necessary precautions for hazards
- promote ownership by seafarers
- use familiar language and be easily understood by everyone on board
- take into account potential differences in language ability
- use consistent terminology
- be in a suitable format
- be accessible to all on board.

Appendix 1

Regulations, marine notices and guidance issued by the Maritime and Coastguard Agency

This appendix lists all the regulations, marine notices and other guidance referred to in this Code.

Statutory instruments (regulations) are available on www.legislation.co.uk

TSO publications are available from TSO, PO Box 29, Norwich, NR3 1GN; www.tsoshop.co.uk

Copies of Maritime and Coastguard Agency (MCA) marine notices and forms can be downloaded from www.gov.uk/government/organisations/maritime-and-coastguard-agency

There are three different types of marine notice:

- **Merchant Shipping Notices (MSNs)** are used to convey mandatory information that must be complied with under UK legislation. These MSNs relate to statutory instruments and contain the technical detail of such regulations.
- **Marine Guidance Notes (MGNs)** give significant advice and guidance relating to the improvement of the safety of shipping and of life at sea, and to prevent or minimise pollution from shipping.
- **Marine Information Notes (MINs)** are intended for a more limited audience, such as training establishments or equipment manufacturers, or they contain information which will only be of use for a short period of time, such as timetables for MCA examinations. MINs are numbered in sequence and have a cancellation date (which will typically be no more than 12 months after publication).

These notices publicise to the shipping and fishing industries important safety, pollution prevention and other relevant information.

Within each series of marine notices suffixes are used to indicate whether documents relate to merchant ships or fishing vessels, or to both. The suffixes following the number are:

- (M) for merchant ships
- (F) for fishing vessels
- (M+F) for both merchant ships and fishing vessels.

Copies of MCA published leaflets are available from EC Group, Europa Park, Grays, Essex, RM20 4DN; email: mca@ecgroup.uk.com

The Code reference is shown in **bold** and the information is arranged in chapter order.

Chapter 1: Managing occupational health and safety

Regulations
1.1; 1.2.5 SI 1997/2962 The Merchant Shipping and Fishing Vessels (Health and Safety at Work) Regulations 1997 (as amended).

Marine notices
1.2.5 MSN 1838 (M) Amendment 1 Maritime Labour Convention, 2006: Minimum age.

MSN 1890 (M+F) Amendment 3 The Merchant Shipping and Fishing Vessels (Health and Safety at Work) Regulations 1997 and the Merchant Shipping (Maritime Labour Convention) (Medical Certification) Regulations 2010: New and Expectant Mothers.

1.2.8; Annex 1.2 MGN 484 (M) Amendment 3 Maritime Labour Convention, 2006: Health and safety: Published accident statistics – information and advice.

Annex 1.2 MGN 636 (M) Amendment 2 Merchant Shipping and Fishing Vessels (Health and Safety at Work) Regulations 1997.

MGN 587 (F) Amendment 1 International Labour Organization Work in Fishing Convention (No. 188), Health and safety: responsibilities of fishing vessel owners, managers, skippers and fishermen.

Guidance
1.1 Maritime and Coastguard Agency (MCA) (2020) *Wellbeing at Sea: A guide for organisations*, London: TSO (ISBN 978-0-11-553608-3).

1.2.3 MCA (2014) *Leading for Safety: A practical guide for leaders in the maritime industry:* www.gov.uk/government/publications/leading-for-safety/leading-for-safety-a-guide-for-leaders-in-the-maritime-industry

Chapter 2: Safety induction for personnel working on ships

Regulations
2.1.2 SI 1997/2962 The Merchant Shipping and Fishing Vessels (Health and Safety at Work) Regulations 1997 (as amended).

2.4.1 SI 1999/2722 The Merchant Shipping (Musters, Training and Decision Support Systems) Regulations 1999.

2.8.1 SI 2020/621 The Merchant Shipping (Prevention of Pollution by Garbage from Ships) Regulations 2020.

Marine notices
2.1.2 MGN 636 (M) Amendment 2 Merchant Shipping and Fishing Vessels (Health and Safety at Work) Regulations 1997.

2.4.1 MGN 71 (M) Musters, drills, on-board training and instructions, and decision support systems.

2.6.2 MGN 652 (M+F) Amendment 1 Infectious disease at sea.

2.8.1 MGN 632 (M+F) Amendment 1 The Merchant Shipping (Prevention of Pollution by Garbage from Ships) Regulations 2020.

Guidance
2.2.1 Maritime and Coastguard Agency (MCA) (2020) *Wellbeing at Sea: A guide for organisations*, London: TSO (ISBN 978-0-11-553608-3).

Chapter 3: Living on board

Regulations
None

Marine notices

3.2.1; 3.4.1 MSN 1886 (M+F) Amendment 1 Maritime Labour Convention, 2006: Work in Fishing Convention, 2007 (ILO No. 188) Medical Examination System: Appointment of Approved Doctors and Medical and Eyesight Standards.

3.2.1 MSN 1815 (M+F) Amendment 6 Countries whose seafarer medical certificates are accepted as equivalent to the UK seafarer medical fitness certificate (ENG1).

3.5.1; 3.5.3 MGN 652 (M+F) Amendment 1 Infectious disease at sea.

3.6.4 MGN 505 (M) Amendment 1 Human element guidance – Part 1 Fatigue and fitness for duty: Statutory duties, causes of fatigue and guidance on good practice.

3.13.3 MGN 357 (M+F) Night-time lookout – Photochromatic lenses and dark adaptation.

3.15.7 MGN 299 (M+F) Interference with safe navigation through inappropriate use of mobile phones.

MGN 520 (M) Human element guidance – Part 2 The deadly dozen: 12 significant people factors in maritime safety.

MGN 638 (M+F) Human element guidance – Part 3 Distraction: The fatal dangers of mobile phones and other personal devices when working.

Guidance

3.1.1 Maritime and Coastguard Agency (MCA) (2020) *Wellbeing at Sea: A guide for organisations*, London: TSO (ISBN 978-0-11-553608-3).

MCA (2020) *Wellbeing at Sea: A pocket guide for seafarers*, London: TSO (ISBN 978-0-11-553787-5).

3.1.1; 3.5.3; 3.15.7 MCA, *The Ship Captain's Medical Guide*, London: TSO.

Chapter 4: Emergency drills and procedures

Regulations

4.1.1; 4.1.7; 4.4.12; 4.4.23 SI 1999/2722 The Merchant Shipping (Musters, Training and Decision Support Systems) Regulations 1999.

4.2.15 SI 1998/2514 The Merchant Shipping (Passenger Ship Construction: Ships of Classes I, II, and II(A)) Regulations 1998.

4.8.2 SI 2022/96 The Merchant Shipping and Fishing Vessels (Entry into Enclosed Spaces) Regulations 2022.

Marine notices

4.1.1; 4.1.9; 4.1.10; 4.2.14; 4.4.12 MGN 71 (M) Musters, drills, on-board training and instructions, and decision support systems.

MSN 1579 (M) Minimum training requirements for personnel nominated to assist passengers in emergency situations.

4.2.14 MGN 276 (M+F) Amendment 1 Fire protection – maintenance of portable fire extinguishers.

4.3.1 MGN 653 (M) Amendment 1 Electric vehicles onboard passenger roll-on/roll-off (ro-ro) ferries.

4.4.13; 4.4.21 MGN 560 (M) Amendment 2 Requirements for life-saving appliances.

4.4.15 MSN 1722 (M+F) Guidelines for training crews for the purpose of launching lifeboats and rescue boats from ships making headway through the water.

4.4.23 MGN 540 (M+F) Life-saving appliances – lifeboats and rescue boats – fitting of 'fall preventer devices' to reduce the danger of accidental hook release.

4.13.5 MGN 558 (M) Amendment 1 Life-saving appliances – marine evacuation systems (MES) – servicing and deployments.

Chapter 5: Fire precautions

None

Chapter 6: Security on board

Regulations
None

Marine notices
None

Guidance
6.1.1 Maritime and Coastguard Agency (MCA) (2020) *Wellbeing at Sea: A guide for organisations*, London: TSO (ISBN 978-0-11-553608-3).

MCA (2020) *Wellbeing at Sea: A pocket guide for seafarers*, London: TSO (ISBN 978-0-11-553787-5).

Chapter 7: Workplace health surveillance

7.1.1; 7.3.1 SI 1997/2962 The Merchant Shipping and Fishing Vessels (Health and Safety at Work) Regulations 1997 (as amended).

SI 2007/3100 The Merchant Shipping and Fishing Vessels (Health and Safety at Work) (Carcinogens and Mutagens) Regulations 2007.

SI 2007/3075 The Merchant Shipping and Fishing Vessels (Control of Noise at Work) Regulations 2007.

SI 2007/3077 The Merchant Shipping and Fishing Vessels (Control of Vibration at Work) Regulations 2007.

SI 2010/330 The Merchant Shipping and Fishing Vessels (Health and Safety at Work) (Chemical Agents) Regulations 2010 (as amended).

SI 2010/332 The Merchant Shipping and Fishing Vessels (Health and Safety at Work) (Work at Height) Regulations 2010.

SI 2010/2984 The Merchant Shipping and Fishing Vessels (Health and Safety at Work) (Asbestos) Regulations 2010 (as amended).

SI 2010/2987 The Merchant Shipping and Fishing Vessels (Health and Safety at Work) (Artificial Optical Radiation) Regulations 2010.

7.1.1; 7.2.3 SI 2010/323 The Merchant Shipping and Fishing Vessels (Health and Safety at Work) (Biological Agents) Regulations 2010.

7.3.1 SI 2014/1616 The Merchant Shipping (Maritime Labour Convention) (Health and Safety) (Amendment) Regulations 2014.

Marine notices
7.2.3 MSN 1889 (M+F) Amendment 3 The Merchant Shipping and Fishing Vessels (Health and Safety at Work) (Biological Agents) Regulations 2010 (as amended).

7.3.4 MSN 1850 (M) Amendment 1 Maritime Labour Convention, 2006: Health and Safety Reporting of Occupational Diseases.

7.3.5 MSN 1888 (M+F) Amendment 3 The Merchant Shipping and Fishing Vessels (Health and Safety at Work) (Chemical Agents) Regulations 2010 (as amended), Annex 3.

Forms
7.3.4 MSF 4159 Occupational disease report form for UK registered merchant ships.

Chapter 8: Personal protective equipment

Regulations
8.1.1; 8.1.2; 8.1.4; 8.1.5; 8.1.6; 8.1.7; 8.1.8; 8.1.9; 8.2.2 SI 1999/2205 The Merchant Shipping and Fishing Vessels (Personal Protective Equipment) Regulations 1999.

8.5.1 SI 2007/3075 The Merchant Shipping and Fishing Vessels (Control of Noise at Work) Regulations 2007.

Marine notices
8.1.1; 8.1.2; 8.1.6 MSN 1870 (M+F) Amendment 5 The Merchant Shipping and Fishing Vessels (Personal Protective Equipment) Regulations 1999.

8.5.1 MGN 658 (M+F) The Merchant Shipping and Fishing Vessels (Control of Noise at Work) Regulations 2007.

Guidance
8.5.1 Maritime and Coastguard Agency (MCA) (2009) *Code of Practice for Controlling Risks Due to Noise on Ships*, London: TSO (ISBN 978-0-11-553075-3).

Chapter 9: Safety signs and their use

Regulations
9.1.1; Annex 9.1 SI 2001/3444 The Merchant Shipping and Fishing Vessels (Safety Signs and Signals) Regulations 2001.

Marine notices
9.1.1; Annex 9.1 MGN 556 (M+F) Amendment 1 The Merchant Shipping and Fishing Vessels (Safety Signs and Signals) Regulations 2001.

9.3.9 MSN 1676 (M) Amendment 1 The Merchant Shipping (Life-saving Appliances and Arrangements) Regulations 2020 and The Merchant Shipping (Life-saving Appliances for Ships of Classes III to VI(A)) Regulations 1999.

9.8.1 MSN 1665 (M) Amendment 1 The Merchant Shipping (Fire Protection) Regulations 2023 and The Merchant Shipping (Fire Protection: Small Ships) Regulations 1998: Fire-fighting equipment.

Chapter 10: Manual handling

Regulations

10.1.3; 10.1.7 SI 1998/2857 The Merchant Shipping and Fishing Vessels (Manual Handling Operations) Regulations 1998.

Marine notices

10.1.3 MGN 90 (M+F) Amendment 3 The Merchant Shipping and Fishing Vessels (Manual Handling Operations) Regulations 1998.

Guidance

10.1.2 Maritime and Coastguard Agency (MCA) (2020) *Wellbeing at Sea: A pocket guide for seafarers*, London: TSO (ISBN 978-0-11-553787-5).

Chapter 11: Safe movement on board ship

Regulations

11.1.1; 11.11.7 SI 1997/2962 The Merchant Shipping and Fishing Vessels (Health and Safety at Work) Regulations 1997 (as amended).

11.1.3 SI 2001/3444 The Merchant Shipping and Fishing Vessels (Safety Signs and Signals) Regulations 2001.

11.9.1 SI 1988/1638 The Merchant Shipping (Entry into Dangerous Spaces) Regulations 1998.

SI 2022/96 The Merchant Shipping and Fishing Vessels (Entry into Enclosed Spaces) Regulations 2022.

Marine notices

11.1.1 MGN 532 (M) Amendment 2 Health and safety at work. Safe movement on board ship.

11.1.3 MGN 556 (M+F) Amendment 1 The Merchant Shipping and Fishing Vessels (Safety Signs and Signals) Regulations 2001.

11.6.1; 11.6.11 MGN 35 (M+F) Amendment 1 Accidents when using power-operated watertight doors.

Annex 11.2 MSN 1844 (M) Maritime Labour Convention, 2006: Crew Accommodation.

MGN 481 (M) Amendment 1 Maritime Labour Convention, 2006: Crew Accommodation, Supplementary Guidance.

Chapter 12: Noise, vibration and other physical agents

Regulations

12.5.1; Annex 12.3 SI 2007/3075 The Merchant Shipping and Fishing Vessels (Control of Noise at Work) Regulations 2007.

12.9.1; 12.14.7; 12.15.3 SI 2007/3077 The Merchant Shipping and Fishing Vessels (Control of Vibration at Work) Regulations 2007.

12.18.1 SI 2010/2987 The Merchant Shipping and Fishing Vessels (Health and Safety at Work) (Artificial Optical Radiation) Regulations 2010.

Annex 12.3 SI 1999/2205 The Merchant Shipping and Fishing Vessels (Personal Protective Equipment) Regulations 1999.

Marine notices

12.1.1; 12.15.3 MGN 436 (M+F) Amendment 3 Whole body vibration, guidance on mitigating against the effects of shocks and impacts on small vessels.

12.1.1; 12.14.7 MIN 588 Amendment 2 Codes of practice for controlling risks due to noise and vibration on ships.

12.5.1; 12.9.1; Annex 12.2; Annex 12.3 MGN 658 (M+F) The Merchant Shipping and Fishing Vessels (Control of Noise at Work) Regulations 2007.

12.9.1; 12.17.1 MGN 353 (M+F) Amendment 2 The Merchant Shipping and Fishing Vessels (Control of Vibration at Work) Regulations 2007.

12.18.1 MGN 428 (M+F) Amendment 2 The Merchant Shipping and Fishing Vessels (Health and Safety at Work) (Artificial Optical Radiation) Regulations 2010.

Guidance

12.1.1 Maritime and Coastguard Agency (MCA) (2020) *Wellbeing at Sea: A guide for organisations*, London: TSO (ISBN 978-0-11-553608-3).

MCA (2020) *Wellbeing at Sea: A pocket guide for seafarers*, London: TSO (ISBN 978-0-11-553787-5).

12.1.1 MCA (2009) *Code of Practice for Controlling Risks Due to Noise on Ships*, London: TSO (ISBN 978-0-11-553075-3).

12.14.7 MCA (2009) *Code of Practice for Controlling Risks Due to Hand-transmitted Vibration on Ships*, London: TSO (ISBN 978-0-11-553090-6).

12.15.3 MCA (2009) *Code of Practice for Controlling Risks Due to Whole-body Vibration in Ships*, London: TSO (ISBN 978-0-11-553076-0).

Chapter 13: Safety officials

Regulations

13.1; 13.2; 13.3; 13.4; 13.8.1 SI 1997/2962 The Merchant Shipping and Fishing Vessels (Health and Safety at Work) Regulations 1997 (as amended).

13.3.7.1; 13.3.8.4; 13.8 SI 2012/1743 The Merchant Shipping (Accident Reporting and Investigation) Regulations 2012.

Marine notices

13.8 MGN 564 (M+F) Amendment 1 Marine casualty and marine incident reporting.

13.8.2 MGN 520 (M) Human element guidance – Part 2 The deadly dozen: 12 significant people factors in maritime safety.

Chapter 14: Permit to work systems

None

Chapter 15: Entering enclosed spaces

Regulations

15.1.1; 15.5.1 SI 2022/96 The Merchant Shipping and Fishing Vessels (Entry into Enclosed Spaces) Regulations 2022.

15.2.7 SI 1997/1713 The Confined Spaces Regulations 1997.

15.11.2 SI 1988/1638 The Merchant Shipping (Entry into Dangerous Spaces) Regulations 1988.

Chapter 16: Hatch covers and access lids

None

Chapter 17: Work at height

Regulations
17.1.1. SI 2010/332 The Merchant Shipping and Fishing Vessels (Health and Safety at Work) (Work at Height) Regulations 2010.

17.2 SI 2006/2183 The Merchant Shipping and Fishing Vessels (Provision and Use of Work Equipment) Regulations 2006.

Marine notices
17.1.1; 17.3.1; 17.4.1; 17.5.1; 17.7.1; Annex 17.1 MGN 410 (M+F) Amendment 2 The Merchant Shipping and Fishing Vessels (Health and Safety at Work) (Work at Height) Regulations 2010.

17.1.2 MSN 1838 (M) Amendment 1 Maritime Labour Convention, 2006: Minimum age.

17.2 MGN 331 (M+F) Amendment 2 The Merchant Shipping and Fishing Vessels (Provision and Use of Work Equipment) Regulations 2006.

MGN 532 (M) Amendment 2 Health and safety at work. Safe movement on board ship.

MGN 533 (M) Amendment 2 Means of access.

17.2.2; 17.4.11 MGN 578 (M) Amendment 1 Use of overside working systems on commercial yachts, small commercial vessels and loadline vessels.

Chapter 18: Provision, care and use of work equipment

Regulations
18.1.2; 18.2.2; 18.3.2; 18.4.3; 18.5.1; 18.6.1; 18.7.1; 18.8.2; 18.9.2; 18.10.5; 18.11.3; 18.12.4; 18.13.4; 18.21.4; 18.22.1; 18.23.1; 18.24.1; 18.25.1; 18.26.1; 18.27.1; Annex 18.3 SI 2006/2183 The Merchant Shipping and Fishing Vessels (Provision and Use of Work Equipment) Regulations 2006.

18.1.2 SI 2016/1025 The Merchant Shipping (Marine Equipment) Regulations 2016 (as amended).

18.2.2 SI 1999/2205 The Merchant Shipping and Fishing Vessels (Personal Protective Equipment) Regulations 1999.

18.13.3 2014/1512 The Merchant Shipping (International Safety Management (ISM) Code) Regulations 2014.

Marine notices
18.1.2 MGN 331 (M+F) Amendment 2 The Merchant Shipping and Fishing Vessels (Provision and Use of Work Equipment) Regulations 2006.

MSN 1874 (M+F) Amendment 7 Marine equipment – United Kingdom conformity assessment procedures for marine equipment, other approval and standards.

18.2.2 MSN 1870 (M+F) Amendment 5 The Merchant Shipping and Fishing Vessels (Personal Protective Equipment) Regulations 1999.

18.6.1 MGN 556 (M+F) Amendment 1 The Merchant Shipping and Fishing Vessels (Safety Signs and Signals) Regulations 2001.

18.20.1 MSN 1838 (M) Amendment 1 Maritime Labour Convention, 2006: Minimum age.

Chapter 19: Lifting equipment and operations

Regulations
19.1.3; 19.3.2; 19.7.1; Annex 19.4 SI 2006/2184 The Merchant Shipping and Fishing Vessels (Lifting Operations and Lifting Equipment) Regulations 2006.

Marine notice
19.1.3; 19.17.5; Annex 19.4 MGN 332 (M+F) Amendment 2 The Merchant Shipping and Fishing Vessels (Lifting Operations and Lifting Equipment) Regulations 2006.

Chapter 20: Work on machinery and power systems

Regulations
20.3.1 SI 2006/2183 The Merchant Shipping and Fishing Vessels (Provision and Use of Work Equipment) Regulations 2006.

20.3.6 SI 2019/42 The Merchant Shipping (Prevention of Oil Pollution) Regulations 2019.

Marine notices
20.3.1 MGN 331 (M+F) Amendment 2 The Merchant Shipping and Fishing Vessels (Provision and Use of Work Equipment) Regulations 2006.

20.17.6 MGN 452 (M) Electrical – potential hazards of arc flash associated with high and low voltage equipment.

Chapter 21: Hazardous substances and mixtures

Regulations
21.4.1 SI 2007/3100 The Merchant Shipping and Fishing Vessels (Health and Safety at Work) (Carcinogens and Mutagens) Regulations 2007.

21.5.1 SI 2010/2984 The Merchant Shipping and Fishing Vessels (Health and Safety at Work) (Asbestos) Regulations 2010 (as amended).

21.6.1 SI 2010/330 The Merchant Shipping and Fishing Vessels (Health and Safety at Work) (Chemical Agents) Regulations 2010 (as amended).

21.7.1 SI 1999/336 The Merchant Shipping (Carriage of Cargoes) Regulations 1999.

21.8.1 SI 2010/323 The Merchant Shipping and Fishing Vessels (Health and Safety at Work) (Biological Agents) Regulations 2010.

Marine notices
21.4.1 MGN 624 (M+F) Amendment 2 The Merchant Shipping and Fishing Vessels (Health and Safety at Work) (Carcinogens and Mutagens) Regulations 2007 (as amended).

21.5.1 MGN 669 (M+F) Amendment 1 The Merchant Shipping and Fishing Vessels (Health and Safety at Work) (Asbestos) Regulations 2010 (as amended).

21.6.1 MSN 1888 (M+F) Amendment 3 The Merchant Shipping and Fishing Vessels (Health and Safety at Work) (Chemical Agents) Regulations 2010 (as amended).

21.7.1 MSN 1718 (M) The safe use of pesticides in ships.

MGN 576 (M) Guidance for those undertaking fumigation operations alongside.

21.8.1 MSN 1889 (M+F) Amendment 3 The Merchant Shipping and Fishing Vessels (Health and Safety at Work) (Biological Agents) Regulations 2010 (as amended).

Chapter 22: Boarding arrangements

Regulations
22.1.3; 22.3.2 SI 1997/2962 The Merchant Shipping and Fishing Vessels (Health and Safety at Work) Regulations 1997 (as amended).

22.7.6; Annex 22.1 SI 2002/1473 The Merchant Shipping (Safety of Navigation) Regulations 2002.

22.7.6; 22.8.1; Annex 22.1, 4 SI 2020/0673 The Merchant Shipping (Safety of Navigation) Regulations 2020.

Marine notices
22.2.8 MGN 533 (M) Amendment 2 Means of access.

22.7.6 MGN 432 (M+F) Amendment 1 Safety during transfers of persons to and from ships.

22.8.1; Annex 22.1, 4 MSN 1874 (M+F) Amendment 7 Marine equipment – United Kingdom conformity assessment procedures for marine equipment, other approval and standards.

22.8.5 MGN 301 (M+F) Manoeuvring information on board ships.

Chapter 23: Food preparation and handling in the catering department

Regulations
23.5.1 SI 2020/621 The Merchant Shipping (Prevention of Pollution by Garbage from Ships) Regulations 2020.

Marine notices
23.1.2 MSN 1845 (M) Amendment 1 Maritime Labour Convention, 2006: Food and catering: Provision of food and fresh water.

MSN 1846 (M) Amendment 1 Maritime Labour Convention, 2006: Food and catering: Ship's cooks and catering staff.

23.5.1 MGN 632 (M+F) Amendment 1 The Merchant Shipping (Prevention of Pollution by Garbage from Ships) Regulations 2020.

23.8.1; 23.8.3 MGN 280 (M) Small vessels in commercial use for sport or pleasure, workboats and pilot boats – Alternative construction standards (Annex 5).

Guidance

23.1.2 Maritime and Coastguard Agency (MCA) (2020) *Wellbeing at Sea: A guide for organisations*, London: TSO (ISBN 978-0-11-553608-3).

MCA (2020) *Wellbeing at Sea: A pocket guide for seafarers*, London: TSO (ISBN 978-0-11-553787-5).

Chapter 24: Hot work

None

Chapter 25: Painting

None

Chapter 26: Anchoring, mooring and towing operations

Regulations
None

Marine notices
26.6.10 MGN 592 (M+F) Amendment 1 Anchoring, mooring, towing or hauling equipment on all vessels: Safe installation and safe operation.

26.8.7 MGN 533 (M) Amendment 2 Means of access.

MGN 591 (M+F) Amendment 1 Provision of safe means of access to fishing vessels and small vessels in ports.

Chapter 27: Roll-on/roll-off ferries

Regulations
27.2.1 SI 1998/1011 The Merchant Shipping (Fire Protection: Small Ships) Regulations 1998.

SI 1998/1012 The Merchant Shipping (Fire Protection: Large Ships) Regulations 1998.

Marine notices
27.1.1; 27.6.2; 27.10.1 MGN 341 (M) Ro-ro ships' vehicle decks: Accidents to personnel, passenger access and the carriage of motor vehicles.

27.3.3; 27.10.2; 27.10.8 MGN 653 (M) Amendment 1 Electric vehicles onboard passenger roll-on/roll-off (ro-ro) ferries.

27.4.3 MGN 621 (M+F) Roll-on/roll-off ships – guidance for the stowage and securing of vehicles.

27.10.1 MGN 545 (M+F) Guidance on the transport of dangerous goods as defined by the International Maritime Dangerous Goods (IMDG) Code when carried in a private vehicle not in commercial use or by a foot passenger on a Ro-Ro ship.

MGN 552 (M) Cargo stowage and securing – safe stowage and securing of specialised vehicles.

Guidance
27.1.1 Maritime and Coastguard Agency (MCA) (2009) *Code of Practice for Controlling Risks Due to Noise on Ships,* London: TSO (ISBN 978-0-11-553075-3).

27.4.2 MCA (2003) *Code of Practice on the Stowage and Securing of Vehicles on Roll-on/Roll-off Ships*, London: TSO: www.gov.uk/government/uploads/system/uploads/attachment_data/file/292367/ro-ro_stowage_securing_of_vehicles_cop.pdf

Chapter 28: Dry cargo

Regulations

28.2.4 SI 1999/336 The Merchant Shipping (Carriage of Cargoes) Regulations 1999.

28.2.6 SI 1998/2241 The Merchant Shipping (Load Line) Regulations 1998.

28.3.1 SI 1997/2367 The Merchant Shipping (Dangerous Goods and Marine Pollutants) Regulations 1997.

Marine notices

28.2.4 MGN 107 (M) The Merchant Shipping (Carriage of Cargoes) Regulations 1999.

28.3.1 MGN 340 (M) International Maritime Dangerous Goods (IMDG) Code and cargoes carried in cargo transport units.

28.3.5 MSN 1706 Amendment 1 The carriage of military and commercial explosives – amendment no. 1.

28.3.7 MSN 1870 (M+F) Amendment 5 The Merchant Shipping and Fishing Vessels (Personal Protective Equipment) Regulations 1999.

Chapter 29: Tankers and other ships carrying bulk liquid cargoes

Regulations

29.1.1 SI 2022/1342 The Merchant Shipping (Standards of Training, Certification and Watchkeeping) Regulations 2022.

29.1.12 SI 2022/96 The Merchant Shipping and Fishing Vessels (Entry into Enclosed Spaces) Regulations 2022.

29.2.3 SI 1998/1561 The Merchant Shipping (International Safety Management (ISM) Code) Regulations 1998.

29.2.6 SI 2014/1512 The Merchant Shipping (International Safety Management (ISM) Code) Regulations 2014.

29.3.1 SI 1994/2464 The Merchant Shipping (Gas Carriers) Regulations 1994.

29.4.4 SI 1996/3010 The Merchant Shipping (Dangerous or Noxious Liquid Substances in Bulk) Regulations 1996.

Marine notices

29.1.1 MSN 1866 (M) Amendment 1 Training and certification requirements for seafarers on tankers, ships subject to the IGF Code, ships that operate in polar waters and passenger ships.

Chapter 30: Port towage industry

Regulations

None

Marine notices

30.3.4 MGN 592 (M+F) Amendment 1 Anchoring, mooring, towing or hauling equipment on all vessels: Safe installation and safe operation.

30.8.3 MGN 199 (M) Dangers of interaction.

Chapter 31: Ships serving offshore oil and gas installations

None

Chapter 32: Ships serving offshore renewables installations

Regulations

32.1.3 SI 2002/1587 The Merchant Shipping (Diving Safety) Regulations 2002.

32.1.4 SI 2015/51 The Construction (Design and Management) Regulations 2015.

32.3.2 SI 2005/735 Work at Height Regulations 2005.

SI 2007/114 Work at Height (Amendment) Regulations 2007.

Marine notices

32.1.2 MGN 372 (M+F) Amendment 1 Safety of navigation: guidance to mariners operating in the vicinity of UK offshore renewable energy installations (OREIs).

32.1.3 MSN 1762 (M+F) Amendment 1 The Merchant Shipping (Diving Safety) Regulations 2002.

32.4.1 MGN 497 (M+F) Dangerous goods – including chemicals and other materials – storage and use on board ship.

MGN 280 (M) Small vessels in commercial use for sport or pleasure, workboats and pilot boats – Alternative construction standards.

Guidance

32.4.1 Maritime and Coastguard Agency (MCA) (2014) *Workboat Code: Industry Working Group Technical Standard*: www.gov.uk/government/publications/workboat-code

MCA (2018) *The Workboat Code: Edition 2 Amendment 1*: www.gov.uk/government/publications/the-workboat-code-edition-2-amendment-1

Chapter 33: Ergonomics

Regulations

33.1.1 SI 1997/2962 The Merchant Shipping and Fishing Vessels (Health and Safety at Work) Regulations 1997 (as amended).

Marine notices

33.1.1 MGN 636 (M) Amendment 2 Merchant Shipping and Fishing Vessels (Health and Safety at Work) Regulations 1997.

Guidance

Maritime and Coastguard Agency (MCA) (2020) *Wellbeing at Sea: A guide for organisations*, London: TSO (ISBN 978-0-11-553608-3).

MCA (2020) *Wellbeing at Sea: A pocket guide for seafarers*, London: TSO (ISBN 978-0-11-553787-5).

Appendix 2

Other sources of information

This appendix lists the sources of the other documents referred to in this Code.

This list has been compiled by the MCA, with support from industry. It provides a non-comprehensive list of suggested reading references and publication titles, with the sole purpose of providing further information on the topics covered by the Code. The MCA is not recommending the purchase of any publications nor endorsing the guidance within, unless MCA involvement is specified within the resource.

International Maritime Organization (IMO) publications are available from IMO Publishing, 4 Albert Embankment, London, SE1 7SR, email sales@imo.org or telephone +44 (0)20 7735 7611.

Registration is required for access to some IMO publications. Links have been provided as available at time of publication.

The Code reference is shown in **bold** and the information is arranged in chapter order.

Chapter 1: Managing occupational health and safety

References

National Health Service (NHS), Every Mind Matters: www.nhs.uk/every-mind-matters

NHS, Mental health: www.nhs.uk/conditions/stress-anxiety-depression

NHS, Live Well: Sleep and tiredness: www.nhs.uk/live-well/sleep-and-tiredness

1.2.1 *Code of Conduct for the Merchant Navy*: www.nautilusint.org/en/news-insight/resources/partnership-publications/code-of-conduct-for-the-merchant-navy/

1.2.7 National Maritime Occupational Health and Safety Committee (NMOHSC), *Guidelines to Shipping Companies on Behavioural Safety Systems.*

Chapter 2: Safety induction for personnel working on ships

References

2.1.1 International Maritime Organization (IMO), International Convention on Standards of Training, Certification and Watchkeeping (STCW) for Seafarers, 1978, as amended.

International Labour Organization (ILO), Maritime Labour Convention, 2006: www.ilo.org/global/standards/maritime-labour-convention/lang--en/index.htm

Chapter 3: Living on board

References

National Health Service (NHS), Mental health: www.nhs.uk/mental-health/NHS, Mental health conditions: www.nhs.uk/mental-health/conditions/

3.2.4 National Health Service (NHS) Live Well: Eat well: www.nhs.uk/live-well/eat-well/

3.3.2 NHS Better Health, Quit smoking: www.nhs.uk/better-health/quit-smoking

3.3.4 NHS Live Well: Quit smoking, 'Using e-cigarettes to stop smoking': www.nhs.uk/live-well/quit-smoking/using-e-cigarettes-to-stop-smoking

Public Health England, 'Evidence review of e-cigarettes and heated tobacco products 2018: executive summary': www.gov.uk/government/publications/e-cigarettes-and-heated-tobacco-products-evidence-review/evidence-review-of-e-cigarettes-and-heated-tobacco-products-2018-executive-summary#poisonings-fires-and-explosions

3.6 NHS Live Well: Sleep and tiredness: www.nhs.uk/live-well/sleep-and-tiredness/

3.6.1 International Maritime Organization (IMO), MSC/Circ 813 The Role of Human Element, List of Human Element Common Terms.

3.7.2 Health and Safety Executive (HSE), 'Thermal comfort': www.hse.gov.uk/temperature/thermal

3.8.1 Canadian Centre for Occupational Health and Safety (CCOHS), 'Cold environments: working in the cold': www.ccohs.ca/oshanswers/phys_agents/cold/cold_working.html

3.8.2 NHS, Frostbite: www.nhs.uk/conditions/frostbite

NHS, Hypothermia: www.nhs.uk/conditions/hypothermia

Chapter 4: Emergency drills and procedures

References

4.1.1 International Maritime Organization (IMO), International Convention for the Safety of Life at Sea (SOLAS), 1974: Chapter II-1 Construction – Subdivision and stability, machinery and electrical installations.

IMO, SOLAS, 1974: Chapter II-2 – Fire protection, fire detection and fire extinction.

IMO, SOLAS, 1974: Chapter III – Life-saving appliances and arrangements.

IMO, SOLAS, 1974: Chapter V – Safety of navigation.

4.1.9; 4.1.11; 4.4.13; 4.8.2; 4.4.21; 4.13.5 IMO, SOLAS, 1974: Chapter III – Life-saving appliances and arrangements, Regulation 19 Emergency training and drills: Drills.

4.1.10; 4.13.1 IMO, SOLAS, 1974: Chapter III – Life-saving appliances and arrangements, Regulation 30 Drills.

4.2.1 IMO, SOLAS, 1974: Chapter II-2 – Fire protection, fire detection and fire extinction, Regulation 15 Instructions, onboard training and drills.

4.2.15 IMO, SOLAS, 1974: Chapter II-2 – Fire protection, fire detection and fire extinction, Regulation 10 Fire fighting.

4.4.7; 4.4.21 IMO, MSC 1/Circ 1206/Rev 1, Annex 2 Guidelines on safety during abandon ship drills using lifeboats.

4.4.21 IMO, MSC 1/Circ 1206/Rev 1, Appendix Guidelines for simulated launching of free-fall lifeboats.

4.6.1 International Chamber of Shipping (ICS) (2022) *Bridge Procedures Guide:* www.ics-shipping.org/publication/bridge-procedures-guide-sixth-edition/

4.11.2 IMO Resolution A.716(17) The International Maritime Dangerous Goods Code (IMDG Code).

Emergency Procedures for Ships Carrying Dangerous Goods (EmS Guide): www.imo.org/en/OurWork/Safety/Pages/EmS-Guide.aspx

4.13.1; 4.13.5 IMO, SOLAS, 1974: Chapter II-1 Construction – Subdivision and stability, machinery and electrical installations, Regulation 19-1 Damage control drills for passenger ships.

4.13.5 IMO, SOLAS, 1974: Chapter V – Safety of navigation, Regulation 26 Steering gear: testing and drills.

Chapter 5: Fire precautions

References
5.5.4 International Maritime Organization (IMO), MSC 1/Circ 1321 Guidelines for Measures to Prevent Fires in Engine Rooms and Cargo Pump Rooms.

Chapter 6: Security on board

References
6.2.1; 6.3 International Maritime Organization (IMO), The International Ship and Port Facility Security (ISPS) Code.

6.2.1 Regulation (EC) No 725/2004 of the European Parliament and of the Council of 31 March 2004 on enhancing ship and port facility security: www.legislation.gov.uk/eur/2004/725

Chapter 7: Workplace health surveillance

References
National Health Service (NHS), Mental health conditions: www.nhs.uk/mental-health/conditions/

NHS, Mental health: www.nhs.uk/mental-health/

NHS Live Well: Sleep and tiredness, 'Why am I tired all the time?': www.nhs.uk/live-well/sleep-and-tiredness/why-am-i-tired-all-the-time/

7.1.2 Health and Safety Executive (HSE), 'Record keeping': www.hse.gov.uk/health-surveillance/record-keeping.htm

7.1.8 HSE, 'Manage performance and act on results': www.hse.gov.uk/health-surveillance/manage-performance.htm

Chapter 8: Personal protective equipment

None

Chapter 9: Safety signs and their use

References
9.2 Health and Safety Executive (HSE) (2015) *Safety Signs and Signals: The Health and Safety (Safety Signs and Signals) Regulations 1996.* HSE Books (3rd edn, ISBN 978 0 7176 6598 3).

9.3.6 International Maritime Organization (IMO) Resolution A.952(23) Graphical Symbols for Shipboard Fire Control Plans: wwwcdn.imo.org/localresources/en/KnowledgeCentre/IndexofIMOResolutions/AssemblyDocuments/A.952(23).pdf

(For ships constructed before 1 January 2004, IMO Resolution A.654(16) Graphical Symbols for Fire Control Plans: wwwcdn.imo.org/localresources/en/KnowledgeCentre/IndexofIMOResolutions/AssemblyDocuments/A.654(16).pdf)

9.3.9 IMO Resolution A.1116(30) Escape Route Signs and Equipment Location Markings: wwwcdn.imo.org/localresources/en/KnowledgeCentre/IndexofIMOResolutions/AssemblyDocuments/A.1116(30).pdf

(For ships constructed before January 2019, IMO Resolution A.760(18) Symbols Related to Life-saving Appliances and Arrangements: wwwcdn.imo.org/localresources/en/KnowledgeCentre/IndexofIMOResolutions/AssemblyDocuments/A.760(18).pdf)

9.4.2 IMO Resolution A.1021(26) Code on Alerts and Indicators, 2009: wwwcdn.imo.org/localresources/en/KnowledgeCentre/IndexofIMOResolutions/AssemblyDocuments/A.1021(26).pdf

9.4.4 IMO Resolution A.918(22) Standard Marine Communication Phrases, 2002 (IMO SMCP 2001) (IMO sales no. IA987E) and IMO SMCP on CD-ROM 2004, Standard Marine Communication Phrases: A pronunciation guide, London: IMO (IMO sales no. D987E): wwwcdn.imo.org/localresources/en/KnowledgeCentre/IndexofIMOResolutions/AssemblyDocuments/A.918(22).pdf

Chapter 10: Manual handling

None

Chapter 11: Safe movement on board ship

References

Annex 11.2 Health and Safety Executive (HSE), HSG38 *Lighting at work*: www.hse.gov.uk/pubns/books/hsg38.htm

Convention on the International Regulations for Preventing Collisions at Sea (COLREG), 1972 (as amended).

Chapter 12: Noise, vibration and other physical agents

References

12.1.1; 12.6.3; Annex 12.2 International Maritime Organization (IMO) Resolution MSC 337(91), Code on Noise Levels on Board Ships.

12.9.1 Health and Safety Executive (HSE), HSG260 *Sound advice: Control of noise at work in music and entertainment:* www.hse.gov.uk/pubns/books/hsg260.htm

12.12.6 HSE, 'Vibration at work': www.hse.gov.uk/vibration

12.16.2 HSE, 'Vibration risk assessment': www.hse.gov.uk/vibration/hav/advicetoemployers/assessrisks.htm

HSE, 'Providing health surveillance' (hand–arm vibration): www.hse.gov.uk/vibration/hav/advicetoemployers/healthsurveillance.htm

12.16.3 HSE, 'Health monitoring and review' (whole-body vibration): www.hse.gov.uk/msd/wbv.htm

12.18 *Non-binding guide to good practice for implementing Directive 2006/25/EC Artificial Optical Radiation:* op.europa.eu/en/publication-detail/-/publication/556b55ab-5d1a-4119-8c5a-5be4fd845b68/language-en/format-PDF/source-261805315

Non-binding guide to good practice for implementing Directive 2013/35/EU Electromagnetic Fields: op.europa.eu/en/publication-detail/-/publication/c6440d35-8775-11e5-b8b7-01aa75ed71a1

Annex 12.3 HSE, 'Accounting for "real world" factors': www.hse.gov.uk/noise/goodpractice/hearingrealworld.htm

Chapter 13: Safety officials

References

13.3.8.4 Marine Accident Investigation Branch (MAIB): www.gov.uk/government/organisations/marine-accident-investigation-branch

Chapter 14: Permit to work systems

None

Chapter 15: Entering enclosed spaces

References

15.2; 15.5.4; 15.5.12 International Maritime Organization (IMO) Resolution A.1050(27) Revised Recommendations for Entering Enclosed Spaces Aboard Ships: wwwcdn.imo.org/localresources/en/KnowledgeCentre/IndexofIMOResolutions/AssemblyDocuments/A.1050(27).pdf

15.5.2 MSC 1/Circ 1477 Guidelines to Facilitate the Selection of Portable Atmosphere Testing Instruments for Enclosed Spaces as Required by SOLAS Regulation XI-1/7.

15.5.12 Health and Safety Executive (HSE), 'EH40/2005 Workplace exposure limits': www.hse.gov.uk/pubns/books/eh40.htm

Chapter 16: Hatch covers and access lids

None

Chapter 17: Work at height

References

17.3.4 Health and Safety Executive (HSE) and The Ladder Association, 'LA455 Safe Use of Ladders and Stepladders: a brief guide': ladderassociation.org.uk/la455

17.7 Health and Safety Executive (HSE), 'Tower scaffolds': www.hse.gov.uk/construction/safetytopics/scaffold.htm

Annex 17.1 HSE Research Report RR708, 'Evidence-based review of the current guidance on first aid measures for suspension trauma': www.hse.gov.uk/research/rrpdf/rr708.pdf

Chapter 18: Provision, care and use of work equipment

References

18.28.1 British Tugowners Association (BTA) (2021) *Rope Selection, Procurement and Usage*: britishtug.com/bta-produced-tow-rope-guidance

18.28.19 IMO MSC 1/Circ 1620 Guidelines for Inspection and Maintenance of Mooring Equipment including Lines: wwwcdn.imo.org/localresources/en/Documents/MSC.1-Circ.1620%20-.pdf

Chapter 19: Lifting equipment and operations

References

19.3.2 Health and Safety Executive (HSE), 'INDG422 Thorough examination of lifting equipment: A simple guide for employers': www.hse.gov.uk/pubns/indg422.pdf

19.18.1 HSE, 'INDG339 Thorough examination and testing of lifts: Simple guidance for lift owners': www.hse.gov.uk/pubns/indg339.pdf

Chapter 20: Work on machinery and power systems

References

20.3.2 International Maritime Organization (IMO), International Convention for the Safety of Life at Sea, 1974 (SOLAS) II-2 Reg 4.2.2.6 Construction fire-protection, fire detection and fire extinction: Arrangements for oil fuel, lubrication oil and other flammable oils; Protection of high temperature surfaces: www.imo.org/en/About/Conventions/Pages/International-Convention-for-the-Safety-of-Life-at-Sea-(SOLAS),-1974.aspx

20.3.4 IMO Resolution A.1021 (26) Code on Alerts and Indicators, 2009: wwwcdn.imo.org/localresources/en/KnowledgeCentre/IndexofIMOResolutions/AssemblyDocuments/A.1021(26).pdf

20.3.13 IMO MSC 1/Circ 1321 Guidelines for Measures to Prevent Fires in Engine Rooms and Cargo Pump Rooms.

20.3.15 IMO MSC/Circ 834 Guidelines for Engine-Room Layout, Design and Arrangement.

20.17.3 National Health Service (NHS), 2022, Overview: Burns and scalds, types of burn: www.nhs.uk/conditions/burns-and-scalds

Chapter 21: Hazardous substances and mixtures

References

21.2.6 Control of Substances Hazardous to Health (COSHH) Regulations: www.hse.gov.uk/coshh/index.htm

21.3.1 Health and Safety Executive (HSE), EH40/2005 *Workplace exposure limits:* www.hse.gov.uk/pubns/books/eh40.htm

21.6.2; Annex 21.1 European Regulation (EC) 1272/2008 Classification, Labelling and Packaging of Substances and Mixtures (CLP Regulation): eur-lex.europa.eu/LexUriServ/LexUriServ.do?uri=OJ:L:2008:353:0001:1355:en:PDF

21.7.2 International Maritime Organization (IMO) MSC 1/Circ 1264 Recommendations on the Safe Use of Pesticides in Ships Applicable to the Fumigation of Cargo Holds.

IMO MSC 1/Circ 1358 Recommendations on the Safe Use of Pesticides in Ships.

21.7.4 HSE, HSG251 Fumigation: *Health and safety guidance for employers and technicians carrying out fumigation operations:* www.hse.gov.uk/pubns/books/hsg251.htm

Annex 21.1 HSE (2021) *The Approved List of Biological Agents:* www.hse.gov.uk/pubns/misc208.pdf

Chapter 22: Boarding arrangements

References

22.2.1; 22.6.1; Annex 22.1 International Maritime Organization (IMO) MSC 1/Circ 1331 Guidelines for Construction, Installation, Maintenance and Inspection/Survey of Means of Embarkation and Disembarkation.

22.2.1; 22.8.1; Annex 22.1, 4 IMO, SOLAS V.23 Safety of navigation: Pilot transfer arrangements.

22.2.6 UK Port Skills and Safety, 'SIP014 Guidance on safe access and egress': www.portskillsandsafety.co.uk/resources/sip014-guidance-safe-access-and-egress

22.6.1 IMO, SOLAS II.1/3-9 Construction – Structure, subdivision and stability, machinery and electrical installations: Means of embarkation on and disembarkation from ships.

22.8.1; Annex 22.1, 4 IMO Resolution A.1045(27) Pilot Transfer Arrangements: wwwcdn.imo.org/localresources/en/KnowledgeCentre/IndexofIMOResolutions/AssemblyDocuments/A.1045(27).pdf

22.8.1; Annex 22.1, 4 IMO Resolution A.1108(29) Amendments to the Recommendation on Pilot Transfer Arrangements (Resolution A.1045(27)): wwwcdn.imo.org/localresources/en/KnowledgeCentre/IndexofIMOResolutions/AssemblyDocuments/A.1108(29).pdf

22.8.1; Annex 22.1, 4 IMO MSC 1/Circ 1428 Pilot Transfer Arrangements: Required boarding arrangements for pilots.

22.9.1 International Maritime Pilots' Association, 'Required boarding arrangements for pilots': www.impahq.org/system/files/2021-04/Pilot%20Ladder%20Poster.pdf

22.10.3 UK Port Skills and Safety, 'SIP021 Guidance on safe access to fishing vessels and small craft in ports': www.portskillsandsafety.co.uk/resources/sip-021-guidance-safe-access-fishing-vessels-and-small-craft

Annex 22.1, 3.1 SOLAS III/3.13 Life-saving appliances and arrangements: Definitions. (For further information, see MSN 1676(M) Amendment 1 The Merchant Shipping (Life-saving Appliances and Arrangements) Regulations 2020 and The Merchant Shipping (Life-saving Appliances for Ships of Classes III to VI(A)) Regulations 1999.)

Chapter 23: Food preparation and handling in the catering department

None

Chapter 24: Hot work

References

24.10 Health and Safety Executive (HSE), HSG139 *The safe use of compressed gases in welding, flame cutting and allied processes:* www.hse.gov.uk/pubns/books/hsg139.htm

Chapter 25: Painting

None

Chapter 26: Anchoring, mooring and towing operations

References

26.2.1 International Association of Independent Tanker Owners (INTERTANKO) (2019) *Anchoring Guidelines: A risk-based approach:* www.intertanko.com/info-centre/intertanko-guidance/guidancenotearticle/anchoring-guidelines-a-risk-based-approach

Oil Companies International Marine Forum (OCIMF) (2010a) *Anchoring Systems and Procedures:* www.ocimf.org/publications/books/anchoring-systems-and-procedures

OCIMF (2010b) *Estimating the Environmental Loads on Anchoring Systems:* www.ocimf.org/document-libary/131-estimating-the-environmental-loads-on-anchoring-systems/file

The Swedish Club, 'Mooring and anchoring: Correct care and maintenance of mooring and anchoring equipment': www.swedishclub.com/loss-prevention/ship/mooring-and-anchoring

26.2.5 International Association of Classification Societies (IACS) UR A2 Rev 5 (Sept 2020) Shipboard fittings and supporting hull structures associated with towing and mooring on conventional ships: www.iacs.org.uk/publications/unified-requirements/ur-a/ur-a2-rev5-cln/

IACS Recommendation No 10 Rev 5 (June 2023) Chain anchoring, mooring and towing equipment: www.iacs.org.uk/download/1965

26.3.1 OCIMF (2019) *Effective Mooring* (4th edn): www.ocimf.org/publications/books/effective-mooring

26.8.7 UK Port Skills and Safety, 'SIP005 Guidance on mooring operations': www.portskillsandsafety.co.uk/resources/sip005-guidance-mooring-operations

UK Port Skills and Safety, 'SIP014 Guidance on safe access and egress': www.portskillsandsafety.co.uk/resources/sip014-guidance-safe-access-and-egress

UK Port Skills and Safety, 'SIP021 Guidance on safe access to fishing vessels and small craft in ports': www.portskillsandsafety.co.uk/resources/sip021-guidance-safe-access-fishing-vessels-and-small-craft

Chapter 27: Roll-on/roll-off ferries

References

27.8.2 IMO (2021) *Code of Safe Practice for Cargo Stowage and Securing* (CSS Code).

27.10.3 International Maritime Organization (IMO), International Convention for the Safety of Life at Sea, 1974 (SOLAS) II-2 Reg 20 Protection of vehicle, special category and ro-ro space**s**.

27.10.4 Department for Environment, Food and Rural Affairs (Defra), *Guidance: Animal welfare*: www.gov.uk/guidance/animal-welfare#animal-welfare-during-transport

Chapter 28: Dry cargo

References

28.2.2 International Maritime Organization (IMO) Resolution A.862(20) Code of Practice for the Safe Loading and Unloading of Bulk Carriers: wwwcdn.imo.org/localresources/en/KnowledgeCentre/IndexofIMOResolutions/AssemblyDocuments/A.862(20).pdf

IMO Resolution MSC 23(59) Adoption of the International Code for the Safe Carriage of Grain in Bulk: wwwcdn.imo.org/localresources/en/KnowledgeCentre/IndexofIMOResolutions/MSCResolutions/MSC.23(59).pdf

28.2.3; 28.2.4 IMO Resolution A.714(17) Code of Safe Practice for Cargo Stowage and Securing: wwwcdn.imo.org/localresources/en/KnowledgeCentre/IndexofIMOResolutions/AssemblyDocuments/A.714(17).pdf

28.2.4 IMO Resolution A.1048(27) Code of Safe Practice for Ships Carrying Timber Deck Cargoes, 2011 (2011 TDC Code): https://wwwcdn.imo.org/localresources/en/KnowledgeCentre/IndexofIMOResolutions/AssemblyDocuments/A.1048(27).pdf

28.3.4 IMO Resolution MSC 268(85) Adoption of the International Maritime Solid Bulk Cargoes (IMSBC) Code: wwwcdn.imo.org/localresources/en/KnowledgeCentre/IndexofIMOResolutions/MSCResolutions/MSC.268(85).pdf

28.3.5 Emergency Procedures for Ships Carrying Dangerous Goods (EmS Guide): www.imo.org/en/OurWork/Safety/Pages/EmS-Guide.aspx

IMO MSC/Circ 857 *The Medical First Aid Guide for Use in Accidents Involving Dangerous Goods* (MFAG).

28.3.10 IMO (2007) *Revised Recommendations on the Safe Transport of Dangerous Cargoes and Related Activities in Port Areas.*

28.4.1 UK Port Skills and Safety, 'SIP008 Guidance on the storage of dry bulk cargo': www.portskillsandsafety.co.uk/resources/sip-008-guidance-storage-dry-bulk-cargo

28.4.3 IMO Resolution MSC 355(92) Amendments to the International Convention for Safe Containers (CSC), 1972.

28.5 IMO MSC/Circ 886 Recommendation on Safety of Personnel During Container-Securing Operations.

IMO MSC/Circ 888 Preventing Falls at Corrugated Bulkheads in General Cargo Ships.

Chapter 29: Tankers and other ships carrying bulk liquid cargoes

References

29.1.1; 29.2.3 International Association of Ports and Harbors (IAPH), International Chamber of Shipping (ICS) and Oil Companies International Marine Forum (OCIMF) (2020) *International Safety Guide for Oil Tankers and Terminals (ISGOTT 6):* ocimf.org/publications/books/international-safety-guide-for-tankers-and-terminals-1

29.1.1; 29.3.1 ICS (2018) *Tanker Safety Guide (Liquefied Gas):* ics-shipping.org/publication/tanker-safety-guide-liquefied-gas/

29.1.1; 29.4.3 International Chamber of Shipping (ICS) (2021) *Tanker Safety Guide (Chemicals):* ics-shipping.org/publication/tanker-safety-guide-chemicals/

29.1.5 International Maritime Organization (IMO) MSC 1/Circ 1588 Carriage of Dangerous Goods, International Maritime Dangerous Goods (IMDG) Code, Revised Emergency Response Procedures for Ships Carrying Dangerous Goods (EmS Guide).

29.1.5 IMO MSC/Circ 857 The Medical First Aid Guide for Use in Accidents Involving Dangerous Goods (MFAG).

29.1.9 The International Convention for the Prevention of Pollution from Ships (MARPOL).

29.2.6 IMO Resolution MSC 98(73) Adoption of the International Code for Fire Safety Systems (FSS Code).

29.3.1 IMO Resolution MSC 5(48) Adoption of the International Code for the Construction and Equipment of Ships Carrying Liquefied Gases in Bulk (IGC Code).

29.4.3 IMO International Code for the Construction and Equipment of Ships carrying Dangerous Chemicals in Bulk (IBC Code).

IMO Code for the Construction and Equipment of Ships Carrying Dangerous Chemicals in Bulk (BCH Code).

Chapter 30: Port towage industry

None

Chapter 31: Ships serving offshore oil and gas installations

References

IMCA Guidance: *Guidance on the Transfer of Offshore Personnel to and from Offshore Vessels and Structures:* IMCA SEL 025, Rev 1, IMCA M 202 Rev 1: www.imca-int.com/publications/287/guidance-on-the-transfer-of-personnel-to-and-from-offshore-vessels-and-structures

31.6 Offshore Energies UK (2015) *Best Practice for the Safe Packing and Handling of Cargo to and from Offshore Locations.*

International Maritime Organization (IMO) Resolution A.716(17) The International Maritime Dangerous Goods Code (IMDG Code).

31.6; 31.8.5; 31.9.1; 31.11.1; 31.15 Guidelines for Offshore Marine Operations (G-OMO): www.g-omo.info

Chapter 32: Ships serving offshore renewables installations

References

32.3.2 IMCA Guidance: *Guidance on the Transfer of Offshore Personnel to and from Offshore Vessels and Structures* – IMCA SEL 025 Rev 1, IMCA M 202 Rev 1: www.imca-int.com/publications/287/guidance-on-the-transfer-of-personnel-to-and-from-offshore-vessels-and-structures

32.4.1 National Workboat Association – *Good Practice Guide for Offshore Vessels:* www.workboatassociation.org/news/nwa-publishes-good-practice-guide-for-offshore-energy-service-vessels/

32.6.1 *G+ Integrated Offshore Emergency Response – Renewables (IOER-R): Good Practice Guidelines for Offshore Renewable Energy Developments:* publishing.energyinst.org/topics/renewables/offshore-wind/g-integrated-offshore-emergency-response-g-ioer-good-practice-guidelines-for-offshore-renewable-energy-developments

G+ Good Practice Guideline: The safe management of small service vessels used in the offshore wind industry: publishing.energyinst.org/topics/power-generation/offshore-wind/good-practice-guideline-the-safe-management-of-small-service-vessels-used-in-the-offshore-wind-industry

Health and Safety strategy for renewables: www.renewableuk.com/page/HealthSafety

The National Workboat Association *Good Practice Guide for Offshore Energy Service Vessels:* www.workboatassociation.org/news/nwa-publishes-good-practice-guide-for-offshore-energy-service-vessels

Emergency rescue cooperation planning: www.gov.uk/government/publications/offshore-renewable-energy-installations-orei

Chapter 33: Ergonomics

References

33.1.1 Health and Safety Executive (HSE), HSG48 *Reducing error and influencing behaviour:* www.hse.gov.uk/pubns/priced/hsg48.pdf

33.2.5 HSE, INDG36 *Working with display screen equipment (DSE):* www.hse.gov.uk/pubns/indg36.pdf

Appendix 3

Standards and specifications referred to in this code

NOTE: Copies of standards produced by the British Standards Institution (BSI) can be obtained from BSI Customer Services, 389 Chiswick High Road, London W4 4AL; or online from shop.bsigroup.com

Standards double-prefixed 'BS EN' are the UK version in English of a European harmonised standard. The prefix 'BS EN ISO' appears where an international standard has been adopted by Europe as a European Standard.

The transition from CE marking to UKCA marking in Great Britain occurred from 31 December 2020 and a new UK Conformity Assessed (UK CA) marking system was introduced, with the full transition to this UK CA marking system from 1 January 2023.

All standards are subject to periodic updating; the most recent version should be used.

Copies of EU Directives are available from eur-lex.europa.eu/homepage.html?locale=en

The Code reference is shown in **bold**. Information arranged in chapter order.

Chapter 1: Managing occupational health and safety

None

Chapter 2: Safety induction for personnel working on ships

None

Chapter 3: Living on board

3.13.6 BS EN ISO 12311:2013 Personal protective equipment – Test methods for sunglasses and related eyewear.

BS EN ISO 12312-1:2013+A1:2015 Eye and face protection – Sunglasses and related eyewear. Part 1: Sunglasses for general use.

3.14.8 BS EN ISO 23907-1:2019 Sharps injury protection – Requirements and test methods – Part 1: Single-use sharps containers.

Chapter 4: Emergency drills and procedures

None

Chapter 5: Fire precautions

None

Chapter 6: Security on board

None

Chapter 7: Workplace health surveillance

None

Chapter 8: Personal protective equipment

None

Chapter 9: Safety signs and their use

9.6.1 BS EN 1089-3:2011 Transportable gas cylinders. Gas cylinder identification (excluding LPG). Colour coding.

9.6.3 BS EN ISO 407:2023 Small medical gas cylinders – Pin-index yoke-type valve connections.

9.7.1 BS ISO 14726:2008 Ships and marine technology – identification colours for the content of piping systems.

BS 4800:2011 Schedule of paint colours for building purposes.

9.8.1 BS EN 3-10:2009 Portable fire extinguishers. Provisions for evaluating the conformity of a portable fire extinguisher to EN 3-7.

9.8.4 BS 5306-10:2019 Fire extinguishing installations and equipment on premises. Colour coding to indicate the extinguishing medium contained in portable fire extinguishers. Code of practice.

BS EN 3-7:2004+A1:2007 Portable fire extinguishers: Characteristics, performance requirements and test methods.

Chapter 10: Manual handling

None

Chapter 11: Safe movement on board ship

11.5.5 ISO 15085:2003+A2:2018 Small craft – Man-overboard prevention and recovery.

Chapter 12: Noise, vibration and other physical agents

None

Chapter 13: Safety officials

None

Chapter 14: Permit to work systems

None

Chapter 15: Entering enclosed spaces

15.12.4 EN 137:2006 Respiratory protective devices. Self-contained open-circuit compressed air breathing apparatus with full face mask. Requirements, testing, marking.

BS EN 14593-1:2018 Respiratory protective devices. Compressed air line breathing devices with demand valve. Devices with a full face mask. Requirements, testing and marking.

BS EN 14593-2:2005 Respiratory protective devices. Compressed air line breathing apparatus with demand valve. Apparatus with a half mask at positive pressure. Requirements, testing, marking.

BS EN 14594:2018 Respiratory protective devices. Continuous flow compressed air line breathing devices. Requirements, testing and marking.

BS EN 1146:2005 Respiratory protective devices. Self-contained open-circuit compressed air breathing apparatus incorporating a hood for escape. Requirements, testing, marking.

Chapter 16: Hatch covers and access lids

None

Chapter 17: Work at height

None

Chapter 18: Provision, care and use of work equipment

None

Chapter 19: Lifting equipment and operations

19.1.5 BS 7121-2-4:2013 Code of practice for the safe use of cranes. Inspection, maintenance and thorough examination. Loader cranes.

Annex 19.3 Council Directive 92/58/EEC of 24 June 1992 on the minimum requirements for the provision of safety and/or health signs at work (ninth individual Directive within the meaning of Article 16 (1) of Directive 89/391/EEC).

Chapter 20: Work on machinery and power systems

None

Chapter 21: Hazardous substances and mixtures

None

Chapter 22: Boarding arrangements

22.8.1; Annex 22.1, 4, 4.2 BS ISO 799-1:2019 Ships and marine technology – Pilot ladders – Part 1: Design and specification.

22.8.1; Annex 22.1, 4 BS ISO 799-3:2022 Ships and marine technology – Pilot ladders – Part 3: Attachments and associated equipment.

Annex 22.1 BS ISO 5488:2015 Ships and marine technology – Accommodation ladders.

ISO 7061:2015 Ships and marine technology – Aluminium shore gangways for seagoing vessels.

Annex 22.1, 2.1 BS MA 78:1978 Specification for aluminium shore gangways.

Annex 22.1, 3.2 ISO 7364:2016 Ships and marine technology – Deck machinery – Accommodation ladder winches.

Chapter 23: Food preparation and handling in the catering department

None

Chapter 24: Hot work

24.3.1 BS EN 12941:1998+A2:2008 Respiratory protective devices. Powered filtering devices incorporating a helmet or a hood. Requirements, testing, marking.

24.4.1 BS EN ISO 11611:2015 Protective clothing for use in welding and allied processes.

BS EN 169:2002 *(withdrawn)* Personal eye-protection. Filters for welding and related techniques. Transmittance requirements and recommended use.

BS EN ISO 16321-1:2022 Eye and face protection for occupational use – Part 1. General requirements.

BS EN ISO 16321-2:2021 Eye and face protection for occupational use – Part 2: Additional requirements for protectors used during welding and related techniques.

BS EN 1146:2005 Respiratory protective devices. Self-contained open-circuit compressed air breathing apparatus incorporating a hood for escape. Requirements, testing, marking.

24.6.9 BS EN 60529:1992+A2:2013 Degrees of protection provided by enclosures (IP Code).

24.7.1 BS EN ISO 20345:2022 Personal protective equipment – Safety footwear.

BS EN 50321-1:2018 Live working. Footwear for electrical protection. Insulating footwear and overboots.

24.9.13; Annex 24.3 BS EN 1256:2006 Gas welding equipment. Specification for hose assemblies for equipment for welding, cutting and allied processes.

Annex 24.2 BS EN IEC 60974-1:2022+A11:2022 Arc welding equipment – Welding power sources.

BS EN IEC 60974-11:2021 Arc welding equipment – Part 11: Electrode holders.

Annex 24.3 BS EN ISO 3821:2019 Gas welding equipment. Rubber hoses for welding, cutting and allied processes.

BS EN 16436-1:2014+A3:2020 Rubber and plastics hoses, tubing and assemblies for use with propane and butane and their mixtures in the vapour phase. Hoses and tubings.

BS EN 16436-2:2018 Rubber and plastics hoses, tubing and assemblies for use with propane and butane and their mixture in the vapour phase. Assemblies.

ISO/TR 28821:2012 Gas welding equipment – Hose connections for equipment for welding, cutting and allied processes – Listing of connections which are either standardised or in common use.

BS EN 561:2002 Gas welding equipment. Quick-action coupling with shut-off valves for welding, cutting and allied processes.

BS ISO 7289:2018 Gas welding equipment. Quick-action coupling with shut-off valves for welding, cutting and allied processes.

Chapter 25: Painting

None

Chapter 26: Anchoring, mooring and towing operations

None

Chapter 27: Roll-on/roll-off ferries

None

Chapter 28: Dry cargo

None

Chapter 29: Tankers and other ships carrying bulk liquid cargoes

None

Chapter 30: Port towage industry

None

Chapter 31: Ships serving offshore oil and gas installations

None

Chapter 32: Ships serving offshore renewables installations

None

Chapter 33: Ergonomics

None

Information arranged by Standards number

British Standards (BS)/European Norm (EN)/International Organization Standardization (ISO)

BS EN ISO Standards number	BS/EN/ISO full title	Code reference
BS EN ISO 407:2023	Small medical gas cylinders – Pin-index yoke-type valve connections	9.6.3
BS EN ISO 3821:2019	Gas welding equipment. Rubber hoses for welding, cutting and allied processes	Annex 24.3
BS EN ISO 11611:2015	Protective clothing for use in welding and allied processes	24.4.1
BS EN ISO 12311:2013	Personal protective equipment – Test methods for sunglasses and related eyewear	3.13.6
BS EN ISO 12312-1:2013+A1:2015	Eye and face protection – Sunglasses and related eyewear. Part 1: Sunglasses for general use	3.13.6
BS EN ISO 16321-1:2022	Eye and face protection for occupational use – Part 1. General requirements	24.4.2
BS EN ISO 16321-2:2021	Eye and face protection for occupational use – Part 2: Additional requirements for protectors used during welding and related techniques	24.4.2
BS EN ISO 20345:2022	Personal protective equipment – Safety footwear	24.7.1
BS EN ISO 23907-1:2019	Sharps injury protection – Requirements and test methods – Part 1: Single-use sharps containers	3.14.8

British Standards

BS number	BS full title	Code reference
BS MA 78:1978	Specification for aluminium shore gangways	Annex 22.1, 2.1
BS 4800:2011	Schedule of paint colours for building purposes	9.7.1
BS 5306-10:2019	Fire extinguishing installations and equipment on premises. Colour coding to indicate the extinguishing medium contained in portable fire extinguishers. Code of practice	9.8.4
BS 7121-2-4:2013	Code of practice for the safe use of cranes. Inspection, maintenance and thorough examination. Loader cranes	19.1.5

BS/EN

BS/EN number	BS/EN full title	Code reference
BS EN 3-7:2004+A1:2007	Portable fire extinguishers: Characteristics, performance requirements and test method.	9.8.4
BS EN 3-10:2009	Portable fire extinguishers. Provisions for evaluating the conformity of a portable fire extinguisher to EN 3-7	9.8.1
BS EN 169:2002 (withdrawn)	Personal eye-protection. Filters for welding and related techniques. Transmittance requirements and recommended use	24.4.1
BS EN 561:2002	Gas welding equipment. Quick-action coupling with shut-off valves for welding, cutting and allied processes	Annex 24.3
BS EN 1089-3:2011	Transportable gas cylinders. Gas cylinder identification (excluding LPG). Colour coding	9.6.1
BS EN 1146:2005	Respiratory protective devices. Self-contained open-circuit compressed air breathing apparatus incorporating a hood for escape. Requirements, testing, marking	15.12.4; 24.4.1
BS EN 1256:2006	Gas welding equipment. Specification for hose assemblies for equipment for welding, cutting and allied processes	24.9.13; Annex 24.3
BS EN 12941:1998+A2:2008	Respiratory protective devices. Powered filtering devices incorporating a helmet or a hood. Requirements, testing, marking	24.3.1
BS EN 14593-1:2018	Respiratory protective devices. Compressed air line breathing devices with demand valve. Devices with a full face mask. Requirements, testing and marking	15.12.4

BS EN 14593-2:2005	Respiratory protective devices. Compressed air line breathing apparatus with demand valve. Apparatus with a half mask at positive pressure. Requirements, testing, marking	15.12.4
BS EN 14594:2018	Respiratory protective devices. Continuous flow compressed air line breathing devices. Requirements, testing and marking	15.12.4
BS EN 16436-1:2014+A3:2020	Rubber and plastics hoses, tubing and assemblies for use with propane and butane and their mixtures in the vapour phase. Hoses and tubings	Annex 24.3
BS EN 16436-2:2018	Rubber and plastics hoses, tubing and assemblies for use with propane and butane and their mixture in the vapour phase. Assemblies	Annex 24.3
BS EN 50321-1:2018	Live working. Footwear for electrical protection. Insulating footwear and overboots	24.7.1
BS EN 60529:1992+A2:2013	Degrees of protection provided by enclosures (IP code)	24.6.9

BS EN IEC Standards

BS EN IEC Standard	BS EN IEC full title	Code reference
BS EN IEC 60974-1:2022+A11:2022	Arc welding equipment – Welding power sources	Annex 24.2
BS EN IEC 60974-11:2021	Arc welding equipment – Part 11: Electrode holders	Annex 24.2

BS/ISO Standards

BS/ISO Standards	BS/ISO full title	Code reference
BS ISO 799-1:2019	Ships and marine technology – Pilot ladders – Part 1: Design and specification	22.8.1; Annex 22.1, 4, 4.2
BS ISO 799-3:2022	Ships and marine technology – Pilot ladders – Part 3: Attachments and associated equipment	22.8.1; Annex 22.1, 4
BS ISO 5488:2015	Ships and marine technology – Accommodation ladders	Annex 22.1
BS ISO 7289:2018	Gas welding equipment. Quick-action coupling with shut-off valves for welding, cutting and allied processes	Annex 24.3
BS ISO 14726:2008	Ships and marine technology – identification colours for the content of piping systems	9.7.1

EN Standards

EN number	EN full title	Code reference
EN 137:2006	Respiratory protective devices. Self-contained open-circuit compressed air breathing apparatus with full face mask. Requirements, testing, marking	15.12.4

ISO Standards

ISO number	ISO full title	Code reference
ISO 7061:2015	Ships and marine technology – Aluminium shore gangways for seagoing vessels	Annex 22.1
ISO 7364:2016	Ships and marine technology – Deck machinery – Accommodation ladder winches	Annex 22.1, 3.2
ISO 15085:2003+A2:2018	Small craft – Man-overboard prevention and recovery	11.5.5
ISO/TR 28821:2012	Gas welding equipment – Hose connections for equipment for welding, cutting and allied processes – Listing of connections which are either standardised or in common use	Annex 24.3

Other applicable standards: EEC Directives

Directive number	Full title	Code reference
92/58/EEC	Council Directive 92/58/EEC of 24 June 1992 on the minimum requirements for the provision of safety and/or health signs at work (ninth individual Directive within the meaning of Article 16 (1) of Directive 89/391/EEC)	Annex 19.3

Appendix 4

Acknowledgements

The Maritime and Coastguard Agency (MCA) would like to give special thanks to all those involved in the 2023/4 review of this Code. The following organisations contributed to the development of this update:

Abu Dhabi Maritime Academy – Abu Dubi Ports Group

Anglo Eastern

Bernicia Marine Consultants Ltd

CalMac Ferries Ltd

Clyde & Co/BML Solicitors

Condor Ferries

Dover Harbour Board

Fairbrother Consultancy Services

Fleetwood Nautical College

Fletcher Group

Honourable Company of Master Mariners

InterManager

Isles of Scilly Steamship Group

MSC Cruise Management LTD UK

Nautilus International

Northlink Ferries

Offshore Turbine Services

P&O Ferries

Royal Fleet Auxiliary

Scottish Sea Farms

SeaRegs Training Ltd

SERCO Northlink Ferries

Svitzer Marine Ltd

The National Union of Rail, Maritime and Transport Workers (RMT)

The Shipowners' Club

UK Chamber of Shipping

VShips/DNV

Wightlink Ferries

Williams Shipping

All other private seafarers/industry individuals in the UK and abroad are recognised for their contribution.

Additionally, we would like to acknowledge the individuals and organisations that gave their time to respond to the public consultation carried out from 29 August to 21 November 2023.

The Maritime and Coastguard Agency (MCA) acknowledges the kind permission for the following sources of information and illustrations used in this Code.

Links to the information are provided in Appendix 2, Other sources of information.

The Code reference is shown in **bold** and the information is arranged in chapter order.

Chapter 1: Managing occupational health and safety

1.2.9 Eddie Perkins, Figure 1.2, Knowledge management diagram.

Annex 1.1 John Blaikie, Simple change (management of change).

Chapter 2: Safety induction

None

Chapter 3: Living on board

National Health Service (NHS) website, including Live Well and Better Health sections.

3.7.2 Health and Safety Executive (HSE), 'Controlling thermal comfort'.

3.8.1 York University, Canada, 'Guidelines for working in cold weather'.

Chapter 4: Emergency drills and procedures

None

Chapter 5: Fire precautions

None

Chapter 6: Security on board

None

Chapter 7: Health surveillance

7.1.8 Health and Safety Executive (HSE), Figure 7.1, the health surveillance cycle diagram.

Chapter 8: Personal protective equipment

None

Chapter 9: Safety signs and their use

None

Chapter 10: Manual handling

Annex 10.1 Health and Safety Executive (HSE), Figure 10.7, Guideline weight diagram.

Chapter 11: Safe movement on board ship

None

Chapter 12: Noise, vibration and other physical agents

12.12.6; 12.16.2; 12.16.3 Health and Safety Executive (HSE), 'Vibration at work'; 'Providing health surveillance' (hand–arm vibration); and 'Health monitoring and review' (whole-body vibration).

Chapter 13: Safety officials

None

Chapter 14: Permit to work systems

Annex 14.1 North Star Shipping, Permit to work (general).

Chapter 15: Entering dangerous (enclosed) spaces

None

Chapter 16: Hatch covers and access lids

None

Chapter 17: Work at height

None

Chapter 18: Provision, care and use of work equipment

None

Chapter 19: Lifting equipment and operations

None

Chapter 20: Work on machinery and power systems

None

Chapter 21: Hazardous substances and mixtures

None

Chapter 22: Boarding arrangements

None

Chapter 23: Food preparation and handling in the catering department

None

Chapter 24: Hot work

None

Chapter 25: Painting

None

Chapter 26: Anchoring, mooring and towing operations

26.2.3 Figure 26.1 Anchoring. Image courtesy of Intertanko.

Annex 26.1 Swedish Accident Investigation Administration, Complex snap-back zone. Acknowledged as source of diagram of snap-back zones on the foredeck of a ship.

Chapter 27: Roll-on/roll-off ferries

None

Chapter 28: Dry cargo

None

Chapter 29: Tankers and other ships carrying bulk liquid cargoes

None

Chapter 30: Port towage industry

None

Chapter 31: ships serving offshore oil and gas installations

None

Chapter 32: Ships serving offshore renewables

None

Chapter 33: Ergonomics

33.2.2 Health and Safety Executive (HSE), Figure 33.1, 'Working with display screen equipment (DSE)'.

Glossary

Additional earth	An earth connection applied to apparatus after the application of a circuit main earth. This is normally at the point of work if the equipment is not already fitted with a circuit main earth.
Approved	A type of form sanctioned for use by the superintendent/senior electrical engineer.
Athwartships restraint	Mooring lines leading ashore as perpendicular as possible to the ship's fore and aft line.
Authorising officer	A person appropriately trained and appointed in writing by the superintendent/electrical engineer to carry out work as permitted by these rules.
Breast lines	Lines that restrain the ship in one direction (off the berth).
Caution notice	Conveys a warning against interference with the apparatus to which it is attached.
Chafe guard	An anti-abrasion device.
Chief engineer	A senior engineer on board the vessel who is responsible for all vessel technical operations and maintenance.
Chock	A means of restraint placed against the wheels to prevent movement.
Circuit main earth	An earth connection applied to make apparatus safe to work on before a permit to work or sanction for test is issued, and which is nominated on the document.
Competent person	A person who is appropriately trained and has sufficient technical knowledge or experience to enable them to avoid danger. It is the duty of the authorising officer issuing a permit to work covered by these rules to satisfy themselves that persons are competent to carry out the work involved.
Danger notice	Calls attention to the danger of approach or interference with the apparatus to which it is attached.
Dead	At or about zero voltage and disconnected from all sources of electrical energy.
Double and reverse stoppering	A quick, practical whipping knot; a method of using twine to secure the end of a rope to prevent it fraying. Also known as West Country whipping.
Down-flooding	Water ingress on board that may affect the stability of the vessel.
Dynamic risk assessment	The continuous process of identifying hazards, assessing risk, taking action to eliminate or reduce risk, monitoring and reviewing, in the rapidly changing circumstances of an operational incident.

Earthed	Connected to the general mass of earth in such a manner as will ensure at all times an immediate discharge of electrical energy without danger.
Electro-technical officer	A specialist electronic engineer who is competent to work on high-voltage systems.
Elimination	Taking action to remove exposure to risk.
Girting	The capsizing movement of the tug caused by a ship's sudden movement.
Go and return	In welding, consists of two cables to complete the electric circuit that allows for electrical arc-based welding (eg TIG, MIG/MAG, MMA) to work.
Gog rope (gob rope)	A specific tug rope for the purpose of maintaining safe manoeuvres during towing.
High voltage	A voltage exceeding 1000 volts.
High-voltage apparatus	Any apparatus, equipment or conductors normally operated at a voltage higher than 1000 volts.
Isolate	To disconnect and separate the electrical equipment from every source of electrical energy in such a way that the disconnection and separation are secure.
Just culture	An effective safety culture is necessary for operating safely. It means organisations must be able to learn from their mistakes and continuously improve.
Key safe	A device for the secure retention of keys used to lock means of isolation, earthing or other safety devices.
Lee	The side of the ship not exposed to the prevailing wind.
Limitation of access	A form issued by an authorising officer to a competent person, defining the limits of the work to be carried out in the vicinity of, but not on, high-voltage electrical apparatus.
Live	Electrically charged from a supply of electricity.
Lock out tag out (LOTO)	A safety procedure to ensure that energy and power sources are properly isolated, shut off, locked and tags applied, to notify others that energy or power systems are being worked on and the restarting of these energy and power sources is prohibited, while locks and tags are in place.
Longitudinal restraint	Mooring lines leading in a nearly fore and aft direction to prevent longitudinal movement (surge) of the ship while in berth.
Manual handling	Any operation that includes transporting or supporting a load, lifting, putting down, pushing, pulling, carrying or moving by hand or bodily force.
Monkey's fist	A knot formed in the shape of a fist or ball used to add natural weight to a rope.

Permit to work	A form of declaration signed and given by an authorising officer to a competent person in charge of the work to be carried out on or in close proximity to high-voltage apparatus, making known to the competent person the extent of the work, exactly what apparatus is dead, is isolated from all live conductors, has been discharged and earthed and, insofar as electric hazards are concerned, on which it is safe to work.
Physical agent	An external environmental factor within the working location such as noise, vibration, optical radiation (ultra-violet and infra-red light), heat and electromagnetic fields.
Safety lock	A lock used to secure points of isolation, safety devices and circuit earths, being unique from any other locks used on the system.
Sanction for test	A form of declaration, signed and given by an authorising officer to another authorising officer in charge of testing high-voltage apparatus, making known to the recipient what apparatus is to be tested and the conditions under which the testing is to be carried out.
Shark jaws	A remotely controlled chain and wire stoppers used on board to unshackle lengths of wire on deck when carrying a loaded wire over the stern roller. They are designed to guide wire/chain safely without crew being present and also work as a securing mechanism.
Simultaneous operations	Work activities that involve more than one vessel, or a vessel and an installation.
Snap-back zone	Designated area on board within which a mooring line or rope could break.
Spring lines	Lines that restrain the ship in two directions: headsprings prevent forward motion and backsprings prevent aft motion.
Superintendent/senior electrical engineer	A senior electrical/mechanical engineer suitably qualified and appointed in writing by the company to be responsible for compilation and administration of rules for high-voltage installations and operations.
Switching plan	A plan or programme, developed by the authorised person, which details the intended sequence of switching, isolation and earthing operations required to be carried out to isolate and make dead, or reinstate and make live, high-voltage equipment or installation. The plan must be agreed between the authorised person and the competent persons undertaking the task before executing the plan. If contractors are involved, then their agreement is also required.
Task based risk assessment (TBRA)	A formal process of identifying the hazards associated with each task to be performed, assessing the risk, and providing the safety controls to manage the risk.
Thorough examination	An examination carried out by a competent person who should be sufficiently independent and impartial to allow objective decisions to be made.

Ullaging	The amount of space remaining empty in a tank.
Whelps	The projections which stand out from the barrel of a capstan or winch. They provide extra bite for a rope under strain than if the barrel were smooth.
Whip check	A safety cable used to connect air hoses across the coupling. It secures hoses from movement in case the connection unexpectedly separates.

Index

References are shown by chapter, section and paragraph numbers. The letter A before a chapter number indicates the annex at the end of the chapter.

abandon ship drills 4.4
abandoning ship 4.5
abrasive wheels 18.17
access lids 16.1–16.7 *see also* hatch covers
accident investigation 13.8, A13.2
accident recording 13.3.7, 13.4.5
accidents 2.5
accommodation ladders A22.1.1, A22.1.3, A22.2
accountability 1.2.7
accumulators 20.23.1
acetylene 24.9
adverse weather 11.11
aerials 20.22.8–20.22.9
aerosols 3.10.2
AHTS (anchor handling towing supply) vessels 31.15
alcohol 3.4.2, 3.7.1, 29.4.5
alkaline storage batteries 20.20
aluminium equipment 22.6.3, A22.2
anchor handling towing supply (AHTS) vessels 31.15
anchoring 26.1–26.2
arc-flash 20.17
arc-welding systems A24.2
armed robbery 6.7
asbestos 3.11.4, 21.6
atmosphere testing 15.5
auxiliary machinery 20.7

batteries 20.18–20.20
bench machines 18.16
beryllia 20.24.3–20.24.5
bilges 20.3.6, 20.3.9
biological agents 7.2, 21.9, A21.2
boarding arrangements 22.1–22.10
    ladders 22.5, 22.8–22.9 *see also* accommodation ladders; pilot ladders; portable ladders; rope ladders
    maintenance of equipment 22.6
    pilots 22.8
    positioning of equipment 22.4
    safety nets 22.3
    small craft 22.10
    special circumstances 22.7
    standards A22.1
    transfers between ships at sea 22.7.6–22.7.7
body protection 8.10
boilers 20.6
bosun's chairs 17.5
breast lines 26.3.10
breathing apparatus 8.7.14–8.7.15, 15.12
bulk liquid cargoes 29.1–29.4
bulk ore/oil carriers 29.2
bulkheads, moveable 28.8
bulldog grips A18.2
bump caps 8.4.4
buoys, mooring to 26.4
burns 3.12.7–3.12.8
busbars 20.16.15

carbon dioxide 21.9.6–21.9.7
carcinogens 21.5, A21.1
cardice 21.9.6–21.9.7
cargo gear 19.15, A19.2
cargo spaces, lighting in 28.6
casting off 26.3
casualties 4.10
catering equipment 23.4–23.7, 23.11
change management 1.2.5, A1.1
chemical agents 21.7
chemical carriers 29.4
choppers 23.12
clothing 3.9
cold conditions 3.8
communications 1.2.3
compressed gas cylinders 9.6, 24.8
consultation procedures 2.11
containers, dry cargo 28.4
cradles 17.4
cranes 19.11
critical equipment 20.10
cross-contamination 23.3.2
culture for safe working 1.2.1–1.2.9
cuts 3.12.6

damage control drills 4.13
dangerous goods 27.9, 28.3
dangerous occurrences 13.4.5
dangerous spaces *see* enclosed spaces
dangerous substances, spillage of 28.3.8–28.3.10
dangerous work, duty to stop 13.4.6
davit-launched life rafts 4.4.25–4.4.26
deck work 11.10
deep fat fryers 23.7, 23.9
derricks 19.12–19.13
diesel engines 20.7.2–20.7.5
display screen equipment (DSE) 33.2
domestic passenger craft, mooring 26.7, A26.2
drainage 11.2
Drikold™ 21.9.6–21.9.7
drills *see* emergency drills
drive units 18.27
drowning protection 8.11
dry cargo 28.1–28.8
    containers 28.4
    dangerous goods 28.3
    moveable bulkheads 28.8
    precautions for personnel 28.7
    safe access 28.2.7–28.2.11
    stowage 28.2
    working 28.5
dry ice 21.9.6–21.9.7
dry-cleaning solvents 21.9.8
DSE (display screen equipment) 33.2
dust 3.11.6
dynamic risk assessment 1.2.6

ear defenders 8.5.4, A12.3
earplugs 8.5.2–8.5.3, A12.3
e-cigarettes 3.3.4–3.3.5, 5.4
electric arc welding 24.7
electric welding equipment 24.6
electrical equipment *see also* high-voltage systems
   arc-flash 20.17
   LOTO (lock out tag out) 20.13.4
   precautions 20.13.6–20.13.11
   work on 20.13
electrical hazards 3.12.12–3.12.14
electrical wiring 9.5
electronic equipment 20.22–20.24
emergency drills 4.1–4.15
  abandon ship drills 4.4
  damage control drills 4.13
  emergency steering drills 4.14
  for enclosed spaces 4.8
  fire drills 4.2
  frequency of A4.1
  launching drills A4.2
  leakage drills 4.11
  man overboard drills 4.6
  MES drills 4.15
  spillage drills 4.11
emergency exit signs A9.1
emergency procedures 2.4, 4.1–4.15
emergency steering drills 4.14
enclosed spaces
  atmosphere, suspect/unsafe 15.10
  atmosphere testing 15.5
  breathing apparatus 15.12
  control systems 15.6
  emergencies in 4.9
  emergency drills for 4.8, 15.13
  entry into 11.9, 15.1–15.14
    authorised officers 15.2
    competent persons 15.2
    completion procedures 15.9
    information 15.11
    instruction 15.11
    precautions 15.3, 15.7
    preparing and securing the space 15.4
    procedures and arrangements during 15.8, 15.13
    training 15.11
  flammable vapours 15.4.15–15.4.16
  hazards 15.14
  oxygen deficiency 15.14.1–15.14.2
  oxygen-enriched atmosphere 15.4.3
  permits to work A14.1
  rescue arrangements 15.13.3–15.13.9
  toxic substances 15.4.7–15.4.14
engines 20.7–20.8
environmental responsibilities 2.8
ergonomics 33.1–33.2, A33.1
escape route signs A9.1
escorting 30.9
expectations of seafarers 1.2.2
extension runners 20.21
eye injuries 3.12.4
eye protection 8.6

face protection 8.6
fall protection 8.9
fast-rescue boats 4.4.22–4.4.24
fatigue 3.6
fire drills 4.2

fire extinguishers 9.8, A9.1
fire precautions 2.4, 5.1–5.6
fire procedures 4.3
fire safety/prevention, roll-on/roll-off (ro-ro) ferries 27.3
firefighting symbols A9.1
first-aid signs A9.1
fitness 3.2
fixed installations 18.16
flammable gases and vapours, testing for 15.5.9–15.5.11
flammable vapours 15.4.15–15.4.16
floating work platforms 17.6
food hygiene 23.1.3
food preparation and handling 23.1–23.13
  burns 23.7.3–23.7.9
  choppers 23.12
  deep fat fryers 23.7, 23.9
  electric shocks 23.7.3–23.7.4
  equipment 23.4–23.7, 23.11
  galley stoves 23.7
  knives 23.12
  liquid petroleum gas appliances 23.8
  liquid spills 23.7
  meat saws 23.12
  microwave ovens 23.10
  personal preparation 23.2
  refrigerated rooms 23.13
  slips, falls and tripping hazards 23.6
  steam boilers 23.7
  store rooms 23.13
  surfaces 23.4
  waste disposal 23.5
foot injuries 3.12.3
foot protection 8.8
footwear 8.8
fork-lift trucks 18.24, 19.16
free-fall lifeboats 4.4.20–4.4.21
frostbite 3.8.4

galley stoves 23.7
galleys 5.6
gangways A22.1.2, A22.2
gas cutting 24.2, 24.9
gas cylinders 9.6, 24.8, 27.10.1
gas welding 24.9
gases, testing for 15.5.9–15.5.12
generic risk assessment 1.2.6
gloves 8.8, 12.14.7
guardrails 11.5

hairnets 8.4.5
hand injuries 3.12.2
hand protection 8.8
hand signals A19.3
hand tools 18.14
hand-arm vibration 12.14
harnesses 8.9
hatch covers 16.1–16.7
hatchways 11.5
hazardous substances 3.11, 21.1–21.9
  asbestos 21.6
  biological agents 21.9, A21.2
  carbon dioxide 21.9.6–21.9.7
  carcinogens 21.5, A21.1
  chemical agents 21.7
  dry-cleaning solvents 21.9.8
  health surveillance 21.3
  instruction/training 21.2

hazardous substances *continued*

   mutagens 21.5, A21.1
   pesticides 21.8
   prevention/control of exposure 21.4
head injuries 3.12.5
head protection 8.4
health 2.6, 3.2 *see also* occupational diseases
health and safety
   occupational 2.9
      management of 1.1–1.2.9
   representations 13.3.8
health surveillance 7.1–7.3
   hazardous substances 21.3
   noise 12.8
   vibration 12.16
hearing protection 8.5, A12.3
height, work at *see* work at height
high-pressure jetting equipment 18.18
high-voltage cables 20.16.11
high-voltage systems 20.15–20.16
   busbars 20.16.15
   cables 20.16.11
   earthing 20.16.9
   entry to enclosures 20.16.8
   failure of supply 20.16.7
   insulation testing 20.16.6
   locking off 20.16.4
   notices 20.16.10
   protective equipment 20.16.5
   ring main units 20.16.14
   switchgear 20.16.2
   tags 20.16.10
   transformers 20.16.13
   withdrawn apparatus 20.16.3
hold access, standards for A11.1
hooks 23.13.9, 27.8.13, 28.5.11, 28.5.13
hoses and connections/assemblies A24.3
hot work 3.7, 24.1–24.10
   arc-welding systems A24.2
   compressed gas cylinders 24.8
   electric arc welding 24.7
   electric welding equipment 24.6
   equipment checks 24.5
   gas cutting 24.2, 24.9
   hoses and connections/assemblies A24.3
   lighting up procedures A24.1
   shutting down procedures A24.1
   welding 24.3–24.10, A24.1–24.2
housekeeping 2.7, 3.10, 27.11
hydraulic equipment 20.12
hydraulic jacks 18.19
hydraulic systems 18.18
hygiene 2.6, 3.2
hypothermia 3.8.5

incidents, learning from A1.1
injuries 3.12
insulation testing 20.16.6

just culture policy 1.2.7

knives 23.12
knowledge management 1.2.9

ladders 22.5, 22.8–22.9 *see also* accommodation ladders;
   pilot ladders; portable ladders; rope ladders

launches, mooring 26.7, A26.2
launching drills A4.2
laundry equipment 18.20
lead acid storage batteries 20.19
leadership 1.2.4
leakage drills 4.11
leakages 4.12
lifeboats 4.4.8–4.4.21
lifebuoys 8.11
lifejackets 4.1.13, 8.11
lifting equipment 19.1–19.18
   cargo gear 19.15
   certificates 19.5, A19.1
   controls 19.1.11
   cranes 19.11
   defect reporting 19.4
   derricks 19.12–19.13
   examination 19.3, A19.1
   fork-lift trucks 18.24, 19.16
   hand signals A19.3
   inspection 19.3
   installation 19.7
   maintenance 19.2
   operation 19.8
   operational safety measures 19.10
   personnel-lifting 19.17
   positioning 19.7
   records 19.6
   register of A19.2
   safe working load 19.9
   selection 19.1.10
   standards A19.4
   stoppers 19.14
   testing 19.4, A19.1
   trucks 19.16
   winches 19.11
lifts 19.17–19.18
lighting
   in cargo spaces 28.6
   for safe movement 11.4
   standards A11.2
   and work equipment 18.5
liquefied gas carriers 29.3
liquid cargoes 29.1–29.4
liquid petroleum gas appliances 23.8
living on board 3.1–3.15
lock out tag out (LOTO) 20.13.4

machinery
   auxiliary machinery 20.7
   on the bench 20.21
   boilers 20.6
   critical equipment 20.10
   electrical equipment 20.13
      arc-flash 20.17
      high-voltage systems *see* high-voltage systems
      LOTO (lock out tag out) 20.13.4
      precautions 20.13.6–20.13.11
   electronic equipment 20.22–20.24
   on extension runners 20.21
   hydraulic equipment 20.12
   main engines 20.8
   main switchboards 20.14
   maintenance of 20.5
   pneumatic equipment 20.12
   radio equipment 20.22–20.24
   refrigeration machinery 20.9

machinery *continued*
   spaces 5.5, 20.3
   steering gear 20.11
   storage batteries 20.18–20.20
   thermal oil heaters 20.6
   unmanned spaces 20.4
main engines 20.8
main switchboards 20.14
making fast 26.3
malaria 3.5
man overboard drills 4.6
man overboard procedures 4.7
mandatory safety signs A9.1
manual handling 3.12.10, 10.1–10.2, A10.1
marine evacuation systems *see* MES drills
meat saws 23.12
medical emergencies 2.5
medications 3.4
MES drills 4.15
microwave ovens 23.10
microwave radiation 20.22
mobile phones 3.15
mobile-lifting appliances 11.8
mooring 3.12.11, 26.1, 26.3–26.6
   arrangements, examples A26.3
   to buoys 26.4
   domestic passenger craft 26.7, A26.2
   launches 26.7, A26.2
   to quays 26.7
   self-mooring operations 26.8
moveable bulkheads 28.8
musters 4.1
mutagens 21.5, A21.1

needles 3.14
noise 12.5–12.9
   exposure assessment 12.6
   exposure to A12.1–12.2
   health surveillance 12.8
   levels A12.1–12.2
   limits A12.1–12.2
   from music and entertainment 12.9
   risk assessment 12.7

occupational diseases *see also* health
   reporting of 7.3
occupational health and safety 2.9
   management of 1.1–1.2.9
offshore oil and gas installations 31.1–31.15
   anchor handling 31.15
   bulk cargo 31.7
   cargo 31.6
   cargo-handling operations 31.8
   communications 31.5
   gangway transfers 31.14
   personal protective equipment (PPE) 31.4
   personnel baskets 31.13
   personnel carriers 31.11–31.13
   precautions 31.3
   responsibilities 31.2
   support vessels 31.5
   transfer capsules 31.12
   transfers 31.9
offshore renewable energy installations (OREIs) 32.1–32.6
oil 3.11.5, 15.4.4–15.4.6, 29.2
oil and gas installations *see* offshore oil and gas installations

openings, guarding of 11.5
OREIs (offshore renewable energy installations) 32.1–32.6
oxygen deficiency 15.5.7–15.5.8, 15.14.1–15.14.2
oxygen-enriched atmosphere 15.4.3
painting 25.1–25.4
permit to work systems 14.1–14.3
permits to work A14.1
   enclosed spaces A14.1.1
   general A14.1.3
   at height A14.1.2
   over the side A14.1.2
personal assessment of risk 1.2.6
personal electronic devices 3.15
personal gas monitors 8.7.10–8.7.13
personal injuries 3.12
personal protective equipment (PPE) 8.1–8.11
   body protection 8.10
   drowning protection 8.11
   eye protection 8.6
   face protection 8.6
   fall protection 8.9
   foot protection 8.8
   hand protection 8.8
   head protection 8.4
   hearing protection 8.5, A12.3
   offshore oil and gas installations 31.4
   respiratory protective equipment 8.7
   seafarer duties 8.2
   types of 8.3
   vibration 12.14.7
   welding 24.3–24.4
personal safety 6.9
pesticides 21.8
physical agents 12.1–12.18 *see also* noise; vibration
   consultation 12.3
   control of 12.2
   information on 12.4
   prevention of 12.2
   training 12.4
pilot ladders 10.2.13, 22.2.1, 22.8–22.9, 25.5.4, A22.1.4
pipelines 9.7
piracy 6.7
planning 1.2.5
pneumatic equipment 20.12
pneumatic systems 18.18
port towage industry 30.1–30.9
portable fire extinguishers 9.8
portable ladders 11.7, 16.7.2, 17.3, 22.2.7, 22.5, 22.10.3, 28.7.4, A17.2
positioning techniques A17.3
power systems *see* machinery
power take-off shafts 18.27
power-operated tools 18.15
power-operated watertight doors 11.6
PPE *see* personal protective equipment
prohibitory safety signs A9.1

quays, mooring to 26.7, A26.2

radar equipment 20.22.4
radiation 20.22
radio equipment 20.22–20.24
razor blades 3.14
refrigerated compartments 20.9
refrigerated rooms 23.13
refrigeration machinery 20.9
remote-controlled self-propelled equipment 18.26
rescue boats 4.4.22–4.4.24

respirators 8.7.4–8.7.8
respiratory protective equipment 8.7
resuscitators 8.7.16
rigging accommodation 10.2.13
ring main units 20.16.14
risk assessment 1.2.6, 18.3, A1.2, A1.3
risk awareness 1.2.6
roll-on/roll-off (ro-ro) ferries 27.1–27.11
   cargo, securing of 27.8
   dangerous goods 27.9
   fire safety/prevention 27.3
   gas cylinders 27.10.1
   housekeeping 27.11
   livestock 27.10.4
   safe movement 27.4
   stowage 27.7
   vehicles
      electric 27.10.3, 27.10.5–27.10.7
      inspection of 27.6
      specialised 27.10
   ventilation 27.2
   work equipment 27.5
rope access A17.3
rope ladders 22.2.7, 22.5, A22.1.3, A22.4.1
ropes 18.28, 26.3.9
ro-ro *see* roll-on/roll-off ferries
rubbish collection 3.14.5–3.14.8

safe condition signs A9.1
safe movement on board ship 11.1–11.11
safe weight guidelines A10.1
safe working load (SWL) 19.9
safety caps 8.4.5
safety committees 13.3.4, 13.7
safety culture 1.2.1–1.2.9
safety harnesses 8.9
safety helmets 8.4.1–8.4.3
safety induction 2.1–2.11
safety inspections 13.4.4
safety nets 22.3
safety officers
   accident recording 13.4.5
   appointment of 13.3.2
   compliance advice 13.4.2
   dangerous occurrences 13.4.5
   dangerous work, duty to stop 13.4.6
   duties of 13.4
   inspections 13.4.4
      checklist A13.1
   investigations 13.4.3
   support for 13.3.6
   termination of appointments 13.3.5
safety officials 13.1–13.8 *see also* safety committees; safety officers; safety representatives
safety representatives
   advice to 13.6
   election of 13.3.3
   powers of 13.5
safety signs 9.1–9.8
   duty to display 9.2
   electrical wiring 9.5
   emergency exit A9.1
   escape routes A9.1
   fire extinguishers 9.8
   firefighting symbols A9.1
   first-aid A9.1
   gas cylinders 9.6

   mandatory A9.1
   occasional 9.4
   pipelines 9.7
   portable fire extinguishers 9.8
   prohibitory A9.1
   safe condition A9.1
   standards 9.3
   warning A9.1
sanction to test 14.3, A14.2
scaffolding 17.7, A17.4
scalds 3.12.7–3.12.8
security 6.1–6.9
security levels 6.3
security precautions 6.4
self-mooring operations 26.8
self-propelled work equipment 18.25
semi-conductor devices 20.24
sharps 3.14
ship security plans (SSPs) 6.2
shipboard vehicles 11.8, 19.16
side-launch lifeboats 4.4.8–4.4.19
signs *see* safety signs
smoking 3.3, 3.10.3, 5.4
smoking regulations 2.4.2
smuggling 6.8
snap-back zones 26.3.4, A26.1
solid carbon dioxide 21.9.6–21.9.7
spillage drills 4.11
spillages 4.12
spontaneous combustion 5.3
sprays 3.10.2–3.10.3
spring lines 26.3.10
SSPs (ship security plans) 6.2
stages 17.4
stairways 11.7
steam boilers 23.7
steering gear 20.11
stoppers 19.14
storage batteries 20.18–20.20
store rooms 23.13
stowaways 6.6
sunburn 3.7.5
sunglasses 3.13
supervisory interventions 1.2.7
switchboards 20.14
switchgear 20.16.2
SWL (safe working load) 19.9
syringes 3.14

tankers 29.1–29.4
task-based risk assessments (TBRAs) 1.2.6
terrorism 6.5
thermal oil heaters 20.6
thermal protective clothing 8.10
tiredness 3.6
toolbox talks 1.2.6
tools 18.15
   misuse of 3.12.9
towing 26.1, 26.6, 30.1–30.9
toxic gases, testing for 15.5.12
transformers 20.16.13
transit areas 11.3
tripping hazards 23.6
trucks 19.16 *see also* shipboard vehicles

union purchase 19.13

valves 20.24
ventilation 27.2
vessel familiarisation training 2.2
vibration 12.11–12.16
  effects of 12.10
  exposure to 12.11
  hand-arm vibration 12.14
  health surveillance 12.16
  levels, determination of 12.12
  limits 12.11
  mitigation 12.14
  personal protective equipment (PPE) 12.14.7
  types of 12.10
  whole-body vibration 12.15
voluntary statements A13.2

warning safety signs A9.1
waste disposal 23.5
watertight doors 11.6
weather, adverse 11.11
weighing anchor 26.1–26.2
welding A24.1–24.2
whole-body vibration 12.15
winches 19.11
wires 18.28, 26.3.9
work at height 17.1–17.7
  emergency planning for A17.1
  permits to work A14.1.2
work equipment 18.1–18.28 see also lifting equipment
  abrasive wheels 18.17
  bench machines 18.16
  bulldog grips A18.2
  carrying of seafarers on 18.23
  conformity requirements A18.1
  controls 18.21
  dangerous parts of 18.4
  drive units 18.27
  fixed installations 18.16
  fork-lift trucks 18.24
  hand tools 18.14
  high-pressure jetting equipment 18.18
  hydraulic jacks 18.19
  hydraulic systems 18.18
  information 18.12
  inspection 18.11
  instructions 18.12
  laundry equipment 18.20
  and lighting 18.5
  maintenance 18.10
  markings 18.6
  pneumatic systems 18.18
  power take-off shafts 18.27
  power-operated tools 18.15
  remote-controlled 18.26
  risk assessment 18.3
  roll-on/roll-off (ro-ro) ferries 27.5
  ropes 18.28
  seafarers, duties of 18.2
  self-propelled 18.25
  stability of 18.8
  standards A18.3
  temperatures, extreme 18.9
  training 18.12
  use of 18.22
  warnings 18.7
  wires 18.28
  workers, duties of 18.2

workshop machines 18.16
worker responsibilities for safety 2.10
working aloft see work at height
working clothes 3.9
workplace health surveillance 7.1–7.3
workshop machines 18.16

X-ray radiation 20.22.4